Bioeconomics of Invasive Species

Bioeconomics of Invasive Species

Integrating Ecology, Economics, Policy, and Management

Edited by

Reuben P. Keller
David M. Lodge
Mark A. Lewis
Jason F. Shogren

UNIVERSITY PRESS
2009

OXFORD
UNIVERSITY PRESS

Oxford University Press, Inc., publishes works that further
Oxford University's objective of excellence
in research, scholarship, and education.

Oxford New York
Auckland Cape Town Dar es Salaam Hong Kong Karachi
Kuala Lumpur Madrid Melbourne Mexico City Nairobi
New Delhi Shanghai Taipei Toronto

With offices in
Argentina Austria Brazil Chile Czech Republic France Greece
Guatemala Hungary Italy Japan Poland Portugal Singapore
South Korea Switzerland Thailand Turkey Ukraine Vietnam

Published by Oxford University Press, Inc.
198 Madison Avenue, New York, New York 10016
http://www.oup.com

Library of Congress Cataloging-in-Publication Data
Bioeconomics of invasive species : integrating ecology, economics, policy,
and management / edited by Reuben P. Keller . . . [et al.].
p. cm.
Includes bibliographical references and index.
ISBN 978-0-19-536798-0; 978-0-19-536797-3 (pbk.)
1. Biological invasions. I. Keller, Reuben P.
QH353.B53 2009
577'.18—dc22 2008030333

9 8 7 6 5 4 3 2 1
Printed in the United States of America
on acid-free paper

Foreword

This is a book that all ecologists and economists interested in bioeconomics should read. The master narrative encompassing a bi-disciplinary framework and endogenous risk makes it intuitively and logically appealing. A narrative that can be generalized in such a straightforward manner constitutes a forceful principle for organizing research and for informing policy. The work here should leave even the disciplinary isolationist interested in studying more about what a joint determination framework can offer.

Invasive species are a major environmental policy challenge. They continue to alter, often in undesirable ways, the workings of ecosystems around the globe. This book provides general and species-specific overviews of ecological and economic tools and also consensus propositions for studying interactions of the determinants and behaviors of invasive species. It treats lessons from past attempts to understand and to manage invasive species. It also suggests strategies for understanding and combating the threats to environmental and economic well-being that nonindigenous species pose. Readers will get a thorough treatment of the relevant scientific issues as well as a comprehensive review of the analytical and the empirical tools used by ecologists and economists to research invasive aquatic and terrestrial flora and fauna in North America and around the world.

Pleas for collaboration between ecologists and economists to advance understanding and resolution of environmental problems are so commonplace as to be almost hackneyed. When adherents of each disciplinary personality try to work together, they usually lapse into discord, followed by retreat into remote if not totally separated intellectual pursuits. Most ecologists and economists see only dimly how to clarify assumptions about their respective disciplines. The book provides a master narrative in which ecological and economic expertise complement and make each area more robust than were it to stand alone.

Ecological and economic systems each mediate the behaviors of the opposite system. The appropriate focus is the decision maker working in her or his environment, for in reality, neat separation of natural and human activities does not exist. A species' initial invasion, establishment, spatial spread, and temporal persistence influences and is influenced by abiotic and biotic processes and by individual and institutionalized human decisions. Decision makers adapt to environmental change by changing their personal behaviors as well as by directly changing a particular environment. Interactions and feedback between and among systems and system scales influence the structure, resilience, and dynamics of respective systems. Thus, jointly determined vision encourages individuals from each discipline to consider and understand what the other brings to the table. Each discipline is thereby forced to better scrutinize and document the information needs of the other. Such a vision supplies a framework for fostering sharper questions as well as sharper and smarter answers. This volume makes better-informed outcomes possible.

The focus of this book is on the bioeconomic behavioral roots of invasive species. Evaluation techniques (e.g., energy analysis, benefit–cost analysis) take a secondary role. The authors primarily address what does happen rather than what should happen. They present empirical illustrations demonstrating that the joint determination vision produces different answers from those arising from a framework based solely on either the ecological or the economic system.

Framing the causal relations between the ecology and the economics of invasive species as reciprocating systems does not imply that researchers should reform their entire set of ecological and economic tools or the tenets these tools have uncovered. Similar tools will likely be employed to develop propositions and to extract empirical results, whatever framework is used.

It is possible for model components to become so entangled in a web of interconnectedness, especially when some components are ill defined, that explanatory power is lost rather than gained. Parsimony can trump completeness, implying that there has to be some limit to the reciprocal coupling of the ecological and the economic components of an environmental model. Some intellectual separation is necessary to mark distinctions in system integration and to assure empirical content. This book acknowledges the parsimony–completeness tradeoff. Given limited research budgets and policy goals, this tradeoff immediately brings up the question of those facets of an invasive species model for which accuracy (unbiasedness) and precision (low variability across independent measurements) are especially important. Though the authors offer no firm answers to this question, the background they provide on invasive species will help formulate answers. A key extension of joint determination runs throughout the book. Uncertainty, irreversibilities, and timing issues almost always characterize invasive species problems. Uncertainty about causes or consequences shifts the focus to endogenous risk, a scenario in which decision makers can try to alter the risks (the product of probability and severity, if realized) of the establishment, spread, and persistence of an invasive species. An endogenous-risk focus has the potential to make less costly the tradeoff between model parsimony and completeness. A careful reading of this book strongly conveys this impression.

Whatever the issue, complexity and ambiguity tempt policy makers and even scientific experts to wrap themselves in a cloak of objectivity by picking and choosing the scientific results they deem relevant. The authors are alert to this temptation. Policy makers and experts must often transfer findings from existing original studies to new areas of scientific or policy interest. Several chapters here consider the transfer question. They ponder both the theoretical underpinnings of the question and its statistical and computational treatment. In particular, the authors recognize that combining information from multiple sources and models of a common phenomenon can produce parameter estimates corresponding more closely to a new setting than can any single source.

The book concludes with an appealing human touch. The editors recall and reflect upon the successes and failures of their research and their attempts to communicate and to convince the public and policy makers about the causes and likely consequences of invasive species problems. They view their records of success as mixed. This tentativeness is leavened by the cheery optimism of a young ecologist recounting what inspires him about invasive species research. He nevertheless expresses bewilderment at the frequent reluctance of policy makers and the public to learn about and to accept scientific results.

Thomas D. Crocker
Department of Economics and Finance
University of Wyoming

Preface

Biological invasions can drive global environmental change. Biologists have explained the risks so that both the public and policy makers are now aware of the impacts of invasive species. Economists are also taking greater interest in determining how invasive species interact with economic systems, and in how invaders should be controlled to increase societal welfare. Disciplinary work by ecologists and economists expands our understanding of the drivers and impacts of invasions, but neither ecological nor economic systems operate in isolation. This book provides a greater integration and synthesis of ecological and economic concepts and tools—a bioeconomic approach to understanding and managing invasive species. Such an approach can help policy makers and the public determine optimal expenditures, for example, on preventing and controlling invasive species.

The Integrated Systems for Invasive Species (ISIS) team is a multi-institution collaboration among ecologists, economists, and mathematical biologists. The team came together as a project funded by the U.S. National Science Foundation and has met annually since 2000 (and conducted much research between meetings) to identify and address key questions about the bioeconomics of invasive species. The questions and our best responses are presented here. Our framework blends the work of the ISIS project with results from other researchers working on both disciplinary and interdisciplinary frontiers. Our group's composition ensures analytical and empirical rigor, as well as ecological and economic realism.

As society becomes more aware of global environmental change, people are demanding that policy makers address broader biological and economic realities. This book has two related goals. The first is to reinforce the role of bioeconomic research as the best approach to design policy and management systems for invasive species. The second is to show how bioeconomic research can be conducted to generate realistic invasive species policy recommendations. Throughout the ISIS project, we aim to place our bioeconomic research approach into a context that is useful to researchers and society.

The book's structure follows the linked economic and ecological processes that lead to invasion—starting with the vector of introduction, through establishment and spread, to the impacts of successful invaders. Our main thesis throughout is that a bioeconomic approach is required both to understand and to manage invasions. The first two chapters introduce the study of species invasions and give the rationale for this thesis. The next four chapters track the invasion process, including risk assessment tools to predict the identity of likely invaders and methods to identify the extent of suitable habitat for non-native species, also treating model approaches for predicting species establishment and dispersal. We consider throughout how the science can inform management and policy actions to reduce total impacts. Next we explore general issues, addressing uncertainty in models and methods for economic valuation, then tie it all together in an integrated bioeconomic model for determining appropriate management decisions in response to particular species invasions. The final four chapters include case studies based on ISIS research and a discussion of the possibilities and challenges for future bioeconomic research.

We gratefully acknowledge the funding agencies that have supported the ISIS project. These include the U.S. National Science Foundation, the Economic Research Service of the U.S. Department of Agriculture, the U.S. Environmental Protection Agency, the U.S. National Oceanographic and Atmospheric Agency (both directly and through SeaGrant), and the Natural Sciences and Engineering Research Council of Canada. We thank the Banff International Research Station for providing us a retreat where we edited the book.

Contents

Contributors

Caroline J. Bampfylde
Alberta Environment
Alberta Government
Edmonton, Alberta TG5 1G4 Canada

Angela M. Bobeldyk
Department of Biological Sciences
University of Notre Dame
Notre Dame, IN 46556 USA

Jonathan M. Bossenbroek
Department of Environmental Sciences and the Lake Erie Center
University of Toledo
Toledo, OH 43606 USA

John M. Drake
Odum School of Ecology
University of Georgia
Athens, GA 30602 USA

Kevin J. Egan
Department of Economics
University of Toledo
Toledo, OH 43606 USA

David C. Finnoff
Department of Economics and Finance
University of Wyoming
1000 E. University Avenue
Laramie, WY 82071 USA

Leif-Matthias Herborg
BC Ministry of the Environment
Victoria, British Columbia Canada

Christopher L. Jerde
Center for Aquatic Conservation
Department of Biological Sciences
University of Notre Dame
Notre Dame, IN 46556 USA

Reuben P. Keller
Center for Aquatic Conservation
Department of Biological Sciences
University of Notre Dame
Notre Dame, IN 46556 USA

David W. Kelly
Landcare Research
764 Cumberland Street
Private Bag 1930
Dunedin 9054
New Zealand

Martin Krkošek
Department of Mathematical and Statistical Sciences
CAB 545B
University of Alberta
Edmonton, Alberta T6G 2G1
Canada

Gary A. Lamberti
Department of Biological Sciences
University of Notre Dame
Notre Dame, IN 46556 USA

Mark A. Lewis
Department of Mathematical and Statistical Sciences
CAB 545B
University of Alberta
Edmonton, Alberta T6G 2G1
Canada

David M. Lodge
Center for Aquatic Conservation
Department of Biological Sciences
University of Notre Dame
Notre Dame, IN 46556 USA

Hugh J. MacIsaac
Great Lakes Institute for Environmental Research
University of Windsor
Windsor, Ontario N9B 3P4 Canada

Christopher R. McIntosh
Department of Economics
University of Minnesota
Duluth, MN 55812 USA

Jim R. Muirhead
Department of Mathematical and Statistical Sciences
University of Alberta
Edmonton, Alberta T6G 2G1 Canada

Jody A. Peters
Center for Aquatic Conservation
Department of Biological Sciences
University of Notre Dame
Notre Dame, IN 46556 USA

Alexei B. Potapov
Department of Mathematical and Statistical Sciences
University of Alberta
Edmonton, AB, T6G 2G1 Canada

John D. Rothlisberger
Center for Aquatic Conservation
Department of Biological Sciences
University of Notre Dame
Notre Dame, IN 46556 USA

Chad Settle
Department of Economics
University of Tulsa
800 Tucker Drive
Tulsa, OK 74104 USA

Jason F. Shogren
Department of Economics and Finance
University of Wyoming
1000 E. University Avenue
Laramie, WY 82071 USA

John Tschirhart
Department of Economics and Finance
University of Wyoming
1000 E. University Avenue
Laramie, WY 82071 USA

Travis W. Warziniack
Department of Economics and Finance
University of Wyoming
1000 E. University Avenue
Laramie, WY 82071 USA

Bioeconomics of Invasive Species

1

Introduction to Biological Invasions: Biological, Economic, and Social Perspectives

David M. Lodge, Mark A. Lewis, Jason F. Shogren, and Reuben P. Keller

In a Clamshell

Invasive species are now recognized worldwide as a serious side effect of international trade. They often spread irreversibly, and damages increase over time. To reduce such damages, private and public investments are increasing in an effort to prevent the arrival of species or eradicate them early in an invasion, control their local abundance once they have become established, or slow their spread. Most often, however, the damages of invasive species are accepted as a new cost of doing business, and humans change their behavior to minimize the impact. In this chapter, we argue that integrating ecological and economic analyses is essential to guide policy development in support of more cost-effective management. A key goal is to describe quantitatively the feedbacks between economic and ecological systems and to provide answers to such questions as how many dollars should be invested in prevention versus control, and what benefits are derived from such investments. This chapter describes the impacts of some high-profile invasive species, explains the extent to which ecological and economic systems are integrated, and looks to epidemiology for a model of how research and management could be better integrated to inform policy.

In the last two decades, experts and the public have recognized two important things about many anthropogenic environmental changes: first, these changes are increasingly global in scope, and second, they are hard to reverse. These characteristics apply with special force to harmful nonindigenous species, which we refer to as "invasive species" throughout this book. Both the global scope and the difficulty of reversing invasions impart considerable urgency to increasing our understanding of this problem. Invading organisms reproduce and spread, even if we cease introducing more individuals. The problem of harmful invasive species gets worse without management.

Research to better understand invasions comes naturally to scientists and social scientists, especially to those of us in universities. We also, however, believe it is urgent to focus our research on questions important to natural resource managers and policy makers, given society's explicit desire to reduce the current and future damages caused by invasive species. We want our research and its implementation to increase social welfare. Using the perspectives and tools of economists is appropriate because invasive species are, by definition, driven by human activities, usually commercial enterprises. Solutions will derive from changes in industry practices and consumer behavior.

Humans are as much the target of our study as the species that humans move around the globe. If research is to inform natural resource management and policy, it must be conducted collaboratively by natural and social scientists, and in the context of possible management and policy responses to invasive species. We elaborate on these general points after considering some specific examples of invasive species, their environmental and economic costs, and societal responses to them.

CAULERPA: SUCCESSFUL ERADICATION

Aquarium keepers, like owners of all sorts of plants and animals, sometimes tire of the organisms under their care and release them. In 2000, populations of the invasive seaweed *Caulerpa taxifolia* were discovered in two Southern California coastal embayments. This species, including a very invasive strain, has been sold widely in aquarium shops because it is fast growing, hardy, and beautiful (Walters et al. 2006). Some of these same characteristics have caused a well-documented history of harmful invasions. In various invaded marine ecosystems, including the Mediterranean Sea, commercial and recreational fishing, recreational activities like scuba diving, and tourism have all suffered (Meinesz 1999). When the species was discovered in California, a consortium of private and government agencies launched a concerted eradication effort using chlorine applications under anchored tarps. The effort cost at least $3.7 million over 5 years (Woodfield and Merkel 2005), and it was successful.

Without policy responses to prevent additional *Caulerpa* introductions, however, the need for many similarly expensive management situations would probably occur in the future as other naive aquarium owners dispose of unwanted plants (Walters et al. 2006). The U.S. Department of Agriculture (USDA) used its authority under the Plant Protection Act of 2000 to declare the Mediterranean aquarium strain of *C. taxifolia* a federal noxious weed. Such a designation gives the USDA authority to prohibit importation, exportation, or movement of the species in interstate commerce. In 2001, the state of California went a step further and made it illegal to possess *C. taxifolia* and nine other *Caulerpa* species. Nevertheless, various species and strains of *Caulerpa* remain easy to purchase in all states (Walters et al. 2006). The story of *Caulerpa* eradication near San Diego, then, is a success story. It is an example of successful implementation of a strategy referred to as "early detection, rapid response, and eradication," supported by additional efforts (of minimal success thus far) to prohibit future introductions.

SEA LAMPREY: SUCCESSFUL CONTROL

Across the continent and about a century earlier, the construction of the Welland Canal by-passed Niagara Falls and allowed sea lamprey (*Petromyzon marina*), along with ships and barges, access to the upper Great Lakes. Despite the fact that most sea lamprey previously lived their adult lives in the Atlantic Ocean, large and self-sustaining populations soon thrived in the upper lakes. While the increased navigation fostered commercial activities that were beneficial to humans, the invasion by sea lamprey was not. Adult sea lamprey are parasitic on other fish species, using their rasping and suckerlike mouth to feast on the blood of commercially valuable species such as lake trout (*Salvelinus namaycush*) and whitefish (*Coregonus* spp.). The result was declining fisheries and a public outcry.

Fortunately, larval sea lamprey are confined to the tributaries of the Great Lakes, where they reside for about 7 years before assuming their adult bloodsucking habits. The larvae are easy to locate and are highly susceptible to TFM (3-trifluoromethyl-4-nitrophenol), a chemical discovered in 1955. When applied at appropriate concentrations in tributaries, TFM kills sea lamprey larvae with acceptably low effects on other species. Since 1956 the United States and Canada have together spent about $15 million annually on monitoring and poisoning sea lamprey. Sea lamprey populations plummeted, and harm to the fisheries is kept tolerably low with these continuous expenditures. The management efforts directed at sea lamprey constitute a remarkably successful "control" effort, the ongoing expense of which is justified by even larger benefits in the protection of Great Lakes fisheries.

GYPSY MOTH: SUCCESSFULLY SLOWING THE SPREAD

In 1869, gypsy moth (*Lymantria dispar*), which had been imported from its native range in Europe, escaped an unsuccessful attempt at silk production in Massachusetts. Thus began an invasion of North America that is ongoing today. Gypsy moth infestations can completely defoliate vast forests of oak and other trees and can achieve such abundance that their excrement and bodies are sometimes a serious nuisance in urban areas. Outbreaks of gypsy moths are often controlled with an aggressive integrated pest management program. In areas where the gypsy moth is now a permanent resident, expenditures to keep their populations acceptably low are very high when the periodic population outbreaks are treated with pesticides. As for sea lamprey, the best that can be hoped for in these areas is successful control, not eradication. Therefore, for every acre that becomes infested as the invasion progresses, future control costs will be high (perhaps forever) if pesticide treatments are chosen. Otherwise, humans must simply adapt (*sensu* economics, not evolution) to the periodic damage to urban and natural forests.

Because of the damage and/or control costs once gypsy moths become established, the USDA and states from Wisconsin south to North Carolina spend about $12 million annually to slow the southwestward march of gypsy moths across the country. A combination of trapping, aerial spraying of insecticides, and mating-disrupting pheromones has slowed by 50% the advance of the invasion front, from about 13 miles per year to about 6 miles per year (Sharov et al. 2002). Although this effort is expensive, it is cost-effective because damages are avoided, at least for a year, in the area in advance of the invasion front—an area of roughly 9,000 square miles (1,500 miles × 6 miles). The avoided damages are much higher than the costs of the slow-the-spread program (Sharov 2004). Preventing long-distance, especially human-mediated, dispersal ahead of the advancing invasion front remains a challenge for this program, but overall the scientific and management responses to the gypsy moth are a successful example of a slow-the-spread strategy.

MOST OTHER INVASIVE SPECIES: UNCONTROLLED DAMAGES AND UNCHECKED SPREAD

Stories that end in at least some level of success—eradication of *Caulerpa*, control of sea lamprey, slowing the spread of the gypsy moth—are rare and unfortunately are vastly outnumbered by harmful invasions that proceed apace to a grim and often irreversible outcome. Some of the most visible, dramatic, and widespread examples come from forests.

In the United States, a combination of nonindigenous insects, fungi, and other parasites and pathogens have essentially extirpated American chestnut (*Castanea dentata*) and American elm (*Ulmus americana*), previously two of the dominant trees in eastern natural and urban forests, respectively (Burnham 1988; Gilbert 2002). Many other beloved and valuable species seem likely to face a similar demise from ongoing invasions: flowering dogwood (*Cornus florida*), destroyed by the anthracnose pathogen, has declined in abundance by more than 90% in some forest types over the last two decades (Holzmueller et al. 2006); American beech (*Fagus grandifolia*) is succumbing to beech bark blister; Eastern hemlock (*Tsuga canadensis*) is declining as the hemlock wooly adelgid spreads across the East and Midwest; butternut (*Juglans cinerea*) invariably dies after infection by butternut canker, which is common and spreading in the Northeast and Midwest (Ostry and Woeste 2004); mortality of ashes (*Fraxinus* spp.) hovers near 100% as the emerald ash borer advances across the Midwest (BenDor et al. 2006); and several species of oak (*Quercus* spp.) are vulnerable to sudden oak death, the spread of which has only recently begun but has already jumped from the West Coast to the East Coast in the nursery trade (Gilbert 2002). All the responsible pests and pathogens are nonindigenous, with many arriving in the United States as hitchhikers in shipments of plants, wood products, or wood packing material.

It is not just accidentally introduced pests and pathogens that damage forestry production and damage natural and urban forests. Deliberately introduced plants, such as the kudzu vine (*Pueraria lobata*), are also outcompeting native vegetation for light, nutrients, and space. And, like the gypsy moth, they can seem like a good thing at first. The American public first saw the fast-growing, attractively purple-flowered kudzu vine from Japan at the 1876 Centennial Exposition in Philadelphia (Forseth and Innis 2004). For decades thereafter, particularly in the southeastern United States, it served well as an ornamental plant that also provided summer shade under overgrown porches. Later, especially during the first half of the twentieth century, as justifiable concerns grew about the severe soil erosion and nutrient depletion that accompanied intensive cotton agriculture, the U.S. government distributed 85 million seedlings, paying southern farmers to plant them (Forseth and Innes 2004). As for so many introduced species, only later did the downsides to kudzu become apparent, especially as other economic forces caused the decline of row cropping and livestock operations that had included management of kudzu. Millions of kudzu plants began to escape control altogether (Forseth and Innes 2004).

By mid-century, the costs of kudzu had become painfully obvious. Kudzu now occurs from Texas to Florida and north to New York, covering over 3 million hectares, which increases by about 50,000 hectares per year (Forseth and Innes 2004). Forest productivity losses are between $100 million and $500 million per year, power companies spend about $1.5 million annually to control kudzu, and a 6-year effort was required to eradicate kudzu from the Chickamauga and Chattanooga National Military Park. The best that can be hoped for is locally successful eradication efforts, whose long-term success depends on continued monitoring and control, as the species continues to expand its geographic range from the southeastern United States. Unfortunately, the list of deliberately introduced plants like kudzu that have become very harmful to agriculture, livestock, forestry, and natural ecosystems is long, including hundreds of species. It also continues to grow.

In addition to lost productivity and increased expenditures for control efforts in human-managed landscapes, the result of these invasive species is an ongoing shift in the composition of forests that is similar in magnitude to that of a nationwide forest fire, only slower. Large negative consequences exist for industries involving horticulture, landscaping, wood products, recreation, and tourism, as well as for natural ecosystems. Forest ecosystems provide the most obvious examples of damaging, unreversed invasions, but the same patterns characterize other terrestrial, marine, and freshwater ecosystems.

Zebra mussels (*Dreissena polymorpha*) and quagga mussels (*D. bugensis* (= *D. rostriformis bugensis* [Andrusov (1897)])) are the best-documented examples of similar phenomena in freshwater ecosystems in North America. Both are small striped bivalve mollusks. Zebra mussel was discovered in Lake St. Clair, between lakes Erie and Huron, in the mid-1980s, with quagga mussels following within a few years. These mussels were released when ships discharged ballast water that had been taken up in a port in northern Europe, where zebra and quagga mussels had previously invaded from their native ranges around the Black Sea. With those ballast water releases, Lake St. Clair, and quickly other Great Lakes, became the beachhead for ongoing invasions of freshwater ecosystems of North America. From the Great Lakes, two major human-driven vectors of dispersal allowed zebra and quagga mussels to spread. First, the Chicago Sanitary and Ship Canal provided a ready conduit for the mussels to escape Lake Michigan (crossing a former watershed divide) and colonize the Illinois and Mississippi rivers downstream. From the Mississippi River proper, the mussels, especially zebra mussel so far, hitched rides upstream on barges to colonize tributaries, including the Ohio, Tennessee, and Missouri rivers.

Second, recreational boaters, who often visit multiple rivers and lakes, inadvertently carried mussels overland on their boat trailers and boats to inland lakes that are not connected by water to initial sites of infestation. Within a decade, zebra mussel colonized much of the Great Lakes–St. Lawrence River and Mississippi River drainage basins. In 2007 and 2008, colonization of the West Coast by quagga and zebra mussels, respectively, began. Quagga mussel was discovered in Lake Mead, the Colorado River, and the California Aqueduct (Stokstad 2007), while zebra mussel was discovered in a California reservoir. Much suitable habitat for zebra and quagga mussels remains to be colonized east of the Appalachians and in the West, including the Columbia and Sacramento-San Joaquin rivers (Drake and Bossenbroek 2004). While the probability of transport of live mussels to those regions from the Midwest is lower than to waterways in the Midwest, mussels are being transported, and without increased slow-the-spread efforts, these regions almost surely will be colonized and suffer damages in the future (Bossenbroek et al. 2007), especially with new sources of invasion in the western waterways.

Efforts to slow the spread of mussels are occurring at regional, state, and federal levels, but their efficacy is poorly documented, and they are almost certainly underfunded (Leung et al. 2002; Lodge et al. 2006). Additional investments in such efforts are warranted because the damages caused by zebra mussels are large, including at least $150 million annually in the Great Lakes region by clogging up water intake pipes in power plants, municipal water supplies, and industrial facilities that withdraw raw surface water (O'Neill 1996). In addition, sharp zebra mussel shells foul beaches, hinder recreation, extirpate native clam species, increase harmful algal blooms, and likely contribute to botulism outbreaks that devastate migrating waterfowl and fishes in the Great Lakes region (Yule et al. 2006). Zebra mussels are successfully (if expensively) controlled inside industrial facilities, and have been eradicated from one quarry lake in Virginia, but no technique exists to reduce the population of zebra mussel in an entire lake or waterway without killing many other organisms.

The zebra mussel invasion, like those described above for terrestrial ecosystems, will continue, more slowly perhaps if a more effective slow-the-spread campaign is implemented, but humans in North America are stuck with zebra and quagga mussels. Forevermore in North America, they will be abundant, and native clams and many other native species will be less abundant, some perhaps extinct (Strayer and Malcom 2007). The changes in our behavior to cope with these changes, and the expenditures necessary to control them in power plants, will likely grow over time until zebra and quagga mussels occupy all suitable waterways in North America. And many other invasive species already in the Great Lakes are following the mussels across the country.

The invasive species vignettes above bring up a very important question: is prevention a management option? Though prevention is little practiced in North America, the answer is yes, of course, prevention is possible. Slow-the-spread programs show that, on a regional scale, prevention is possible even if only temporary. Prevention is also possible at the continent's borders. Anyone who has returned to North America from a trip abroad knows not to try to bring any fresh fruit, or the insects or pathogens that it might harbor, into the country. And some rare rigorous inspection programs at borders show how much potential damage could be avoided with rigorous screening and interdiction programs. For example, comprehensive inspections of air cargo at Kahului Airport, Hawaii, during 20 weeks in 2000–2001 revealed 279 insect species, 125 of which were not known from Hawaii, and 47 plant pathogen species, 16 of which were not known from Hawaii (Hawaii Department of Agriculture 2002). Most of the time at this and other airports in North America, however, inspections are far fewer. Such organisms ordinarily go undetected and are released into the environment. Some will cause great harm.

Prevention is possible, then, but it is reasonable to wonder how much prevention would cost, and whether it would be cheaper than the damages that occur in the absence of prevention. The vignettes above illustrate how costly invasions can be, either through damages suffered or the expenditures to support eradication or control efforts, but would prevention be equally costly? These sorts of questions motivate much of this book. Despite the slowness of these and many other unfolding invasion disasters, they should be regarded with urgency because the costs are high, grow over time as the populations of harmful species spread, and are too often irreversible. Are we simply stuck with such costs, or are prevention and more aggressive control approaches viable alternatives? In this book, we focus on freshwater examples to illustrate the causes, consequences, and potential management responses to invasive species. We combine ecological modeling with economic modeling to answer questions about management and policy.

HUMAN VALUES

Human values determine both which environmental changes we call damages and what investments in management responses seem appropriate. The positive and negative values that humans assign to species or other characteristics of ecosystems are appropriately informed by various financial, scientific, religious, and ethical considerations, but inescapably it is humans that do the valuing and responding (Hamlin and Lodge 2006). Invasions occurred before humans appeared, but the rate at which global commerce now causes them is orders of magnitude higher than natural background rates (Lodge and Shrader-Frechette 2003). More and increasingly international transportation of goods causes invasions, and human behavior will either continue to increase invasions or rein them in. The combination of natural and social science represented in this book is essential to both diagnose invasions and respond to them.

Invasion Process and Feedbacks between Biological and Economic Systems

Following the vignettes above, we could continue to illustrate the issue of invasive species with thousands of additional examples, replete with idiosyncratic biological details. Such catalogs of examples, however, can obscure the processes that are common to all invasions (figure 1.1, left column). Understanding the processes, in turn, is essential to prescribing appropriate management responses (figure 1.1, right column).

Species are carried in a vector, which transports the species either overtly (e.g., the pet and horticultural trades) or incidentally (e.g., insect pests in lumber shipments, ballast water of ships, viruses carried by humans themselves) (figure 1.2). Depending on the traits of the species, and the conditions and the duration in the vector, some proportion of the organisms may be alive when they are released or escape at a location outside their native range.

Depending on the taxonomic group of organisms, many to most species subsequently go extinct in a new location, but a proportion—on the order of 5% for plants (Keller et al. 2007) and up to 50% for animals (Jeschke and Strayer 2005)—establish a self-sustaining population. While some of these established species remain localized, perhaps not even detected by humans, a proportion, again about 5–50%, spread

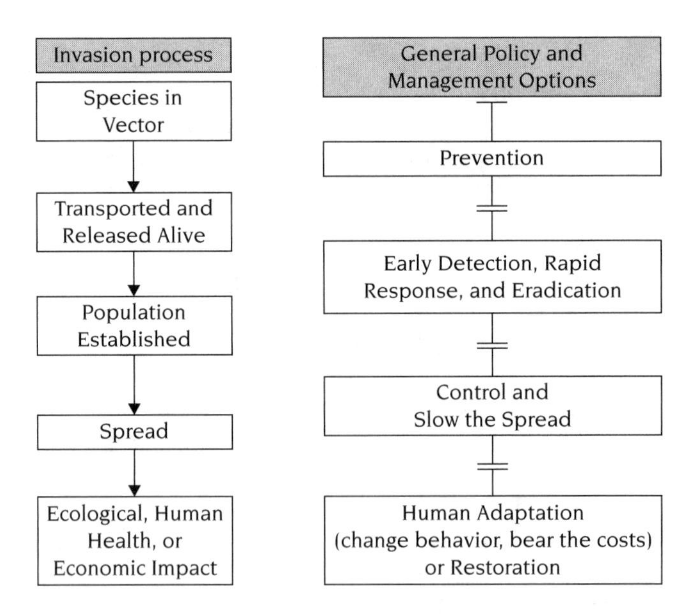

FIGURE 1.1.
The stages of biological invasion (left column) and the management and policy options available to society (right column) at each stage of invasion. The desire to reduce the negative impacts of species (bottom left) motivates the study of biological invasions. Reprinted from Lodge et al. (2006), with permission of the Ecological Society of America.

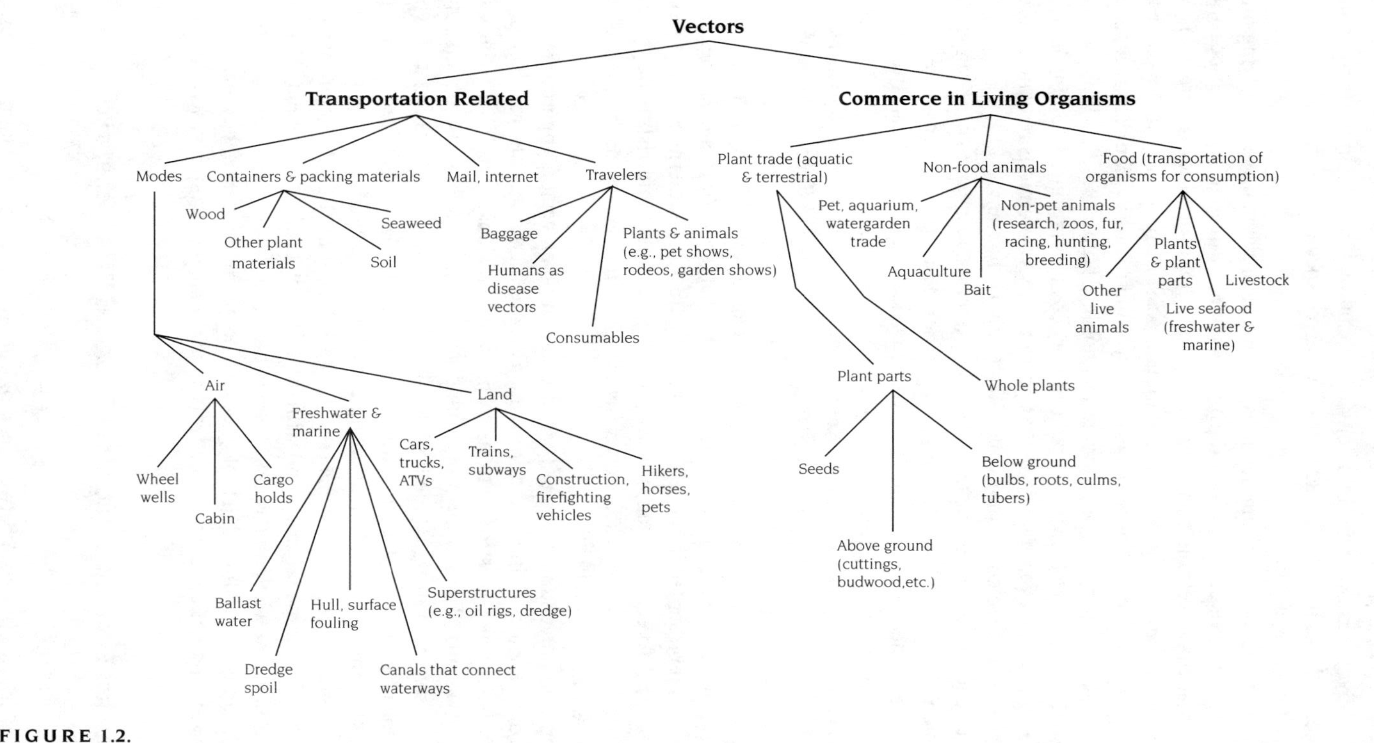

FIGURE 1.2.
Vectors by which nonindigenous species enter the United States and are transported within the United States. Reprinted from Lodge et al. (2006), with permission of the Ecological Society of America.

widely and become abundant at many new locations. Such species—roughly 0.3% of introduced plants and up to 25% of introduced animals (as calculated from the numbers above)—cause undesirable environmental and/or economic changes and are categorized as invasive. By definition, invasive species, which are a subset of nonindigenous species, are bad.

Policy and management implications become clear when these underlying processes and probabilistic transitions during invasion are recognized. The possible human management responses narrow as any invasion progresses (figure 1.1). As illustrated by the above vignettes, prevention is possible only early in the process, before a species arrives in a new range or at the point of entry. Eradication depends on the rapid convergence of appropriate technology, political will, and resources. Once a species is well established, eradication is costly and sometimes impossible. When the opportunity for eradication has passed, only two options remain: control of populations in selected locations, and adaptation by humans.

In most countries, including those in North America, adaptation has been vastly more typical than any other response, except when pests or pathogens have threatened either humans directly or highly valuable agricultural crops. Apart from these exceptions, we passively suffer the consequences of invasions. In the last decade, however, investments in eradication, control, and finally prevention have increased for natural ecosystems, and policy discussions in the United States and elsewhere increasingly feature prevention efforts.

In this book, we assess current scientific capability to forecast the identity, spread, and impact of potential invasive species. In chapters 3–6 we address the series of transitions represented in the left column of figure 1.1. Furthermore, we explore how ecological forecasting can be used in risk assessment and risk management of invasive species, testing especially whether cost-effective approaches, including prevention, can be identified.

Interest in prevention necessarily focuses attention on vectors (figure 1.2). Vectors are commercial activities driven by human desires for the benefits from increased trade. In the absence of strong efforts to prevent invasions, increasing trade will increase invasions. The numbers of nonindigenous plant pathogens, insects, and mollusks discovered in the United States since 1920 are strongly correlated with importation of goods over the same time period (figure 1.3). Trade with many countries is increasing (figure 1.4), and documented invasions are increasing in marine, terrestrial, and aquatic ecosystems (Ricciardi 2006; figure 1.5). Different vectors operate at different spatial scales and with different potential management interventions. Detailed knowledge of vectors, as well as of different taxonomic groups of organisms, must be combined in biological and economic models if they are to guide management and policy to cost-effectively reduce damages from invasive species.

Feedbacks: Economic Activity, Biological Processes, and Damages from Invasion

A circle of feedbacks exists between ecological processes and economic processes (figure 1.6): the economic benefits of trade drive invasions, invasions cause negative

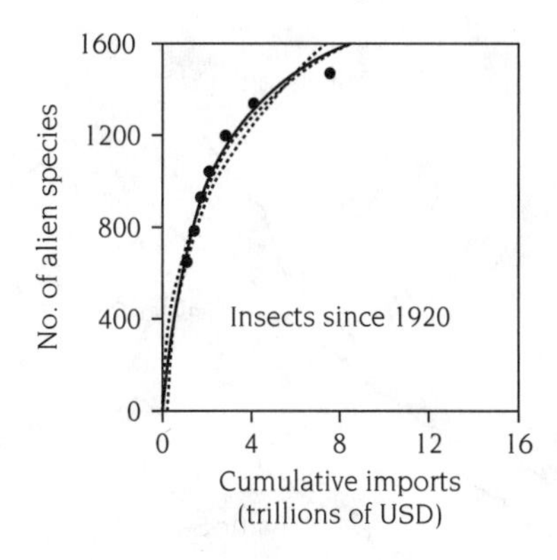

FIGURE 1.3.
Total imports into the United States since 1920 (measured in dollars) as a potential driver of cumulative invasions by terrestrial insects in the United States since 1920. Modified from Levine and D'Antonio (2003).

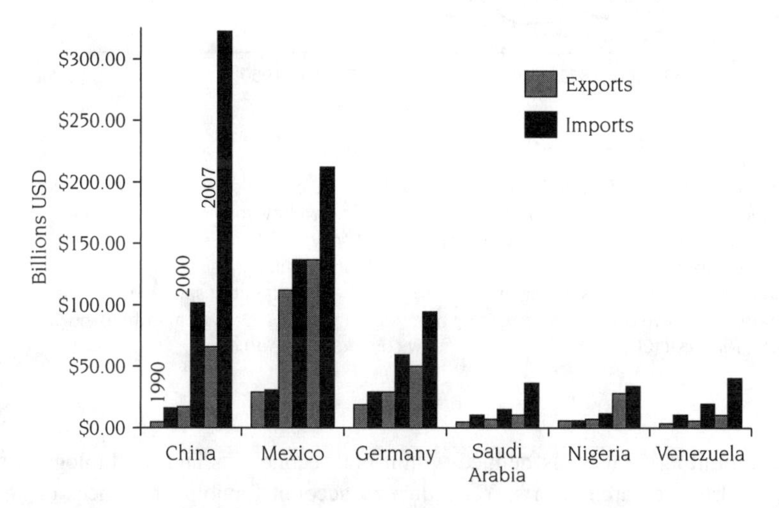

FIGURE 1.4.
Changes in total trade volume between selected countries and the United States, 1900–2007. First bars for each country are imports/exports during 1990; subsequent pairs of bars are for 2000 and 2007, respectively. Data from the U.S. Department of Commerce (2008).

economic and environmental impacts, and human perception of those impacts feeds back as management or policy initiatives to reduce trade or at least reduce the negative side effects of trade. Another way to look at this situation is as an adaptive loop, among risk assessment, risk perception, and risk management, that changes the risks to be assessed. A distinctive strength of this book lies in applying a combination of ecology and economics, with strong mathematical and statistical foundations, to management and policy questions.

Economics and the biological sciences have many similarities. Both are disciplines of limits—both examine how species deal with scarcity. Whether it is a human's reaction to a limited budget and unlimited wants or a squirrel's response to limited food and unlimited appetite for reproduction, all species deal with limits.

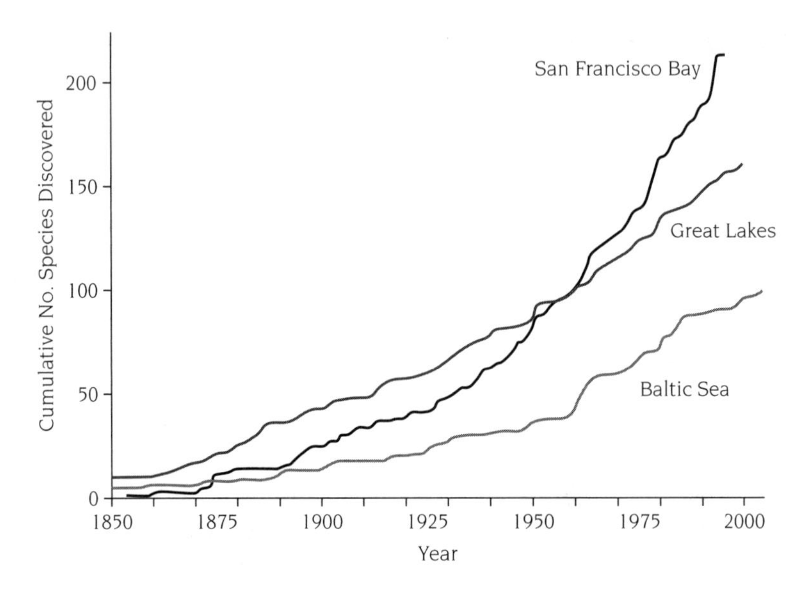

FIGURE 1.5.
Cumulative number of nonindigenous species that have been discovered in three major aquatic ecosystems in the last 150+ years. It is not known how many species remain undiscovered in each ecosystem, or how long the discovered species had been present before they were discovered (Costello et al. 2007). Nevertheless, the data suggest strongly that trade and/or other mechanisms by which humans cause the movement of species (e.g., canal construction) have caused an increasing number of invasions. Data from Cohen and Carlton (1998), Ricciardi (2001), and http://http://www.corpi.ku.lt/nemo/).

These limiting factors, as defined within both economics and the biological sciences, drive research efforts. Yet failure to account for joint influences on these limits in economic systems and biological systems can cause inaccurate perceptions of how each system works and provide misleading policy guidance. The idea of joint determination applies: links between the biological and economic systems create a progression of natural and human actions and reactions, in which a feedback loop emerges. Disturbances in one system set off repercussions in the other system, and these repercussions feed back into the system where the disturbances originated (e.g., Daly 1968; Clark 1990; Crocker and Tschirhart 1992; Sohngen and Mendelsohn 1998; Wilson 1998; Shogren and Crocker 1999; Dasgupta et al. 2000; Finnoff and Tschirhart 2003).

The impact of invasive species is a good example of joint determination. Thresholds for expansion of invasive species are functions of the present distributions and trends of their populations, their interactions with habitats, and the economic circumstances that cause introductions of additional individuals and the quality of potential habitat (e.g., fragmentation). Important economic circumstances include the relative prices of alternative sites for economic development and relative wealth of the landholders in the area. Sites with low relative returns in their "highest and best" use are

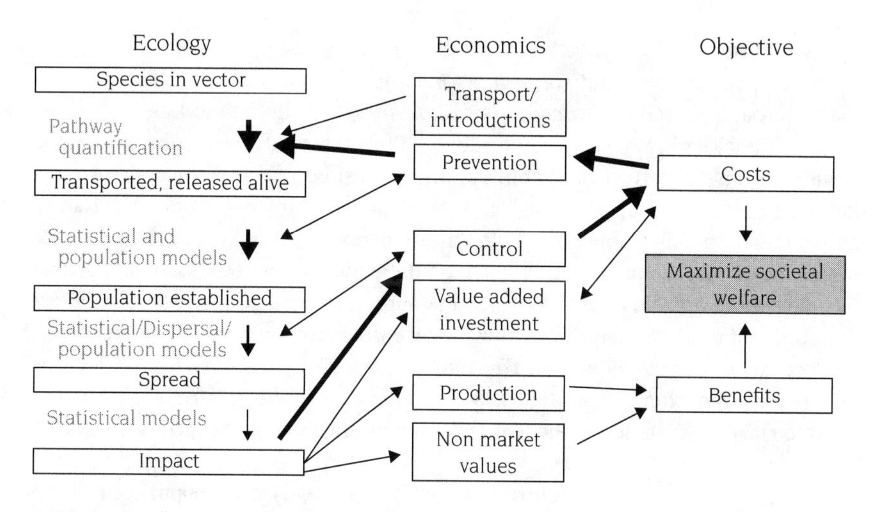

FIGURE 1.6.
Feedback between ecological processes (left column) and economic processes. The lightface text indicates the variety of tools, many recently developed or applied, that we use to model and forecast different stages of invasion. The bold arrows indicate possible feedback pathways in which damages from an invasion cause humans to change investments to reduce future damages: the impact of a species is expressed in increased control costs; in response, humans increase prevention expenditures that reduce the number of organisms entrained in the responsible vector. Modified from Leung et al. (2002).

more likely to be left undisturbed. Moreover, the rich can better afford to set aside undisturbed habitat that may be less susceptible to invasions.

These interactions demonstrate that invasive species establishment and spread are determined by both economic and biological parameters. Effective models of the spread and impact of invasive species require natural and social scientists to integrate their respective tools and their indicators of success and failure. Integration across disciplinary boundaries is especially crucial when a proposed policy may trigger a political feud fueled by misperceptions of benefits and costs imposed on natural and social systems. The resulting challenge is to integrate models, methods, and mind-sets to help researchers and decision makers better understand and manage the delicate balance between private rights of self-determination and social rights to environmental protection.

The most straightforward and pragmatic method is to form a research team that includes both economists and ecologists to construct an explicit model to estimate the trade-offs associated with alternative policy options. Models are always abstractions and must never be mistaken for reality. Nevertheless, the integrative thought process of model construction focuses attention on the most important links between systems. The differences and similarities between economics and ecology can be addressed directly by forcing researchers to construct and link the human and natural sectors of the model. A linked model can then provide informed guidance for pragmatic choices among the trade-offs necessarily involved in policy making.

We illustrate this approach using a model that captures the risks posed by one invasive species, lake trout (*Salvelinus namaycush*), on one endangered species, the cutthroat trout (*Oncorhynchus clarki bouvieri*) in Yellowstone Lake in Yellowstone National Park, Wyoming. Settle et al. (2002) explored how feedbacks between humans and nature affect the likelihood of the desired result—an increased population of cutthroat trout, because many more anglers prefer to catch cutthroat. In a dynamic modeling framework, Settle et al. incorporated both economic and ecological flows and reciprocal flows between the two systems. To test the importance of the economic-ecological feedbacks, the authors compared the modeling results with and without the reciprocal flows between the two systems. They considered two scenarios: (1) a *remove-all-lake-trout* scenario, in which lake trout are immediately removed from Yellowstone Lake; and (2) a *leave-the-lake-trout-be* scenario, in which lake trout are left alone to reach a steady state within the Yellowstone Lake ecosystem.

Under the remove-all-lake-trout scenario, the steady-state population of cutthroat trout is about 2.7 million and 3.4 million, without and with feedbacks. Without feedback between the economic and ecological systems, park visitors continue to fish as before, putting constant pressure on the cutthroat. With feedback, visitors react to declining cutthroat populations by fishing less and visiting other attractions more. This behavioral reaction by park visitors, which reflects an increase in what economists call the shadow price of fishing, now affects the ecosystem because a decline in fishing time produces an increase in the population of cutthroat. Incorporating feedbacks between the economic and ecological system produced estimates of a 26% larger population of cutthroat, the desired species.

Under the leave-the-lake-trout-be scenario, Settle et al. (2002) found a different result. Now a no-feedback model (fishing continued as before) suggested a more desirable outcome than would be likely to occur—almost 1 million cutthroat trout remain versus zero cutthroat trout when feedbacks were included. Without feedbacks, visitors continued to fish and acted as a control on the population of lake trout, even though it is an incidental catch. With feedbacks, visitors shifted away from fishing as the cutthroat trout population declined and the lake trout population increased, leaving the lake trout to take over as cutthroat were extirpated. Without incorporating feedbacks, policy advice might have led park officials to adopt the cheaper leave-the-lake-trout-be policy, satisfied that at least the cutthroat would continue to exist in Yellowstone Lake. According to the model by Settle et al. (2002), such a policy would likely have resulted in the disappearance of cutthroat. The National Park Service currently uses a policy of gill netting lake trout. (See chapter 2 for additional discussion of this example.)

This example illustrates how integrating the feedbacks between economics and ecology can be essential to provide appropriate advice for management and policy for invasive species. Technical integrated models can be a powerful tool to make the linkages among disciplines transparent and workable. Failure to account for the specific links between ecosystems and economic systems might lead to inappropriate management of either the ecosystem or the economic system. Integration of economics and ecology is fundamental both for science and policy. For science, integration implies more accurate estimates of both economic and ecological phenomena. For policy, integration means a better appreciation of the alternative viewpoints that arise when attempting to address a difficult challenge like invasive species management. Societal responses to infectious disease, including research and the way it informs disease management and policy, provide a useful analogy through which to approach the similar intellectual challenges posed by invasive species.

LESSONS FROM EPIDEMIOLOGY

With the spread of such pathogens as SARS and West Nile virus into new continents, such as North America, distinctions between disease and invasive species become blurred; indeed, some diseases are caused by nonindigenous pathogens and parasites. Perhaps due to clear human impacts and well-publicized public health costs, great investments have been made into bioeconomic research, policy, and management of human infectious disease (Roberts 2006). Such responses to infectious disease provide useful parallels for bioeconomic analysis, management, and policy of ecological invasions, which remain in their infancy.

One essential quantity for characterizing dynamics of an infectious disease is the basic reproduction number, the number of secondary infections arising from direct contact with a single infective organism that is introduced into an otherwise susceptible population (Diekmann et al. 1990). This single statistic has proved a convenient metric for assessing methods of disease control. For example, Wonham et al. (2004) estimated that mosquito control that reduced mosquito populations to 30–60% of endemic levels would have prevented the 1999 outbreak of West Nile virus in New York, an outbreak that eventually lead to the spread of this disease across North America. In the context of a biological population, the basic reproduction number is the number of surviving offspring produced during the lifetime of a single individual (Caswell 2001). Although widely applicable to biological invasions, the actual application of this simple statistic to the control of invading populations remains in its infancy (but see de Camino Beck and Lewis 2007).

Infectious diseases may establish in one city and then jump to another, much the same way as aquatic invasive species can spread from one lake to another. One class of models, successful in predicting these jumps in disease contagion, borrows from physics and transportation theory. Here, modifications of the empirical gravity law are used to define the level of attraction of contagion among cities in a network. Cities are like planets—attractiveness is positively related to city size and negatively related to distance between cities. Sets of rules, based on this principle, have been fitted to observed infection data for diseases such as measles (Xia et al. 2004). When incorporated in a network model, the rules can then be used to track or predict spatial spread of infectious disease among cities. As we show in chapter 7, these so-called gravity models have also been used successfully in modeling the spread of invasive species in networks of lakes.

Investment in the modeling and analysis of infectious disease control measures has extended to the realm of livestock and agriculture (Morris 1999), particularly in cases where the diseases can have devastating market impacts. Modeling of the spread of foot and mouth disease in 2001 in the United Kingdom guided the use of different control measures, including culling, prophylactic vaccination, and vaccination strategies that target key spatial transmission foci (Keeling et al. 2003). For this disease, focal units are the individual livestock farms housing infected cattle. Unfortunately, there is no simple nondestructive action analogous to prophylactic vaccination for the control of noninfectious invasive species. Such an action amounts to wholesale manipulation of the biotic resources available to the invader (analogous to decreasing the density of susceptible individuals). For example, tree thinning has been used as a management strategy to stem the spread of the invasive mountain pine beetle into new areas of pine forest (Steeger and Smith 1999). However, for most invasions, such control methods are considered a method of last resort because their costs, both economic and ecological, are so high.

Economic costs of human disease are an area of active current research (Roberts 2006), and an increasing motivation for public health efforts. The economic impact of animal infectious diseases can also be high and a strong motivator for improved management and policy. For example, botulinum infection of Canadian salmon in the 1980s devastated the salmon fish industry. Livestock diseases such as bovine spongiform encephalopathy ("mad cow disease"), found to be capable of crossing species barriers, and foot and mouth, which is capable of very rapid spread, have played havoc with the British beef industry. Methods of economic impact assessment are well developed at the level of the production unit (herd or farm) (Rushton et al. 1999) but are more elusive at national and international levels (Riviere-Cinnamond 2006). For humans, public health costs of infectious disease are typically measured by cost-of-illness studies, which calculate the implications of illness on the use of resources. Many economists prefer to measure disease impacts through surveying the population's willingness to pay for treatment or prevention services (Mangtani and Shah 2006). This easily translates benefits of treatment or prevention into monetary terms. Analogous methods, outlined in chapter 8, are also employed in the study of the economics of invasive species.

As we emphasized in the preceding section, the coupling between biological dynamics and economics is a two-way street: economic conditions can also affect infectious disease outbreaks. Immune status of a person is affected by living conditions, by the quality and quantity of food consumed, and by access to clean water (World Health Organization 2002). Furthermore, trade activity can spread disease from one place to the next (Narasimham 2006). Evaluation of economics of infectious disease can require such two-way coupling (Roberts 2006). In this book, we demonstrate the necessity of a similar two-way coupling between invasion dynamics and economics.

As economists and ecologists, we also learn from the methods of economic analysis applied to disease. The most common is based on cost-benefit analysis. While many cost-benefit analyses employ a static approach, dynamic analyses have been applied to subjects such as HIV intervention policy (Kumaranayake 2006). Even over 5- or 10-year spans, the abundance of an invasive species can increase by orders of magnitude. This means that dynamical models are needed for invasive species, even more than for disease bioeconomics. In this book, we put a premium on the development of dynamical models, illustrated especially in chapter 9, which can be connected directly to policy and management decision making. In the next section, we briefly consider how current policies at various levels are or are not informed by the integration of economic and ecological analyses.

BIOECONOMIC IMPACT OF EXISTING POLICY ON INVASIVE SPECIES

Important arenas in which the feedbacks between the biological and economic systems are adjudicated are international agreements. Movements of species within countries can also cause great damage (Perry et al. 2002), but a large focus of ongoing policy development is international. Once a species is introduced to one country, dispersal to neighboring countries and to countries strongly connected by trade becomes much more likely. Decisions about importation or exportation by one country affect the interests of many other countries. While national policy often focuses on importation, international agreements are the usual venue for more explicitly recognizing steps that should be taken to prevent exportation, as well as importation, of harmful species.

Although more than 50 international and regional legal instruments address invasive species, few of these are binding (Shine et al. 2005). Of these, the binding agreement most directly aimed at preventing environmental harm is the Convention on Biological Diversity (CBD), ratified by more than 170 parties (not including the United States). Under the CBD, however, the obligation for compliance lies with each signatory country, and the repercussions for noncompliance are virtually nonexistent.

In contrast, international trade agreements have exerted the strongest influence over invasive species policy because the costs of noncompliance are high. Globally, the most relevant agreements are those based in the World Trade Organization (WTO), although the following comments apply also to binational and regional agreements, such as the North American Free Trade Agreement. Because the overarching goal of WTO is to increase international trade (which increases the probability of biological invasions), there is an inherent tension between promoting trade and preventing the introduction of invasive species.

Under WTO, the International Plant Protection Convention specifies standards (through the Agreement on the Application of Sanitary and Phytosanitary Measures [SPS Agreement]) that national laws must meet if a nation wishes to reduce the introduction of invasive species (Hedley 2004). These standards apply to invasive species of all kinds, including plants, plant pests, animals, and animal parasites. Any regulations to reduce the introduction of unwanted species must minimize the impact on trade. The initial burden in demonstrating the need for protection is on the importing nation, which must demonstrate via a scientific risk assessment that an import is likely to cause a harmful introduction. While the role of scientific risk analysis appears preeminent in the SPS Agreement, it remains largely unclear what constitutes a scientific risk assessment that can meet the SPS standards. Most of the cases that have been adjudicated have been decided in favor of the exporting country (Pauwelyn 1999). Countries are under pressure to quickly open their borders to imports rather than take precautionary measures to prevent the introduction of invasive species. The difficult balancing act, not yet achieved, is to provide adequate safeguards to prevent invasive species while not unduly hindering the high-speed, high-volume international flow of goods (Jenkins 2002).

Most national policies, including those in the United States, have responded very little to the threat of invasive species, for at least two related reasons. First, while the costs of invasions have been estimated as $120 billion annually for the United States (Pimentel et al. 2005), such aggregate estimates are certainly incomplete and are difficult to parse with respect to policy options for specific vectors, and few specific rigorous economic analyses exist (Lovell et al. 2006; Olson 2006). Second, policy responses aimed at reducing invasions and increasing human welfare could instead lower human welfare and cause unanticipated economic distortions if their costs (in lower trade or shifts in the economy) outweigh their benefits (in decreased damages from invasive species) (Lovell et al. 2006; Olson 2006). Rational policies depend on better quantification of the externalities of trade manifesting as damage costs of invasive species, the vectors by which they move around (otherwise policies might be misdirected), and the costs of alternative policies.

Fortunately, research progress is rapid at this nexus of biology, economics, and policy. For example, a recent analysis demonstrated that the Australian Weed Risk Assessment, under which any plant proposed for importation into Australia is allowed only if it survives a risk assessment, brings net economic benefits to Australia (Keller et al. 2007). In this book, we explore in more detail under which circumstances of costs, benefits, and spatial scales alternative policy and management strategies are warranted.

GOALS OF THIS BOOK

Determining the expected total benefits from a management action or policy is not straightforward and requires the expertise of both ecologists and economists. Additionally, because this work requires extensive modeling of the outcomes from alternative scenarios, mathematicians are needed to synthesize models from ecology and economics into a unified framework. Only when the expertise from these fields is combined does it become possible to answer the questions asked by managers and policy makers (figure 1.7). In the following paragraphs, we present two of these questions, and explain how expertise from the three disciplines can be used to provide solutions.

How do we rationally spend on prevention versus control for a species that is not yet established? When faced with species that are predicted by ecologists and/or economists to be damaging, the possible responses are to prevent its arrival, to

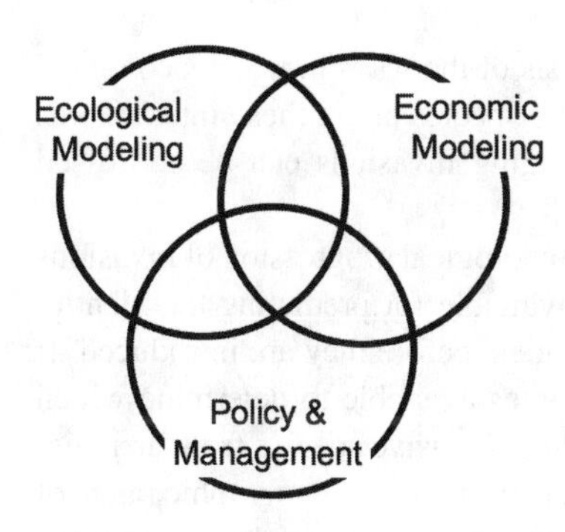

FIGURE 1.7.
Three areas of research that must be jointly considered to understand biological invasions and reduce their impact. The goal of this book is to increase the area of overlap among the three circles.

manage it once it arrives, or to simply live with the impacts. Deciding among these options requires knowledge of the economic costs of prevention (e.g., removing a fish species from the aquarium trade), the costs of managing the species if it arrives (e.g., seasonal pesticide applications to reduce population densities), the benefits from not having the species, and the benefits from controlling the species versus no management. Usually, it will also be necessary to consider multiple methods for both preventing and managing the species. Extracting answers from such potentially complex series of scenarios requires a combination of rigorous economic and ecological models.

What level of resources is it rational to spend on one versus another vector of invasive species transport? Most ecosystems have received invasive species from multiple vectors (see chapter 10 for an analysis of the Great Lakes ecosystem). It is necessary to be mindful of multiple routes of trade and travel when determining how best to reduce the risk from new invasions. Determining the appropriate expenditure on each vector will require knowledge of the total value of that vector, the costs that restrictions on it would cause, and the degree to which the risk of new invasions would be reduced considering any compensatory responses from other vectors. It would also require forecasts of the likely economic and ecological impacts of invasive species from each vector. Scenarios covering the range of management responses would need to be rigorously assessed to determine which approach is optimal.

In the following chapters, we present methods for answering these and related questions. Collectively, the authors of this book come from the disciplines of economics, ecology, and mathematical biology. We have worked together for more than 7 years under the auspices of the Integrated Systems for Invasive Species (ISIS) project, funded by the U.S. National Science Foundation and Canadian Natural Sciences and Engineering Research Council. Over this time, we have identified many important scientific questions and attempted to provide tools for answering them in ways that are relevant to addressing the issues of invasive species. The conceptual frameworks presented in this book have been developed through extensive collaboration, such that all chapters are heavily influenced by each discipline. In each chapter, we critically review the work of both the ISIS project and of the many other researchers working on similar problems.

Chapter 2 presents a more theoretical analysis of the ways in which ecological and economic systems interact, and how invasive species affect these interactions. This chapter shows how the problems of managing invasions can be addressed through a unified bioeconomic framework.

Chapters 3–6 are organized according to the ecological progression of invasions (figure 1.1). Chapter 3 describes the methods available for predicting the identity and the economic and ecological impacts of invaders before they are introduced. If this can be done successfully, information becomes available to determine which species or vectors it is rational to control to prevent invasive species from arriving. Another component in such a decision, however, is the potential geographic range of a species. Chapter 4 reviews recent developments for environmental niche models, tools that describe where a species is able to survive. These models can be used before a species is established to determine the value of prevention and after it is established to determine the value of control, slow-the-spread efforts, and, if possible, eradication.

Because the likelihood that a species will become established is positively related to the number of individuals released—propagule pressure—it may be possible to prevent invasions by controlling the rate and location of releases of individuals. Chapter 5 reviews the ecological and mathematical theory behind this approach and describes the combinations of vectors and species for which management based on it will be appropriate.

Once a species is introduced (e.g., through the pet trade vector), it is often dispersed secondarily by additional human vectors, such as recreational boating. Chapter 6 reviews the available models for predicting such secondary dispersal and illustrates with case studies how these models can inform management responses.

Chapters 7–9 address more general issues for bioeconomic modeling of invasive species. Chapter 7 provides a rigorous analysis of the types of uncertainty that exist and the general issues that they present for modeling. It provides context for many of the models and methods presented in other chapters and suggests research strategies for the future.

Although the market costs of invasive species are generally easy to resolve, the nonmarket costs are extremely difficult to assess, as they are for many other environmental issues. Despite this, there is ample reason to believe that nonmarket costs are often substantial. Chapter 8 describes and reviews methods for determining these costs.

Chapter 9 ties together the models and theory from earlier chapters and presents a framework for integrating economic and ecological data to determine the optimal type and timing of invasive species management. This chapter emphasizes the ways that management efforts affect ecological and economic systems and how the state of those systems feeds back and affects optimal management.

Chapter 10 analyzes for one ecosystem, the North American Great Lakes, the many ways that economic forces and ecology have interacted to create the current state of invasions. This ecosystem is the focus of much economic activity, including canal construction and navigation, commercial and recreational angling, aquaculture, and the ornamental plant and animal trades. Activities such as these have led to the establishment of at least 183 nonindigenous species (Ricciardi 2006), many of which have become invasive.

Chapters 11 and 12 present case studies for two well-known invasive species and show how the bioeconomic framework presented throughout the book has been applied to prescriptions for their management. Chapter 11 focuses on the rusty crayfish (*Orconectes rusticus*), an invader of lakes and streams in the U.S. Upper Midwest with large economic and ecological impacts, and discusses methods for control and eradication. Chapter 12 focuses on the zebra mussel, a well-studied invader across Europe and North America. This species has received much management effort and provides a good case study to demonstrate the necessity for, and effectiveness of, management that is based on a rigorous bioeconomic understanding of an invasive species.

Finally, in chapter 13, we take a step back from the more technical issues and critically assess the contribution that bioeconomic modeling has made to effective management of invasive species (figure 1.7). A number of examples from the authors' own work are described, along with the management responses that have come from them.

References

BenDor, T. K., S. S. Metcalf, L. E. Fontenot, B. Sangunett, and B. Hannon. 2006. Modeling the spread of the emerald ash borer. Ecological Modelling 197:221–236.

Bossenbroek, J. M., L. E. Johnson, B. Peters, and D. M. Lodge. 2007. Westward expansion of the zebra mussel in North America: forecasting a low probability-high impact event. Conservation Biology 21:800–810.

Burnham, C. R. 1988. The restoration of the American chestnut. American Scientist 76:478–487.

Caswell, H. 2001. Matrix population models: construction, analysis, and interpretation, 2nd edition. Sinauer Associates, Sunderland, MA.

Clark, C. W. 1990. Mathematical bioeconomics: the optimal management of renewable resources, 2nd edition. John Wiley and Sons, New York.

Cohen, A. N., and J. T. Carlton. 1998. Accelerating invasion rate in a highly invaded estuary. Science 279:555–558.

Costello, C., J. M. Drake, and D. M. Lodge. 2007. Evaluating the effectiveness of an environmental policy: ballast water exchange and invasive species in the North American Great Lakes. Ecological Applications 17:655–662.

Crocker, T., and J. Tschirhart. 1992. Ecosystems, externalities and economics. Environmental and Resource Economics 2:551–567.

Daly, H. 1968. On economics as a life science. Journal of Political Economy 76:392–406.

Dasgupta, P., S. Levin, and J. Lubchenco. 2000. Economic pathways to ecological sustainability. BioScience 50:339–345.

de Camino Beck, T., and M. A. Lewis. 2007. A new method for calculating net reproductive value from graph reduction with applications to the control of invasive species. Bulletin of Mathematical Biology 69:1341–1354.

Diekmann, O., J. A. P. Heesterbeek, and J. A. J. Metz. 1990. On the definition and the computation of the basic reproduction ratio R0 in models for infectious diseases in heterogeneous populations. Journal of Mathematical Biology 28:365–382.

Drake, J. M., and J. M. Bossenbroek. 2004. The potential distribution of zebra mussels in the United States. BioScience 54:931–941.

Finnoff, D., and J. Tschirhart. 2003. Protecting an endangered species while harvesting its prey in a general equilibrium ecosystem model. Land Economics 79:160–180.

Forseth, I. N., and A. F. Innis. 2004. Kudzu (*Pueraria montana*): history, physiology, and ecology combine to make a major ecosystem threat. Critical Reviews in Plant Sciences 23:401–413.

Gilbert, G. S. 2002. Evolutionary ecology of plant diseases in natural ecosystems. Annual Review of Phytopathology 40:13–43.

Hamlin, C., and D. M. Lodge. 2006. Ecology and religion for a post natural world. Pages 279–309 *in* D. M. Lodge and C. Hamlin, editors. Religion and the new ecology: environmental responsibility in a world in flux. University of Notre Dame Press, South Bend, IN.

Hawaii Department of Agriculture. 2002. Kahului Airport risk assessment. Hawaii Department of Agriculture, Plant Quarantine Division, Honolulu, HI.

Hedley, J. 2004. The international plant protection convention and invasives. Pages 185–202 *in* M. L. Miller and R. N. Fabian, editors. Harmful invasive species: legal responses. Environmental Law Institute, Washington, DC.

Holzmueller, E., S. Jose, M. Jenkins, A. Camp, and A. Long. 2006. Dogwood anthracnose in eastern hardwood forests: what is known and what can be done? Journal of Forestry 104:21–26.

Jenkins, P. T. 2002. Paying for protection from invasive species. Issues in Science and Technology 19:67–72.

Jeschke, J. M., and D. L. Strayer. 2005. Invasion success of vertebrates in Europe and North America. Proceedings of the National Academy of Sciences of the United States of America 102:7198–7202.

Keeling, M. J., M. E. J. Woolhouse, R. M. May, G. Davies, and B. T. Grenfell. 2003. Modelling vaccination strategies against foot-and-mouth disease. Nature 421:136–142.

Keller, R. P., D. M. Lodge, and D. C. Finnoff. 2007. Risk assessment for invasive species produces net bioeconomic benefits. Proceedings of the National Academy of Sciences of the United States of America 104:203–207.

Kumaranayake, L. 2006. Trade and infectious disease outbreaks: ensuring public health without compromising free trade. Pages 341–354 *in* J. A. Roberts, editor. The economics of infectious disease. Oxford University Press, Oxford.

Leung, B., D. M. Lodge, D. Finnoff, J. F. Shogren, M. A. Lewis, and G. Lamberti. 2002. An ounce of prevention or a pound of cure: bioeconomic risk analysis of invasive species. Proceedings of the Royal Society of London Series B Biological Sciences 269:2407–2413.

Levine, J. M., and C. M. D'Antonio. 2003. Forecasting biological invasions with increasing international trade. Conservation Biology 17:322–326.

Lodge, D. M., and K. Shrader-Frechette. 2003. Nonindigenous species: ecological explanation, environmental ethics, and public policy. Conservation Biology 17:31–37.

Lodge, D. M., S. Williams, H. MacIsaac, K. Hayes, B. Leung, S. Reichard, R. N. Mack, P. B. Moyle, M. Smith, D. A. Andow, J. T. Carlton, and A. McMichael 2006. Biological invasions: recommendations for U.S. policy and management. Ecological Applications 16:2035–2054.

Lovell, S. J., S. F. Stone, and L. Fernandez. 2006. The economic impacts of aquatic invasive species: a review of the literature. Agricultural and Resource Economics Review 35:195–208.

Mangtani, P., and A. Shah. 2006. The socio-economic burden of influenza: costs of illness and "willingness to pay" in a publicly funded health care system. Pages 159–180 *in* J. A. Roberts, editor. The economics of infectious disease. Oxford University Press, Oxford.

Meinesz, A. 1999. Killer Algae. University of Chicago Press, Chicago.

Morris, R. S. 1999. The application of economics to animal health programmes: a practical guide. Revue Scientifique et Technique, Office International des Epizooties 18:305–314.

Narasimham, V. 2006. Trade and infectious disease outbreaks: ensuring public health without compromising free trade. Pages 341–354 *in* J. A. Roberts, editor. The economics of infectious disease. Oxford University Press, Oxford.

Olson, L. J. 2006. The economics of terrestrial invasive species: a review of the literature. Agricultural and Resource Economics Review 35:178–194.

O'Neill, C. R. 1996. National zebra mussel information clearinghouse infrastructure economic impact survey—1995. Dreissena! 7(2):1–5.

Ostry, M. E., and K. Woeste. 2004. Spread of butternut canker in North America, host range, evidence of resistance within butternut populations and conservation genetics. Pages 114–120 *in* C. H. Michler, P. M. Pijut, J. W. Van Sambeek, M. V. Coggeshall, J. Seifert, K. Woeste, R. Overton, and F. Ponder, Jr., editors. Black walnut in a new century: proceedings of the 6th Walnut Council Research Symposium. U.S. Department of Agriculture, Forest Service, North Central Research Station. St. Paul, MN.

Pauwelyn, J. 1999. The WTO agreement on sanitary and phytosanitary (SPS) measures as applied in the first three SPS disputes: EC—hormones, Australia—salmon and Japan—varietals. Journal of International Economic Law 2:641–664.

Perry, W. L., D. M. Lodge, and J. L. Feder. 2002. Importance of hybridization between indigenous and nonindigenous freshwater species: an overlooked threat to North American biodiversity. Systematic Biology 51:255–275.

Pimentel, D., R. Zuniga, and D. Morrison. 2005. Update on the environmental and economic costs associated with alien-invasive species in the United States. Ecological Economics 52:273–288.

Ricciardi, A. 2001. Facilitative interactions among aquatic invaders: is an "invasional meltdown" occurring in the Great Lakes? Canadian Journal of Fisheries and Aquatic Sciences 58:2513–2525.

Ricciardi, A. 2006. Patterns of invasion in the Laurentian Great Lakes in relation to changes in vector activity. Diversity and Distributions 12:425–433.

Riviere-Cinnamond, A. 2006. Economics of animal health: implications for public health. Pages 215–236 *in* J. A. Roberts, editor. The economics of infectious disease. Oxford University Press, Oxford.

Roberts, J. A., editor. 2006. The economics of infectious disease. Oxford University Press, Oxford.

Rushton, J., P. K. Thornton, and M. J. Otte. 1999. Methods of economic impact assessment. Revue Scientifique et Technique, Office International des Epizooties 18:315–338.

Settle, C., T. D. Crocker, and J. F. Shogren. 2002. On the joint determination of biological and economic systems. Ecological Economics 42:301–311.

Sharov, A. A. 2004. Bioeconomics of managing the spread of exotic pest species with barrier zones. Risk Analysis 24:879–892.

Sharov A., D. Leonard, A. M. Liebhold, E. A. Roberts, and W. Dickerson. 2002. Slow the spread: a national program to contain the gypsy moth. Journal of Forestry 100(5):30–36.

Shine, C., N. Williams, and F. Burhenne-Guilmin. 2005. Legal and institutional frameworks for invasive alien species. Pages 233–284 *in* H. A. Mooney, R. N. Mack, J. A. McNeely, L. E. Neville, P. J. Schei, and J. K. Waage, editors. Invasive alien species: a new synthesis. Island Press, Washington, DC.

Shogren, J., and T. Crocker. 1999. Risk and its consequences. Journal of Environmental Economics and Management 37:44–51.

Sohngen, B., and R. Mendelsohn. 1998. Valuing the impact of large-scale ecological change in a market: the effect of climate change on U.S. timber. American Economic Review 88:686–709.

Steeger, C., R. Holt, and J. Smith. 1999. Enhancing biodiversity through partial cutting. British Columbia Ministry of Forests report. Pandion Ecological Research, Nelson, British Columbia.

Stokstad, E. 2007. Invasive species—feared quagga mussel turns up in western United States. Science 315:453–453.

Strayer, D. L., and H. Malcom. 2007. Effects of zebra mussels (*Dreissena polymorpha*) on native bivalves: the beginning of the end or the end of the beginning? Journal of the North American Benthological Society 26:111–122.

U.S. Department of Commerce. 2008. TradeStats Express. Available at http://tse.export.gov/.

Walters, L. J., K. R. Brown, W. T. Stam, and Olsen J. L. 2006. E-commerce and *Caulerpa*: unregulated dispersal of invasive species. Frontiers in Ecology and the Environment 4:75–79.

Wilson, E. O. 1998. Consilience. Alfred Knopf, New York.

Wonham, M. J., T. de Camino-Beck, and M. A. Lewis. 2004. An epidemiological model for West Nile Virus: invasion analysis and control applications. Proceedings of the Royal Society of London Series B Biological Sciences 271:501–507.

Woodfield, R., and K. Merkel. 2005. Eradication and surveillance of *Caulerpa taxifolia* within Agua Hedionda Lagoon, Carlsbad, California, Fourth Year Status Report. Report prepared for Southern California Caulerpa Action Team.

World Health Organization. 2002. World health report. World Health Organization, Geneva.

Xia, Y., O. N. Bjørnstad, and B. T. Grenfell. 2004. Measles metapopulation dynamics: a gravity model for pre-vaccination epidemiological coupling and dynamics. American Naturalist 164:267–281.

Yule, A. M., J. W. Austin, I. K. Barker, B. Cadieux, and R. D. Moccia. 2006. Persistence of *Clostridium botulinum* neurotoxin type E in tissues from selected freshwater fish species: implications to public health. Journal of Food Protection 69:1164–1167.

2

Integrating Economics and Biology for Invasive Species Management

David C. Finnoff, Chad Settle, Jason F. Shogren, and John Tschirhart

In a Clamshell

Over the last four decades, many economists and biologists have argued that better integration of the two disciplines would improve both science and policy making. The inspiration behind integration is that explicitly accounting for economic and biological circumstances and the key feedback loops between the two systems would yield better information about the risks faced by humans and the environment. An integrated approach to problems such as invasive species management should account for the ability of humans to respond to changes in their surroundings, and vice versa. Ecosystems, in turn, evoke and subsequently respond to human system changes from management. Concurrently, human systems both evoke and respond to ecosystem changes. Capturing this economic system–ecosystem interaction requires one to account for both how people and nature change and the feedbacks within and among the systems. In this chapter we discuss our organizing analytical framework of endogenous risk, and then we address three questions: (1) What do we gain by integrating the web of life into economic analysis? (2) If integration is worthwhile, how deep should we go? (3) What are some challenges of integration?

An established species is considered invasive when it triggers costs that outweigh any attendant benefits. In the past, many researchers have used an approach that assumes the economic system and the ecosystem affect each other in a one-sided way, which causes them to separate risk assessment from risk management. A change in the economic system is viewed as changing the pressure on the ecosystem, or a change in the ecosystem is viewed as only changing the economic system. But this approach does not address the idea of the two-way interactions and feedbacks between human and natural systems (e.g., Clark 1976; Crocker and Tschirhart 1992; Heal 1998; Barbier 2001; Brown and Layton 2001; Wätzold et al. 2006). Ecosystem changes alter human behavior and productivity in the economic system. People recognize the change in their productivity, and they adapt to this change, either by adapting the environment or by adapting to the environment. When people adapt, they alter the pressure they put on the ecosystem, leading to further changes in the ecosystem (Swallow 1996; Sohngen and Mendelsohn 1998; Perrings 2002).

This idea of a bioeconomic system can be addressed by integrating ecological and economic modeling into a single cohesive framework. The motivation behind integration is to get more precise estimates of invasive species damages on human and natural systems. Integration accounts for interdependencies, or *feedback loops*. Traditionally, economists have captured the notion of feedback loops using dynamic models. With a few exceptions, most standard bioeconomic models consider, at most, one or two feedback loops and operate at a relatively aggregate level. Such models can provide the needed insight into the underlying problem at hand. In other cases, however, more ecological or economic detail is needed to help avoid the unintended consequences of poorly advised policy. This challenge of balancing model tractability with more realism is not new in science, but it matters significantly when addressing the economics of invasive species management.

In this chapter, we address three questions that arise when thinking about building integrated bioeconomic modeling for invasive species management: Why bother integrating economics and biology for invasive species management? How deep should integration go? What challenges exist in integration?

WHY INTEGRATION?

Why bother to go through all the trouble to integrate economics and biology for invasive species management? The straightforward answer is that it will give us "better science" for policy—we can provide more environmental protection at less cost. By integrating we account for the impact of economics on biological systems and vice versa, and we capture the feedback loops between the two systems. If we can do so, (1) we generate better risk assessments—predictions of human behavior and species population densities will be less biased, and (2) we support better risk management through more efficient expenditure of scarce public and private resources for prevention and generate greater net benefits.

For instance, consider the case of fishing pressure or harvest effort. Treating fishing pressure or harvesting effort as a constant—unaffected by feedback among the systems—does not account for how humans adapt to a change in the fishery. With constant fishing effort, as fish populations fall due to an array of biological considerations, the harvest of fish also falls. Integrating economic systems and ecosystems via fishing effort captures this initial change. What it does not capture is how a change in one system can lead to a change in behavior in the other. When the fish species declines, will fishing effort actually be constant? When the fish population declines, many economic factors cause humans to transfer their efforts from one fish species to another or from fishing to other activities. This shift in behavior could lead to a different ecosystem steady state than if no account were taken of these feedbacks.

Consider three examples on why and how integration matters for invasive species management—cutthroat trout in Yellowstone Lake, zebra mussels in the U.S. Midwest, and leafy spurge in grazing and cattle ranching. We first illustrate why integration matters with a detailed discussion on the case of the exotic invader, lake trout, in Yellowstone Lake in Yellowstone National Park, Wyoming. We show how accounting for feedback between humans and nature affects the predicted ecological impacts—the population of a native prey species. We then briefly present the two other examples. The interested reader can consult the research articles for more details.

Box 2.1 presents the formal framework we have used to integrate our bioeconomic models—the theory of endogenous risk. In words, the endogenous risk perspective captures the notion that people invest resources to reduce the risks they confront or create (Ehrlich and Becker 1972; Shogren and Crocker 1991). We know that people routinely act and react to risk—examples abound. Farmers and ranchers purposefully alter cropping and pest control strategies to increase the odds they will not suffer from the invasion of exotic species. Landowners modify land-use plans to either enhance or reduce the likelihood that resident wildlife survive another year. The same logic holds for the case of exotic invaders—the risk of introduction, establishment, and impact is both a biological and economic question. Box 2.2 extends the theory of endogenous risk to address some of the stochastic characteristics of exotic invaders. What is clear is that the "devil is in the details" and the intertwined economic and ecological details cannot be neglected. As such, risk assessment should account for heterogeneous wealth and land prices faced by private citizens and communities, affecting this provision of habitat; otherwise, we will underestimate risks in some regions and overestimate risk in others. The precision of species risk assessment and implementation of collective action can be increased by using both biological and economic parameters as determinants of endangerment (Shogren et al. 1999).

BOX 2.1. Endogenous Risk as an Organizing Framework

Bioeconomic integration requires a framework that can act as a common focal point to guide modeling, data collection, baseline development, and policy evaluation. People protect themselves from risks of invasive species through prevention and control. We prevent risk by curtailing species to lower the likelihood that bad states of nature occur; we control risk by changing production and consumption decisions to reduce the severity of a bad state if it does occur. Together prevention and control jointly determine the risks and the costs to reduce them. Since private citizens have the liberty to adapt on their own accord, a policy maker must consider these responses when choosing the optimal degree of public prevention. Otherwise, policy actions will be more expensive than they need to be, with no additional reduction in invader risk.

Most people would agree with this logic—prevention and control are linked—but the full implication is not always appreciated in invasive species policy. It means economic and environmental systems are jointly determined. Human actions and reactions affect nature; nature affects our actions and reactions. This realization challenges the traditional risk-reduction perspective that habitually and artificially separates risk assessment from risk management. This fragmentation of risk policy essentially presumes that economic and environmental systems are not jointly determined. But this assumption might not be useful for many environmental risks.

Consider the risk to biodiversity a motivating example. Conservation biologists often maintain that establishing the threshold of species endangerment is strictly a biological question, determined by the present sizes, trends, and distributions of its populations and their likely interactions with the stochastic forces of nature. These stochastic events are said to be separable from human actions. This perspective is overly narrow if it does not address economic circumstances. The odds of a species surviving depend on the economic forces of today, as revealed by relative prices and wealth, because these parameters drive land-use decisions today and into the future. Assessing the risk to species and setting a minimum acceptable probability of survival are as much economic questions as they are biological ones. Our choice to create and avoid risk is endogenous (see Shogren and Crocker 1991).

The theory of endogenous risk provides a framework to integrate these different parameters across different disciplines. Shogren (2000) defines an endogenous risk framework to help organize our thinking about controlling the risks posed by invasive species. The structure rests on a benevolent manager who allocates scarce resources to maximize expected social welfare subject to the risk of invasion. The manager selects prevention, Q, and control, x, efforts to maximize expected social welfare, EU, of his country:

$$\text{Max}_{x,Q}\left[\int_a^b (p(Q,\theta)V_0[m - c(x,Q)] + [1 - p(Q,\theta)]V_1[m - D(x;\theta) - c(x,Q)])dF(\theta;\beta)\right]$$

Several details of the manager's problem need to be defined and understood. First, $p(Q;\ \theta)$ represents the probability that a good state of nature is realized; that is, no damage from an invasive species; $1 - p(Q;\ \theta)$ is the probability that a bad state of nature is realized, that is, invasive species damage. The probability is assumed to be a function of both economic behavior and biological factors. Here, both economic and biological factors matter in risk assessment and risk management. Second, $D(x;\ \theta)$ is the money equivalent of realized damages if a bad state of nature is realized. The damages are a function of control, and the stochastic variable. Damages can be nonmonetary ecological damages such as changes to the structure and composition of native vegetation. In this case, the manager would be required to assess the social welfare impacts of a change in the quantity of ecosystem services or functions.

Third, x is the manager's investment in control, and Q is his investment in prevention. Prevention is assumed to be a public good—nonexclusive and nonrival in consumption. Assume prevention is represented by a weak-link public good technology, $Q = \min(Q_1, Q_2, \ldots, Q_n)$, which implies that the least successful mitigation effort drives the odds to reduce the likelihood that the community will realize a bad state of nature (see Perrings et al. 2002). For invasive species, prevention methods take the form of quarantines, trade regulations, and transport regulations to reduce the risk of introduction, and if the species is established, control efforts include poisoning, shooting, trapping, weeding, spraying, uprooting, biocontrol agents, viral disease, and sterile insects.

Fourth, θ is a random variable that reflects the basic scientific uncertainty about the impact of invasive species. This variable is crucial because it reflects the state of knowledge about the causes and effects of exotic invaders. But numerous additional questions arise here, for example, which factors matter the most for establishment and diffusion and transformation into a pest—biotic or abiotic factors? Fifth, $F(\theta; \beta)$ represents the cumulative distribution bounded over the support (a, b) that defines the mean and variance of the random variable, θ. Let β represent exogenous collective investments in research to reduce the uncertainty about the likely impact of an invasive species. This research is a public good represented by a best-shot public good technology, $\beta = \max(\beta_1, \beta_2, \ldots, g\beta_n)$, which implies that the most successful research effort to reduce uncertainty spreads through out the entire scientific and policy community (Hirshleifer 1983).

Sixth, $V_0[m - c(x, Q)]$ is the social value of a good state of nature that depends on net wealth, $w = m - c(x, Q)$, where m is endowed wealth, and $c(x, Q)$ is the cost function for control, x, and prevention, Q, activities. $V_1[m - D(x; \theta) - c(x, Q)]$ is the social value of a bad state of nature. We could also measure the value of pure ecological damages by reforming the value function so that the damage function is its own separate argument, $V_1[m - c(x, Q), D(x; \theta)]$. In this case, we would be trying to measure how a change in damages affects social welfare. We could move this damage function outside the welfare function and make it part of some multicriteria analysis that does not try to put a value on each and every change. Assume $V_{0w} > 0, D_x < 0, F_\theta > 0, F_\beta < 0, c_i(x, Q) > 0 (i = x, Q)$, where subscripts denote the relevant partial derivatives.

The manager maximizes the expected returns on investments by equating the marginal returns per costs across prevention and control. This endogenous risk model allows us to frame the invasive species debate in benefit-cost terms. Such calculations are controversial but necessary.

An advantage of the endogenous risk framework is its malleability. Subsequent chapters in this book make clear that problems of irreversibility and uncertainty over time are all key ingredients in the context of invasive species. The endogenous risk framework can include all of these facets through some basic extensions into the arenas of stochastic dynamic programming and real options analysis. Chapter 9 provides a detailed description of stochastic dynamic programming methodologies, and chapter 8 provides a deterministic equivalent of the timing of policy investments as provided in real options analysis in stochastic settings (see box 2.2).

BOX 2.2. Endogenous Risk and Real Options

Most endogenous risks posed by invasive species can be characterized by (1) irreversibilities in outcomes, their costs, and the costs of alternative policies in response; (2) uncertainties over most components of the problem; and (3) the timing of investments in prevention and control. Many chapters throughout this book speak to the irreversibility of invasions and the uncertainty about their properties. Invasion control is not a now-or-never proposition; rather, the timing of investment also matters to outcomes. In these classes of problems, both the policy investments of prevention and control and their timing matter to outcomes. In fact, as noted by the National Invasive Species Management Plan (National Invasive Species Council 2001), understanding the timing of control measures in an uncertain environment requires an analytical framework for invasive species so control strategies can be prioritized and targeted appropriately.

In short, investments in invasion control possess the core features of the real options examined in detail by Dixit and Pindyck (1994) and in the financial economics literature. The theory of real options provides resource managers with a coherent framework to balance urgency of action and neglect. Real-options models provide an analytical framework for our thinking regarding irreversibility and uncertainty. Real-options theory has revealed that when one is evaluating an investment decision, two critical components should be considered (see, e.g., Dixit and Pindyck 1994): the value of the investment itself, which may be uncertain, and the value from having the opportunity to choose the timing of the investment decision given the uncertainty and irreversibility of its cost. When evaluating policy decisions of whether to do anything about an invasive species and, if so, what, given the irreversibilities and uncertainties of invasions and policy investments, the timing of these decisions is another facet that should be included in the analysis.

Consider one use of real-options theory in invasive species work. Saphores and Shogren (2005) used a real-options framework to examine the decision of when, if ever, to control an invasive species invading a given territory (assuming that the relevant bioeconomic parameters are known) and how policy makers should best expend resources in the research of bioeconomic parameters at the heart of these problems. Their approach generated a closed-form action rule—a threshold set by biological and economic parameters that defines the timing of when to control an invasive species. This threshold is similar to the exercise of a financial option. They considered how this threshold varies with different bioeconomic parameters by conducting a sensitivity analysis. They found that the control threshold varies negligibly with the efficacy of a control measure. This result illustrates the key parameters underlying a manager's timing trade-off—how to save control costs that can be spent on gathering more bioeconomic data so that a better control decision can be made later.

Let us briefly consider the analytic framework developed in Saphores and Shogren (2004). An ecosystem is an asset; a manager invests in this asset by controlling the invasive species population. The manager holds a compound option, which gives him or her the right but not the obligation to apply a control measure—this is called a compound stopping problem. The optimal policy is to control once the pest density X reaches the threshold size x^*. The manager selects x^* to minimize the present value of expected damages and control costs. For "low" values of X, waiting is optimum; for high values, immediate control is best.

The invasive species population has an average density over a given area, X, that varies randomly. The manager invests in control at a cost $C_c(x)$ per action to reduce the population to a fraction ω, that is, the efficacy of control (complete inefficacy, $\omega = 1$; complete eradication, $\omega = 0$). The manager's problem is to minimize a value function $V_c(x|y)$ by choosing x, the level of the pest population at which controls should be applied. The manager minimizes the present value of expected pest damages until the next control, given the current pest density y, the present value of the sum of all future control costs and expected pest damages. In general, the manager balances three effects of a change in the timing of control: the flow of damages until the next control, delaying all future controls, and controlling costs and damages between controls.

Waiting to apply the control delays incurring the control costs but increases the flow of pest damages; how long to wait depends on how the discount factor responds to the control threshold. If waiting longer causes the discount factor to decrease a lot, future control costs are discounted more heavily while the present value of the flow of pest damages does not change much. The manager is more likely to wait to control. In contrast, if a small increase in the threshold causes the discount factor to decrease little, future control costs are almost unchanged whereas the present value of the flow of pest damages increases, so the manager is more likely to control now. The manager's timing decision also depends on the responsiveness of the control and damage costs to the control threshold. If control and damage costs are responsive to waiting, the manager controls now; otherwise, he or she waits. Formulating the timing problem in terms of responsiveness provides a relatively intuitive and consistent general rule for managers that helps formulate their thinking about timing based on empirical or judgmental estimates of the relative magnitudes of elasticity measures.

Example 1: Yellowstone Lake

Yellowstone Lake is an inland fishery for the native Yellowstone cutthroat trout (*Oncorhynchus clarki bouvieri*). Cutthroat trout are popular with anglers and many predators, such as ospreys (*Pandion haliaetus*), white pelicans (*Pelecanus erythrorhynchos*), river otter (*Lutra canadensis*), and grizzly bears (*Ursus arctos*). In 1994, however, an angler caught a lake trout (*Salvelinus namaycush*) in Yellowstone Lake. Lake trout are an exotic species to Yellowstone Lake—they prey upon but do not replace cutthroat trout in the food web. If left unchecked, some biologists have predicted that this voracious exotic species could reduce the catchable-size cutthroat population from 2.5 million to 250,000–500,000 within the near future (Kaeding et al. 1995). Furthermore, grizzly bears, ospreys, eagles, river otters, and the other 40 species that rely on cutthroat as part of their food supply are put at risk.

Traditionally, the specifics of threats to species and ecosystems have been estimated using the "damage function" (DF) approach (see Freeman 1993). The DF approach assumes the economic system and the ecosystem affect each other in a one-sided way. A change in the economic system is viewed as only changing the pressure on the ecosystem, or a change in the ecosystem is viewed as only changing the economic system (e.g., Daily 1997). The DF approach therefore does not address the idea of the two-way interactions between human and natural systems (see Daly 1968). Can an explicit accounting of the specifics of feedback links between the two systems yield different policy-relevant results than does assuming that no joint determination occurs?

Settle et al. (2002) constructed a dynamic modeling framework incorporating many of the details of the flows inside each system and specifics of the reciprocal flows between the two systems. Incorporating the particulars of these links makes the model better reflect trade-offs facing managers. They used Stella 2.0 software to simulate the importance of joint determination of the two species, lake trout and cutthroat trout (Settle and Shogren [2006] provide the full mathematical specification of the model). Given the interactions and feedback loops between predator and prey species and between species and humans, the full Stella specification looks like a ball of yarn.

Figure 2.1 simplifies the model to its key interactions and feedback terms based on the five main components: lake trout, cutthroat trout, grizzly bears, birds of prey, and human interactions. For example, the lake trout population is a state variable indicating the population of lake trout in Yellowstone Lake at any given time. Births and deaths are flows out of the state variable; the growth rate of lake trout declines with increases in the population of lake trout. The density dependence factor relates births of lake trout to its primary food source—cutthroat trout. As the cutthroat trout population declines, fewer lake trout can be supported in the lake. Now lake trout reallocate more time/energy to find food and less time/energy to spawn, which reduces spawning success.

The cutthroat trout population is similar except that we have three species feeding upon cutthroat trout—lake trout, birds of prey, and grizzly bears. Therefore, the causes of cutthroat trout deaths include natural causes and predation by lake trout, birds of prey, and grizzly bears. Each relationship is a function of the cutthroat trout population and the predatory species populations. Lake trout are not likely to replace cutthroat trout in the diet of these predators since lake trout primarily stay in deep water and spawn in the lake instead of in the streams, where predators such as grizzly bears catch cutthroat trout. In addition, a density dependence factor for cutthroat trout enters the model; once again, the population of cutthroat trout is shown with a compensation model. Similar to the lake trout, the density dependence factor limits the births of the species and places a carrying capacity on the cutthroat trout population.

Now consider the human interaction with the ecosystem. Figure 2.1 shows this interaction as captured by a representative visitor to Yellowstone National Park

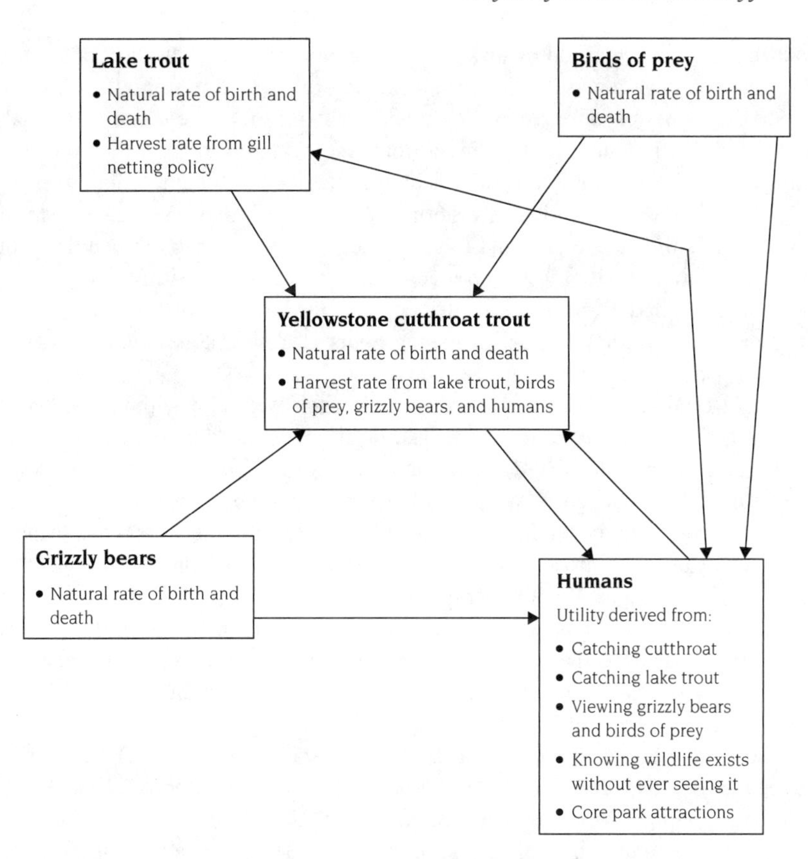

FIGURE 2.1.
Diagram of integrated model of Yellowstone Lake. Reprinted from Settle and Shogren (2002), with permission of the American Agricultural Economics Association.

and its park manager. First, consider the visitor, who gains both direct and indirect benefits from species when visiting Yellowstone. The direct effects include fishing for lake trout and cutthroat trout and visiting the core attractions, such as Old Faithful and other geothermal activity. The indirect effects include seeing birds of prey and grizzly bears while driving or fishing. The number of bird and grizzly bear sightings depends on species populations and time spent fishing and driving.

Our second human interaction is through the park managers. These managers decide how to allocate a fixed budget for the park. They spend this fixed budget on two activities: improving the park public good, and gillnetting the exotic lake trout to reduce pressure on the native cutthroat trout. The success of the gillnetting program, the number of lake trout killed by gillnetting, depends both on the budget spent gillnetting and on the lake trout population.

This human interaction with the ecosystem allows us to capture the feedback links between the economic system and the ecosystem for Yellowstone Lake. These feedback links are captured by how the introduction of lake trout into Yellowstone Lake leads to changes in the ecosystem in and around Yellowstone Lake. Representative visitors respond to these changes by altering their behavior. When lake trout reduce the population of cutthroat trout, anglers find it more difficult to catch cutthroat trout. Since they have less fishing success—the shadow price of fishing increases relative to the prices of alternative activities—they reallocate this time away from fishing at the lake and toward other activities inside or away from the park where productivities have not declined, for example, visiting Old Faithful. (Recall that a "shadow price" for fishing captures the implicit costs of losing fishing productivity.) The total number of fishing hours at Yellowstone Lake falls as the cutthroat population declines.

Settle et al. (2002) consider three cases, each with and without feedbacks. First is the best-case scenario where lake trout are immediately eliminated from Yellowstone Lake without cost. Cutthroat trout return to the world they had before lake trout were introduced. While infeasible in reality, the best-case scenario defines the upper baseline on our indicators of well-being. Second, the worst-case scenario occurs when lake trout are left alone without any interference from the National Park Service. Park managers do not attempt to help cutthroat trout survive. Instead, lake trout and cutthroat trout are left to reach their own steady-state equilibrium. Third, our policy scenario has the National Park Service reducing the risk to cutthroat trout by gillnetting lake trout. Assume that the current level of expenditures on gillnetting lake trout by the National Park Service is continuous and perpetual.

We use the population of cutthroat trout as a yardstick. Table 2.1 summarizes the results for each scenario. Under the best-case scenario without feedbacks, the steady-state population of cutthroat trout is about 2.7 million. With feedbacks, the steady-state population is about 3.4 million. This 700,000 gap is an example of the bias in risk assessment. The bias arises from treating the two systems as a one-way street—when the feedback loops between the two systems are not addressed, one generates biased estimates of the population of fish, which in turn would generate biased estimates of the social net benefits from protecting the fish.

The difference arises from angler's behaviors. Without feedback, anglers continue to fish as before, putting constant pressure on the cutthroat. With feedback, anglers exploiting declining cutthroat populations adapt by fishing less and visiting other attractions more. Reduced human pressure on the cutthroat allows its population to increase by an amount greater than with constant fishing pressure. The resulting population of cutthroat trout is greater with feedback from human control.

TABLE 2.1. Resulting cutthroat trout populations with and without feedbacks.

	No feedbacks	Feedbacks
Best-case scenario	2,700,000	3,400,000
Policy scenario	1,900,000	2,300,000
Worst-case scenario	900,000	0

The results are similar for the policy scenario. The worst-case scenario tells a similar story but with a policy conclusion opposite that for the best-case and the policy scenarios.

What matters is that when people shift their time away from fishing as the cutthroat trout population declines and the lake trout population increases, the incidental catch for anglers of lake trout, an increasingly important control of the lake trout population, declines. While one usually thinks of fishing pressure as pressure on cutthroat trout, in this case the more critical pressure is on lake trout, which were a higher proportion of fishermen's catch. A no-feedback model suggests a healthier outcome than might actually exist—nearly 1 million cutthroat trout versus none is a significant difference however one measures it. Feedbacks yield both different magnitudes and different survival rates, suggesting that one discounts the importance of feedback loops at one's own risk.

The implications for better risk-management outcomes is that one should now assess how much worse off society might be if we apply the optimal policy assuming no feedbacks versus the policy prescription if feedbacks are considered. The bias in fish population is of interest itself, but it is ultimately a stepping stone to better understand how the bias could undermine the net benefits of informed policy decisions. Policy decisions on gillnetting based on 2.7 million fish could well differ from those decisions based on 3.4 million fish—depending on the preferences of the general public, the experts, and the weighting scheme a decision maker might use to balance the two groups.

Example 2: Zebra Mussels

Here we briefly consider our original work, described in Leung et al. (2002) and extended in Finnoff et al. (2005), examining an economic system composed of a Midwest lake ecological system experiencing a zebra mussel invasion, with a resource manager and a power plant, to determine whether integrating the systems is worth the effort. In any period, the power plant takes as given the current state as defined by zebra mussel abundances. Abundances cause damages to the firm as zebra mussels clog coolant systems. Monetized damages diminish the profits of the power plant. In response, the power plant can adapt to the invader. In economics terminology, adaptation is a strategy that accepts the direct damages and compensates in response to reduce the consequences of the damage. The power plant adapts to the damage inflicted by zebra mussel by, for example, operating longer hours or burning more fuel than otherwise necessary. The power plant can also control the invader, which affects the probability of population growth.

In the model, the resource manager can prevent future invasions (if none have occurred) and can control the population growth of the existing population of the invader. If the system has not been invaded at any point in time, prevention reduces the probability of invasion during the transition to the next period. If prevention is effective, no damage occurs; if ineffective, invaders may establish themselves and cause damages in the next period. In the invaded state, population growth increases the magnitude of damages. The resource manager can also use control to affect the probability of population growth in the same fashion as power plant control. (Some details of a simpler version of the model as developed in Leung et al. [2002] are provided in chapter 9.)

In our investigation of whether the integration is worth the effort, we considered two feedbacks—one between the biological system and power plants based on the abundance of zebra mussels, and one between the power plants and the manager based on the manager's expectations of plant behavior. The first is a basic between-system feedback loop and represents whether the power plant takes into account changes in the abundance of zebra mussels as it makes its day-to-day production and control decisions. The second is more subtle and a within-system feedback between the resource manager and the power plant. It represents whether (or not) the resource manager takes into account how the power plant might respond to changes in the abundance of the invader. For both loops, each decision maker's (the resource manager and power plant) beliefs about invasions are central. If the first feedback loop is neglected (the link between the biological system and power plant), the plant behaves as if there is never a change in the biological system—that is, it possesses incomplete knowledge (or beliefs) about the nature of the system. The consequences depend on whether there is an invasion in the initial period, and whether the power plant acknowledges the presence of the invader in the initial period (such that it initially expended some resources on control). For example, with no initial invasion, the power plant neither controls nor adapts. As the biological system changes, the power plant either uses too few or too many inputs relative to what would be optimal as the biological system changes and all feedback loops were included. In turn, output correspondingly either under- or overshoots its targeted level; either way, this results in opportunity cost losses from production shortages or surplus, determined ex post.

The second dimension is the feedback between the resource manager and power plant. Removing the feedback causes the resource manager to act as if the power plant does not respond to changes. For example, following a successful invasion, the manager ignores the private control and adaptation actions of the power plant. This has direct welfare consequences because resources may not be allocated efficiently, but the magnitude of the consequence depends on the actual response of the power plant. The results suggest that feedbacks can matter for this case—but not in every dimension, and in varying degrees. Both biological and economic consequences of not addressing feedbacks are sensitive to the initial environment, behavioral perceptions about the state of the environment, and completeness of beliefs. (Details on the modeling and simulations can be found in Finnoff et al. [2005].)

Example 3: Leafy Spurge

Now consider a model of grazing and cattle ranching given two invasive species—leafy spurge and cheatgrass. Finnoff et al. (2008) developed an integrated model of a grazing land ecosystem and cattle ranching. The ecosystem consists of native grasses, leafy spurge, and cheatgrass. This model considers the stocks of each plant and cattle. Plants in these three species are assumed to behave as if they are maximizing their photosynthetic energy intake minus energy lost to respiration. To photosynthesize, they grow green biomass that provides them access to light; however, the plants are competitors for space. Over time, one species eventually will win the competition by driving out the other two.

The results show that, without humans, the native grasses are most likely to win. When humans enter and introduce cattle to the grazing ecosystem, the native grasses are placed at a competitive disadvantage, and leafy spurge generally becomes dominant, depending on grazing intensity. The model illustrates the importance of accounting for grazing decisions when forecasting the further spread of leafy spurge.

HOW DEEP SHOULD INTEGRATION GO?

Integrating ecological detail into economic models raises many issues on different levels. The fundamental issue is deciding how deep the integration should go within and between the economic and ecological systems. The tradition in economics is to represent ecological systems as a technical constraint, usually in the form of population growth for a single or aggregate species. The influence of all other species and other components of the ecological system are represented by a fixed carrying capacity. Common examples include the Lotka-Volterra model and its variants, where the structures of ecological systems are removed and replaced with lumped parameters that are presumed to be beyond human interference. If the policies prescribed by these models do not affect other components of the ecological system, this representation may be appropriate. But if the policies do affect other components of the ecological system, the system can be "bumped" to different results, with unintended consequences (Crocker and Tschirhart 1992). Models not addressing these other components may miss important linkages between humans and nature and provide misguided policy prescriptions.

Deciding just how deep to dig within and between the economic and ecological systems depends on the number of contact points between the systems and the indirect effects within the systems. For cases with one or two points of contact, a shallow, or abridged, form of integration might suffice. But in cases with multiple contacts or important indirect effects, deep integration is necessary. But in doing so, it is necessary to make other simplifying assumptions. Such deeply integrated models may not be more realistic if the feedback loop or other representations do not conform accurately to reality. Addressing the challenge of adding more realism and being forced to solve a problem computationally rather than analytically require one to work with a solid theoretical framework that guides the depth of integration.

We illustrate the depth of integration challenge by using an example based on Finnoff and Tschirhart (2008) that examines the Alaskan economy and a marine ecosystem comprising Alaska's Aleutian Islands and the Eastern Bering Sea. Figure 2.2 shows the ecosystem and economic interactions and illustrates the 13 key ecological descriptors and the feedback loops. The economy consists of Alaskan

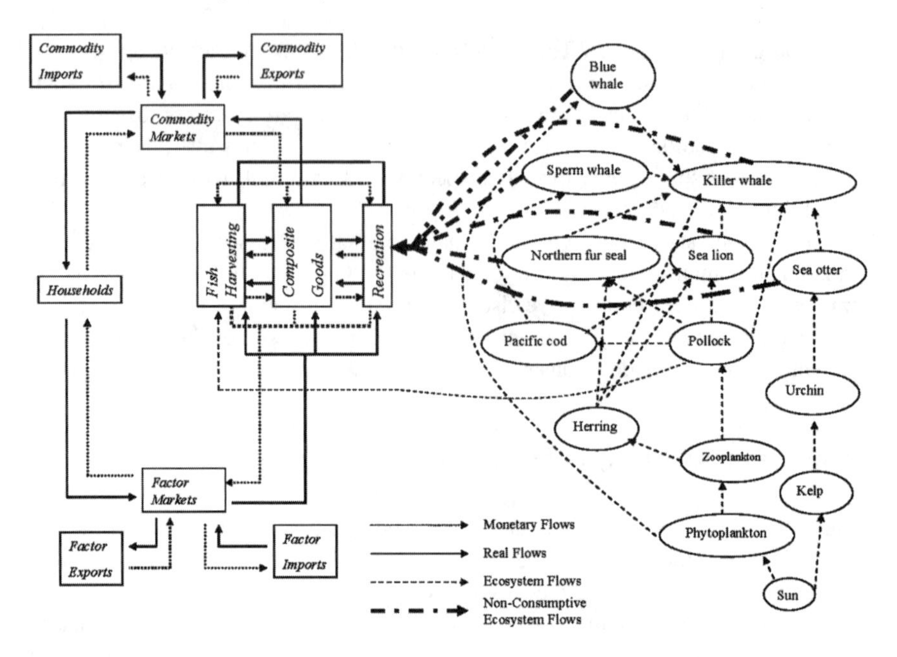

FIGURE 2.2.
Bering Sea web of life. Reprinted from Finnoff et al. (2006), with permission of the American Agricultural Economics Association.

households and producing sectors linked to one another and the rest of the world through commodity and factor markets. All species in the food web are linked through predator–prey relationships, and several species provide inputs to economic production. The prominent groundfish of the system, pollock, support a substantial fishery and marine mammals, including Steller sea lions (an endangered species), killer whales, blue whales, sperm whales, northern fur seals, and sea otters. All of these species provide nonuse inputs to the state's recreation sector. For a policy issue, we focus on the endangered Steller sea lion recovery via alternative pollock harvest quotas.

The first level of analysis is to understand the behavior of the actors in figure 2.2. Economists study the behavior of individual consumers and producers. Consumer behavior has people within the household sector making choices regarding combinations of goods and services. Producer behavior is likewise captured by individual firms within the fish harvesting sector choosing both their optimal mix of inputs and their optimal output level. Alternative quota levels are interpreted as changes in the prices faced by the households or producers. Similarly, ecologists study the behavior of individual animals; they would consider an individual pollock's optimal foraging behavior and how foraging changes affect pollock populationsThe alternative quotas would be interpreted as changes in the pollock populations.

The next level of analysis is to integrate all economic and ecological agents directly affected by pollock quotas through a bioeconomic harvesting model. In the economic system, individual consumer demands for pollock are aggregated to derive market demand, required for producers' decisions. Producer supplies are in part determined by the availability of pollock, which is derived from the aggregation of individual pollock behavior and population dynamics. Therefore, this level requires integration across the household, fish harvester, and pollock components. Linking these three components allows the derivation of market demand and pollock supply, which allows an assessment of how alternative quota policies affect the whole system.

But this level of integration is insufficient if we are interested in how the repercussions of the policies affect all of Alaska. In this case, deepening the analysis a further step within the economic system is necessary to include the other producing sectors of the Alaskan economy (recreation and composite goods in figure 2.2), all other household demands, and trade flows into and out of the region. A complication arises, however, because the recreation industry depends on the marine mammals. Still further depth of integration is needed to increase depth within the ecosystem to account for the predator–prey relationship shown in figure 2.2.

Finally, another level of integration is needed with nonmarket valuation. We discuss this more in chapter 8 (see also Shogren et al. 2006). Nonmarket valuation involves assessing the total values (e.g., existence values) associated with scenarios of reduced human and environmental damages posed by some invasive species so we can better understand the net benefits of policy. The idea is that valuation work needs integration models to develop credible valuation scenarios. In turn, integration models need the parameters as defined by valuation work to capture the full range of benefits associated with the web of life. For instance, in the Yellowstone Lake case, Settle and Shogren (2006) integrated a valuation experiment within their bioeconomic model. They developed the Yellowstone Interactive Survey to ask people to value alternative scenarios designed to inform their integrated model. They determined the value for seeing/catching each species and used these estimates to parameterize the value to see/catch each species in measuring the visitor's welfare from Yellowstone National Park. The disquieting result that people preferred fixing the roads to protecting the native cutthroat trout emerged directly from this integration. Both valuation and bioeconomic modeling can likely be more relevant for policy if the scenarios people are asked to value are valid and if the scenarios created were informed by values stated by actual people. There are gains from joint production of values and feedback loops between economic and ecological systems.

WHAT ARE THE DISCIPLINARY IMPEDIMENTS OF ECOLOGICAL-ECONOMIC MODELING?

Ecological-economic modeling is necessary and feasible in principle and is accompanied by challenges and pitfalls. Ecological-economic modeling combines the knowledge and concepts of two disciplines using a particular methodological approach—modeling. Combining these two disciplines requires (1) an in-depth knowledge of both disciplines by the researchers involved, (2) adequate identification and framing of the problem to be investigated, and (3) a common understanding of modeling and scales between economists and ecologists. In fulfilling these requirements, impediments and pitfalls, which are typical for ecological-economic modeling, are likely to come up. As stated succinctly by a colleague, learn what other disciplines can do for you, and what you can do for them.

Deep Knowledge of the Two Disciplines

The average economist's awareness of what ecologists do, and vice versa, is not well developed. Economics is sometimes confused by ecologists with business or finance. Some economists think ecologists are solely interested in collecting and studying plants and animals for their own sake, failing to appreciate that land management is a major issue in such subdisciplines as landscape ecology and conservation biology. Such confusions or prejudices are probably restricted to researchers who have limited or no experience with the other discipline. But even scientists who closely work with colleagues from the other discipline often do not have a profound understanding of this discipline, for numerous reasons, including the benefits of specialization.

Limited knowledge of the other discipline becomes an issue when a scientist lets his or her own narrow focus assume that simplified views represent a complete picture of the other discipline's concepts, ideas, and methods. Then, she or he misses the opportunity to make full use of the richness of knowledge that exists in the other discipline. Ecologists who assume that, by integrating costs of conservation measures into their models, the full knowledge available in economics has been automatically incorporated may miss essential aspects of a problem. Examples include transaction costs, asymmetric information between policy makers and land users, property rights, and risk aversion of economic agents. Similarly, economists are often unaware of the knowledge ecologists have about the spatial, temporal, and functional structure of ecosystems and restrict themselves to simplified—spatially homogeneous, static or scalar—descriptions of ecological systems and processes.

In addition, ecologists and economists are taught to examine real-world phenomena in different ways. When looking at the same biodiversity management problem, they identify different factors they consider to matter, formulate different research questions, and set up different research projects.

Also, some researchers in both disciplines will acknowledge the depth of the other discipline, and they will learn from each other. Some will agree on the overall aim of the research, and they will cooperate during the course of the entire project. But a key issue remains that strains communication and integration of ecological and economic knowledge: how to handle questions of spatial and temporal scales that frequently differ across the two disciplines. Drechsler et al. (2007) surveyed 60 models related to biodiversity conservation that were randomly selected from eight ecological and economic journals. They found that economic models are conceptual, are formulated and solved analytically, and are static and do not address uncertainty; most ecological models are solved numerically or through simulation. Ecological models are more specialized—specific to a particular species and a geographic region, constructed on rules simulated step by step to model the dynamics of the system and consider various uncertainties. We are aware that computational models are applied in some fields of economics; they are less common relative to ecology, and rare in the economic analysis of biodiversity management.

CONCLUDING REMARKS

In this chapter, we have addressed three key questions: Is integration worth the cost? How much integration should be promoted? What are the challenges of integration? First, we have argued that integration is worth the cost, given the potential opportunity costs associated with inaccurate policy recommendations based on nonintegrated models. That said, we do appreciate the distinction between scientific "truth" and policy "truth," and between accuracy (unbiasedness) and precision (nonvariability). Policies will be made with varying degrees of precision given the existing state of knowledge. But invasive species management challenges all have both an ecological and a socioeconomic dimension. These challenges can be addressed with more accuracy and precision by integrating ecological and economic knowledge that captures both the key elements of the two systems and the feedbacks between the two systems. While the costs of bioeconomic modeling can be sizable today, and the precision can without doubt be improved, as we learn by doing, the costs will fall, and the nonvariability will increase. We are also creating a culture of communication that promotes the trading of ideas. Each discipline has its comparative advantage, and there are most certainly gains from trading ideas and information about what parameters matter and why, and which feedback loops deserve high priority in any modeling exercise. We present the theory of endogenous risk as one potential organizing framework to reduce these costs. In addition, the relative costs of not attempting to advance bioeconomic modeling—as measured by underproductive and cost-ineffective management recommendations—are probably much greater than short-run costs of trying to integrate.

Second, how much integration we should undertake depends on the problem at hand. Over the years, traditional bioeconomic modeling has improved environmental and natural resource decision making. Today, researchers are exploring the next level of integration by expanding the number of feedback loops within and among systems and by making a better link to nonmarket valuation work. This message applies in general to natural resource economics, and in particular to invasive species economics. The open question is how to determine the appropriate level of integration. Where and at what invasion stage is economic precision more important than biological precision, and vice versa? Is a traditional damage function approach sufficient? Does a one- or two-state variable optimal-control model provide enough guidance, or do we require an even deeper integration among and within disciplines that may only be solved computationally? How do biological and economic systems interact across different spatial and temporal scales? What is the influence of cross-scale interactions on system dynamics? A solid theoretical framework provides the needed foundation for the empirical work to rest upon—the question becomes finding the appropriate numerical and computational approaches to complement the theory when closed-formed analytical solutions become unfeasible. One now must judge the array of methods based on results. Our decisions on the depth of integration in invasive species economics will evolve from our experience in pushing the boundaries of both rigorous theory and the observational pattern recognition.

Third, the future challenge for ecologists is to not regard anthropogenic environmental impacts as exogenous parameter shocks. Likewise, the future challenge for economists is to not view the environment as an aggregate or macro form in which interactions with the form can be dismissed. Explicit recognition of joint determination has an additional advantage: both ecology and economics can be better motivated to account for the complexities of both ecosystems and economic systems. To reiterate our adage, learn what other disciplines can do for you, and what you can do for them.

Acknowledgment This chapter draws on material written with co-authors that appears in Settle et al. (2002), Finnoff et al. (2006), and Wätzold et al. (2006). We thank Tom Crocker for influencing our thinking on integration in more ways than is evident throughout this chapter. Two reviewers provided many helpful comments and suggestions.

References

Barbier, E. 2001. A note on the economics of biological invasions. Ecological Economics 39:197–202.

Brown, G., and D. F. Layton. 2001. A market solution for preserving biodiversity: the black rhino. Pages 32–50 *in* J. F. Shogren and J. Tschirhart, editors. Protecting endangered species in the United States: biological needs, political realities, economic choices. Cambridge University Press, New York.

Clark, C. 1976. Mathematical bioeconomics: the optimal management of renewable resources. Wiley, New York.

Crocker, T. D., and J. Tschirhart. 1992. Ecosystems, externalities, and economies. Environmental and Resource Economics 2:551–567.

Daily, G., editor. 1997. Nature's services: societal dependence on natural ecosystems. Island Press, Washington, DC.

Daly, H. 1968. On economics as a life science. Journal of Political Economy 76:392–406.

Dixit, A. K., and R. S. Pindyck. 1994. Investment under uncertainty. Princeton University Press, Princeton, NJ.

Drechsler, M., V. Grimm, J. Mysiak, and F. Wätzold. 2007. Differences and similarities between ecological and economic models for biodiversity conservation. Ecological Economics 62:232–241.

Ehrlich, I., and G. Becker. 1972. Market insurance, self-insurance, and self-protection. Journal of Political Economy 80:623–648.

Finnoff, D., C. Settle, J. Shogren, and J. Tschirhart. 2006. Invasive species and the depth of bioeconomic integration. Choices: The Magazine of Food, Farm and Resource Issues 21:147–151.

Finnoff, D. C., J. Shogren, B. Leung, and D. M. Lodge. 2005. The importance of bioeconomic feedback in invasive species management. Ecological Economics 52:367–381.

Finnoff, D., A. Strong, and J. Tschirhart. 2008. A bioeconomic model of cattle stocking on public land threatened by invasive plants and nitrogen deposition. American Journal of Agricultural Economics, in press.

Finnoff, D., and J. Tschirhart, J. 2008. Linking dynamic economic and ecological general equilibrium models. Resources and Energy Economics 30:91–114.

Freeman, A. M., III. 1993. The measurement of environmental and resource values: theory and methods. Resources for the Future, Washington, DC.

Heal, G. 1998. Valuing the future: economic theory and sustainability. Columbia University Press, New York.

Hirshleifer, J. 1983. From weakest-link to best-shot: the voluntary provision of public goods. Public Choice 41:371–386.

Kaeding, L., G. Boltz, and D. Carty. 1995. Lake trout discovered in Yellowstone Lake. Pages 4–11 *in* J. Varley and P. Schullery, editors. The Yellowstone Lake crisis: confronting a lake trout invasion. A report to the director of the National Park Service. Yellowstone Center for Resources, National Park Service, Yellowstone National Park, WY.

Leung, B., D. M. Lodge, D. Finnoff, J. F. Shogren, M. A. Lewis, and G. Lamberti. 2002. An ounce of prevention or a pound of cure: bioeconomic risk analysis of invasive species. Proceedings of the Royal Society Series B Biological Sciences 269:2407–2413.

National Invasive Species Council. 2001. Management plan: meeting the invasive species challenge. National Invasive Species Council, Washington, DC.

Perrings, C. 2002. Modelling sustainable ecological-economic development. Pages 179–201 *in* H. Folmer and T. Tietenberg, editors. The international yearbook of environmental and resource economics. Edward Elgar, Cheltenham, UK.

Perrings, C., M. Williamson, E. B. Barbier, D. Delfino, S. Dalmazzone, J. Shogren, P. Simmons, and A. Watkinson. 2002. Biological invasion risks and the public good: an economic perspective. Conservation Ecology 6(1):1 (www.consecol.org/vol6/iss1/art1).

Saphores, J. D. M., and J. F. Shogren. 2005. Managing exotic pests under uncertainty: optimal control actions and bioeconomic investigations. Ecological Economics 52:327–339.

Settle, C., T. D. Crocker, and J. F. Shogren. 2002a. On the joint determination of biological and economic systems. Ecological Economics 42:301–311.Settle, C., and J. F. Shogren. 2002. Modeling native-exotic species within Yellowstone Lake. American Journal of Agricultural Economics 84:1323–1328.

Settle, C., and J. Shogren. 2006. Does the integration of biology and economics matter for policy? The case of Yellowstone Lake. Topics in Economic Analysis and Policy 6(1) (www.bepress.com/bejeap/topics/vol6/iss1/art9).

Shogren, J. F. 2000. Risk reduction strategies against the "explosive invader." Pages 56–69 *in* C. Perrings, M. Williamson, and S. Dalmazzone, editors. The economics of biological invasions. Edward Elgar, Cheltenham, UK.

Shogren, J., and T. Crocker. 1991. Risk, self-protection, and ex ante economic value. Journal of Environmental and Economic Management 20:1–15.

Shogren, J., D. Finnoff, C. McIntosh, and C. Settle. 2006. Integration-valuation nexus in invasive species policy. Agricultural and Resource Economics Review 35:11–20.

Shogren, J., J. Tschirhart, T. Anderson, A. Ando, S. Beissinger, D. Brookshire, G. Brown, Jr., D. Coursey, R. Innes, S. Meyer, and S. Polasky. 1999. Why economics matters for endangered species protection. Conservation Biology 13:1257–1267.

Sohngen, B., and R. Mendelsohn. 1998. Valuing the impact of large-scale change in a market: the effect of climate change on U.S. timber. American Economic Review 88:686–710.

Swallow, S. 1996. Resource capital theory and ecosystem economics: developing nonrenewable habitats with heterogeneous quality. Southern Economic Journal 63:106–123.

Wätzold, F., M. Drechsler, C. W. Armstrong, S. Baumgärtner, V. Grimm, A. Huth, C. Perrings, H. Possingham, J. Shogren, A. Skonhoft, J. Verboom-Vasiljev, and C. Wissel. 2006. Ecological-economic modeling for biodiversity management: potential, pitfalls, and prospects. Conservation Biology 20:1034–1041.

3

Trait-Based Risk Assessment for Invasive Species

Reuben P. Keller and John M. Drake

In a Clamshell

Preventing the arrival of invasive species is the only sure way to avoid their impacts. Scientists have long believed that invasive species have different biological and ecological traits compared with species that are not invasive. Recently, trait identification has been formalized as a risk assessment tool for predicting the impact of species before they are introduced. These predictions can be used to justify policy and management actions aimed at preventing the arrival of species that are deemed to pose a high risk. In this chapter, we present the ecological and economic justification for the use of trait-based risk assessment and review the history of the science and associated policy. Although risk assessments are presently mandated by only a small number of countries, recent methodological improvements and the demonstration of large financial benefits mean that research and policy efforts in this area are likely to grow.

The political justification of environmental policy generally begins from one of two points. On the one hand, a strict biocentric view supposes that society has an obligation to protect the environment. A pragmatic biocentrist would amend this assertion to include the qualification that this protection can come at some cost—but not too severe a cost—to society. On the other hand, an anthropocentric view supposes that the obligation to protect the environment only obtains insofar as it serves the long view aimed at protecting the safety and security of citizens, options on society's future productivity, and the general welfare of humanity. The premise of this book—a book about invasive species and human economic systems—is that only the pragmatic biocentric and anthropocentric views are acceptable to society. Thus, in either case we are concerned about human welfare and costs to society as the bottom-line arbiter of acceptable policy: there are some benefits we wish to maximize, and similarly, we wish to minimize costs. Environmental policy is therefore a process of weighing costs and benefits.

Risk analysis is an approach to incorporating both the cost-benefit trade-off that seems a natural basis for decision making and the fact that at any point in time the state of scientific understanding—the ability to predict future events—can only be asserted probabilistically. At its core, risk analysis consists of defining possible future states of the environment, assigning probabilities to these states (such probabilities may depend on events we control), and assigning costs and benefits to these states. By multiplying the costs associated with conditionally independent future states and their probabilities of occurrence, we obtain a distribution of future costs. Rules (i.e., a decision theory) can later be introduced to determine what actions should be taken to maximize welfare over the predicted distribution of future costs.

There are a number of reasons to prefer risk assessment at the introduction stage over other available options. First, risk assessment informs efforts to prevent the introduction of invasive species, making it a more efficient way to reduce impacts than either eradicating species once established or adapting socially and economically to pests. Indeed, eradication is often impossible, meaning that preventing introduction is usually the only option for avoiding the economic and environmental costs of invaders. Second, accurate risk assessments will permit introduction of low-risk species while preventing the arrival of high-risk species, and can thus present a lower detriment to commerce than other methods for preventing the arrival of invaders (e.g., eliminate entire trades). Third, risk assessment allows nations to make a scientifically defensible decision about whether a species should be allowed for trade. This meets World Trade Organization standards for regulating trade (World Trade Organization 2005) and thus ensures that producers have access to international markets as long as their products do not pose a high risk.

In this chapter, we discuss the application of risk analysis methods to the introduction of nonindigenous species. For as long as human societies have engaged in trade and travel, they have intentionally and accidentally transferred species. As humans come to occupy more of the globe, and as societies become more connected, the number of species being introduced increases (Lodge et al. 2006). A proportion of these species go on to become established, and a subset of these causes undesirable impacts. Because trade brings great benefits to society, including the potential to alleviate poverty (Sachs 2005), these invaders have generally been regarded as an unavoidable externality of activities that increase social welfare. Two groups of species have been aggressively managed, however. These are known pests of agriculture, such as foot-and-mouth disease, and human parasites and diseases. Efforts to prevent the proliferation of such species are warranted by large potential costs of invasion without the potential for benefits. In contrast, species that are intentionally introduced, such as garden plants, are chosen for their potential to produce economic benefits for the importer. In these cases, society must weigh the benefits against the costs that will accrue if the species becomes invasive. The risk assessment techniques discussed in this chapter are designed to identify the risk presented by these species, and thereby to determine whether they should be allowed for introduction.

An example where society needs scientific guidance on the likely fate of species introductions concerns fish in the North American Great Lakes. Fully 45 fish species have been introduced to this ecosystem; 24 of these have become established, of which 8 have large impacts and are considered invasive (Kolar and Lodge 2002). New fish species are often introduced to the Great Lakes basin, for example, through the aquarium and live food trades (Keller and Lodge 2007). If we could reliably identify which of these pose a risk of becoming invasive, these could reasonably be targeted for management or exclusion from trade. We refer to this example throughout the chapter to illustrate the concepts involved in risk analysis for invasive species and how it can be used to prevent the introduction of undesirable species.

Although the idea of preventing invader introductions is simple, we must throw into this mix the state of scientific understanding, which is necessarily uncertain. Which species will invade, in what sequence, when, and what the ecological and economic consequences will be are all plausibly open for empirical investigation, but highly uncertain. Thus, risk assessment relies on probabilities that must provide sufficient guidance for the risk assessment to meet two related standards. First, the assessment must discriminate between invasive species and those species that can be introduced with little risk of negative impacts. The first group must be accurately predicted so that the basic function of risk assessment—to reduce the number of invasive species—is met. Noninvaders must also be accurately identified because intentionally introduced species are imported for economic and other benefits. This leads to the second standard that a risk assessment tool should meet: it should produce net economic benefits for the importing country. That is, the total financial benefits (i.e., benefits from trade minus costs of invaders minus the cost of developing and implementing the risk assessment) must be greater under a policy of risk assessment than under either the policy of allowing all species or none.

Even if we cannot exhaustively identify future states or precisely assign probabilities to each, the exercise of risk analysis may still be valuable. We may find, for instance, that the chance of invasion for just one species (one which we can study intensively) is intolerably high. Even if we cannot add to this the future expected costs of all the other species, we may know that a policy is necessary. Alternatively, it may be the case that rates of introduction, establishment, and conversion of naturalized species to invasive ones are so low that when the product of all three transitions is considered, methods to discriminate invasive species from noninvasive species must be unreasonably accurate to be of any practical use (Smith et al. 1999). Recent research has shown that, at least for weeds in Australia, the real world is closer to the first case (Keller et al. 2007a).

RISK ANALYSIS BASED ON SPECIES TRAITS

On what basis, then, should preintroduction invasive species risk analysis be conducted? One possibility is to focus on pathway-based risk analysis, which aims to prevent all introductions from pathways identified as high risk. This approach is used by the U.S. Department of Agriculture, among others (Hennessey 2004). A limitation of the pathway-based approach, however, is that some organisms might be transported by multiple pathways, so eliminating introductions from any pathway may not prevent establishment of any particular invader. Also, probably all pathways are a risk for some group of species, but it is not possible to eliminate all pathways of introduction. Hence, it is unclear if pathway-based regulation is an effective and/or efficient mechanism for controlling introductions. Another approach is to focus on individual release events. This approach recognizes that species will become invasive when their ecologies and natural histories are matched to those of the receiving environment, and when they are released in sufficient numbers to become established (see chapter 5). Limitations of this approach are that rarely will enough information be available about the particular circumstances of an introduction (current state of ecosystem, characteristics of released organisms) to reliably guide the decision to allow or prohibit an introduction. Furthermore, the number of individual introductions is astronomical, so it is not feasible to regulate them individually. Finally, introductions are probably not independent, so the rescue effects of subsequent introductions should be considered when determining the chance of establishment (Drury et al. 2007).

As a complement to pathway and individual-based risk assessment, this chapter is about trait-based risk assessments that aim to determine whether introductions of particular species are risky, rather than the level of risk posed by multiple species through a single pathway or the risk associated with individual introduction events. Trait-based risk assessments assume that the characteristics associated with species becoming invasive are constant across time. Thus, if we can find patterns in traits that discriminate between the groups of species that have and have not become established or invasive in the past, we can apply those patterns to species that may be introduced in the future to determine the risk they pose. These patterns can be represented as a rubric, dichotomous key, computer algorithm, or some other classification instrument, and we can assign species to categories of risk using information about phenotypic, taxonomic, and ecological characters that are relatively well conserved across individuals within a species but likely to be good discriminators between invasive and noninvasive species. This approach differs from controlling individual introductions by using statistical approaches that exploit the cumulative information from many introductions of many species.

Previous Work on Traits of Invasive Species

Trait-based risk assessment aims to discriminate invasive nonnative species from noninvasive nonnative species (reviewed in Lodge 1993; Mack 1996; Kolar and Lodge 2001; Rejmánek et al. 2005; Garcia-Berthou 2007; Pyšek and Richardson 2007; Hayes and Barry 2008). Whether the fate of specific introductions will ever be predictable has been controversial (Daehler and Strong 1993; Lodge 1993; Smith et al. 1999; Williamson 1999; Keller et al. 2007a), and this ongoing debate may be an important factor in explaining why only a small number of countries use risk assessment to determine which species should be allowed for import. To be successful, however, a risk assessment does not need to correctly assign all species to either invasive or noninvasive categories. It does, however, need to be accurate enough to meet the two conditions explained above: it must reduce invasions while not causing undue impacts to trade, and it must produce net economic benefits. In the following, we review a subset of the studies that have investigated the traits associated with transitioning steps in the invasion process.

Rarely have thorough comparative studies been made of introduced species that become invasive and those that do not. In one case, Rejmánek and colleagues (Rejmánek and Richardson 1996; Grotkopp et al. 2002) have built a body of work to predict invasiveness of *Pinus* species. Rejmánek and Richardson (1996) used data from the published literature to discriminate 12 invasive and 12 noninvasive pine species. Their model was 100% correct in predicting invasiveness using just three traits: mean seed mass, minimum juvenile period, and mean interval between large seed crops. Similarly, Grotkopp et al. (2002) found that relative growth rate of pine seedlings was positively correlated with invasiveness, while seed mass and generation time were negatively correlated with invasiveness. In another case, Reichard and Hamilton (1997) compiled data on nonnative woody plants introduced to the United States and Canada prior to 1930. Of 349 species in the data set, 235 are now classified as invasive. An analysis of 149 of 207 species for which all attributes were available correctly classified the invasive status of 86.2% of the remaining 58 species. The strongest predictor was whether the species was invasive elsewhere in the world.

A variation of this approach is to identify predictors of degrees of invasiveness, looking only at nonnative species. For instance, Hamilton et al. (2005) sought to identify traits associated with abundance of 150 nonnative plant species in eastern Australia. Controlling for time since introduction, they identified small seed mass and high specific leaf area as predictors of invasiveness across the group. Hamilton et al. (2005) conclude that predictions can be useful if targeted to an appropriate spatial extent in a particular place; that is, predictive traits will not be universal.

More commonly, differences in trait suites of native species and alien species have been studied. For instance, Baruch and Goldstein (1999) compared field-based measurements of leaf characteristics of 30 nonnative plants and 34 native species in Hawaii. They found that nonnative plants have higher specific leaf area, photosynthetic rates, nitrogen concentration, phosphorus concentration, and nitrogen use efficiency than do native species. Crawley et al. (1996) compared native and nonnative species in the contemporary British flora. Controlling for phylogenetic confounding, they found that nonnative plant species were taller, had larger seeds and reduced duration of seed dormancy, flowered either early or late, and were more likely to be pollinated by insects than were native species. Similarly, Thompson et al. (1995) found that nonnative species are more likely to be clonal, polycarpic perennials with erect leafy stems and a transient seed bank, and Williamson and Fitter (1996) found that distribution and morphology (e.g., plant size) were more important than life history (e.g., reproductive modes).

Finally, Goodwin et al. (1999) identified variables discriminating 165 pairs of European species in which only one of the pair had invaded Canada. Since it is not known if each species of their pair had equal opportunity to establish (i.e., equal propagule pressure), this analysis conflates factors that predispose species to introduction and factors that predispose species to establishment. Using 110 species pairs for model estimation, Goodwin et al. (1999) found that only geographic range was significantly associated with invasive species and could correctly predict invasive status of the remaining 55 pairs 70% of the time.

From Traits Associated with Invasiveness to Trait-Based Risk Assessment

It is perhaps unsurprising that the studies reviewed above ultimately come up with very different results: they focus on different groups of species in different times and places and at different spatial extents (an issue highlighted, e.g., by Hamilton et al. 2005). This emphasizes the need to construct risk assessments at taxonomic and geographic scales for which it is reasonable to expect that the traits associated with invasiveness will be conserved. It is also important to limit risk assessment to just one step in the invasion process (see chapter 1), thus recognizing that different traits may be associated with the transition from introduced to established than with the transition from established to invasive. For fish in the Great Lakes, for example, Kolar and Lodge (2002) defined their taxonomic group as fish, their geographical range as the Great Lakes basin, and their invasion step as the transition from introduced to established.

Once the scale of the risk assessment has been set, the logic is that traits associated with previously introduced species passing through the invasion step will be the same for species introduced in the future. Thus, the type of patterns found by the research reviewed above can be applied to determine the risks posed by species proposed for introduction in the future. Ideally, the species characters used in trait-based risk assessment would be precisely those that confer on species the ability to be invasive in particular habitats and would therefore be the same traits posited by ecological theories of extinction/colonization, r- versus K-selection, and so on. However, in practice, either these theories often fail to hold up, or the data are extremely difficult to obtain for the large number of species that need to be screened and so surrogates, identified through statistical or machine-learning analysis or through causal intuition, may also be used.

Thus, the selection of species characteristics for analysis is a critical part of any risk analysis. First, characteristics should be used only if they can be reasonably expected to be related to invasiveness in that taxon. This step requires a good knowledge of the ecology of the taxonomic group in question and of the receiving environment. Second, characteristics should be chosen only if data are available for all or most species within the taxa that have or have not passed through the invasion step in question and that are likely to reach that invasion step in the future. An important area for research is methodology to handle missing data. Additionally, to be practical, a risk assessment should only require data that are available from the native range of the species being assessed. Much data, such as speed of range expansion, can be gathered only from the nonnative range of species, and these data are often useful for predicting how a species will act if it becomes established elsewhere. Including these data in a risk assessment, however, would mean that only species already established, and well studied, outside their native range could be assessed.

Returning to the example of fish introductions to the Great Lakes, Kolar and Lodge (2002) followed the trait-based approach outlined above to find a rule for predicting which species that may be introduced in the future are likely to become established. They gathered data on 13 life-history traits (e.g., fecundity), five habitat requirements (e.g., minimum temperature tolerance), six variables related to prior invasion success (e.g., species history of invasion elsewhere), and the economic benefit that humans derive from the species. Statistical techniques were used to compare species that have become established to those that did not persist (see below), and a decision tree was created to predict the chance that fish species introduced in the future will successfully establish.

Once the characteristics are chosen and the data gathered, it is necessary to decide on a method for discriminating between species that have and have not passed through the invasion step. A range of quantitative statistical and machine learning tools are available for this, and it is also possible to use qualitative methods. The most prominent risk assessment tool yet created is the Australian Weed Risk Assessment (WRA) (Pheloung 1995), which uses answers to 49 questions (table 3.1) to predict whether the species being assessed poses an unacceptably high risk of becoming invasive. This risk assessment was developed based on a list of 139 known "serious invaders," 147 "moderate invaders," and 84 "noninvader " plants (Pheloung 1995). A list of plant characteristics believed to be associated with invasiveness was then drawn up, and data were gathered for each plant and its characteristics. These data were converted to scores, and the scoring system was calibrated to give the best discrimination among the three groups. Based on responses to the risk assessment questions (table 3.1) a plant receiving a score > 7 is blacklisted (i.e., banned) from introduction, a plant receiving a score < 0 is whitelisted (i.e., allowed), and species in between require "further study" before a decision is made. The 49 questions are divided into categories (e.g., reproductive traits), and although not all questions need to be answered for a decision to be made, a minimum number of questions from each category do need to be answered.

The Australian WRA was one of the first risk assessments created for invasive species, but more recent risk assessments have generally been closer to either of the qualitative or quantitative poles. An example of a qualitative risk assessment is the Generic Nonindigenous Aquatic Organisms Risk Assessment (Orr 2003), which requires assessors to gather information about the potential of the species to establish, spread, and cause impacts, to make a judgment about the risk posed, and thus to decide whether the species is safe for introduction (table 3.2). Although this approach obviously relies on data that will often be quantitative, the ultimate decision is based on the judgments of the assessor rather than strict quantitative criteria. Indeed, for this risk assessment there is no standard list of species traits about which information must be collected.

At the other end of the spectrum is the increasing number of risk assessments involving strict quantitative criteria and where the selection of those criteria is made using statistical or machine learning models. The most commonly used statistical tools include logistic regression (e.g., Keller et al. 2007b) and discriminant analysis (e.g., Rejmánek and Richardson 1996; Kolar and Lodge 2002). These tools analyze

TABLE 3.1. Questions to be answered for the Australian Weed Risk Assessment (modified from Pheloung 1995).

Category		Questions about plant species
Domestication / cultivation	1.01	Is the species highly domesticated? If no, go to question 2.01.
	1.02	Has the species become naturalized where grown?
	1.03	Does the species have weedy races
Climate and distribution	2.01	Species suited to Australian climates?
	2.02	Quality of climate match data?
	2.03	Broad climate suitability (environmental versatility)?
	2.04	Native or naturalized in regions with extended dry periods?
	2.05	Does the species have a history of repeated introductions outside its natural range?
Weed elsewhere	3.01	Naturalized beyond native range?
	3.02	Garden/amenity/disturbance weed?
	3.03	Weed of agriculture/horticulture/forestry?
	3.04	Environmental weed?
	3.05	Congeneric weed?
Undesirable traits	4.01	Produces spines, thorns, or burrs?
	4.02	Allelopathic?
	4.03	Parasite?
	4.04	Unpalatable to grazing animals?
	4.05	Toxic to animals?
	4.06	Host for recognized pests and pathogens?
	4.07	Causes allergies or is otherwise toxic to humans?
	4.08	Creates a fire hazard in natural ecosystems?
	4.09	Is a shade tolerant plant at some stage of its life cycle?
	4.10	Grows on infertile soils?
	4.11	Climbing or smothering growth habit?
	4.12	Forms dense thickets?
Plant type	5.01	Aquatic?
	5.02	Grass?
	5.03	Nitrogen-fixing woody plant?
	5.04	Geophyte?
Reproduction	6.01	Evidence of substantial reproductive failure in native habitat?
	6.02	Produces viable seed?
	6.03	Hybridizes naturally?
	6.04	Self-fertilization?
	6.05	Requires specialist pollinators?
	6.06	Reproduction by vegetative propagation?
	6.07	Minimum generative time (years)?
Dispersal mechanisms	7.01	Propagules likely to be dispersed unintentionally?
	7.02	Propagules dispersed intentionally by people?
	7.03	Propagules likely to disperse as a produce contaminant?
	7.04	Propagules adapted to wind dispersal?
	7.05	Propagules buoyant?
	7.06	Propagules bird dispersed?
	7.07	Propagules dispersed by other animals (externally)?
	7.08	Propagules dispersed by other animals (internally)?
Persistence attributes	8.01	Prolific seed production?
	8.02	Evidence that a persistent propagule bank is formed (>1 year)?
	8.03	Well controlled by herbicides?
	8.04	Tolerates or benefits from mutilation, cultivation, or fire?
	8.05	Effective natural enemies present in Australia?

TABLE 3.2. Factors considered by the Generic Nonindigenous Aquatic Organisms Risk Assessment (modified from Orr 2003).

1. Estimate probability of the nonindigenous species being transported within a vector
2. Estimate probability individuals will survive passage in vector
3. Estimate probability of individuals becoming established where introduced
4. Estimate probability the population will spread
5. Estimate magnitude of economic impacts if species establishes
6. Estimate magnitude of environmental impacts if species establishes
7. Estimate impacts to social and/or political values

The assessors gather data that they consider relevant to address the seven factors listed. Each factor is scored, according to the opinion of the assessor, as being high, medium, or low. Scores are then aggregated to give a final risk rating for the species of negligible, low, medium, or high.

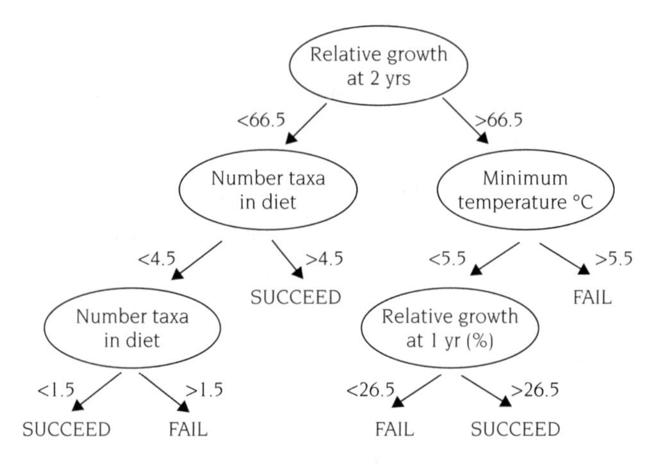

FIGURE 3.1.
Categorical and Regression Tree (CART)-based risk assessment for predicting whether fish species introduced to the Laurentian Great Lakes will succeed or fail to establish. Data for each species needs to be collected only for the questions asked. Reprinted from Kolar and Lodge (2002), with permission of the American Association for the Advancement of Science.

trait data to find the best possible combination and weighting of characteristics to discriminate between species that have and have not become invasive. Categorical and Regression Tree (CART) analysis, a machine learning approach, has also been applied in a number of cases (e.g., Reichard and Hamilton 1997; Kolar and Lodge 2002; Caley and Kuhnert 2006; Keller et al. 2007b). For example, Kolar and Lodge (2002) used CART to analyze the data for fish introductions to the Great Lakes. Their results (figure 3.1) show that, from the 25 species characteristics used, only four are required to correctly classify 42 of 45 introduced species as either established or not. To predict whether a species introduced in the future will establish, it is necessary to gather data on these four characteristics and then work through the decision tree.

CART begins by finding the split in any of the predictor variables that maximizes within-group homogeneity of the two groups produced (De'ath and Fabricius 2000).

Additional splits are made in the resulting groups, again using any of the available variables, to further homogenize the groups of species at the resulting nodes. This continues until a user-defined limit is reached (e.g., minimum number of species at each node). Species proposed for introduction can then be put through the same tree and an assessment made of the likelihood that they will become invasive. In contrast to CART, statistical discrimination techniques are often less intuitive, but may be more accurate. In particular, multivariate statistical model results are not easily visualized, unless few traits are included in the risk assessment.

Risk assessments produced to date vary in the type, quality, and volume of data to be correlated with species invasiveness. While variable in their success, there are some general deficiencies. First, many of these analyses use logistic regression as the classification workhorse. However, logistic regression is quite inflexible with respect to the types of models it fits. Indeed, logistic regression is probably better suited to hypothesis testing on binary observations than for creating a decision tree for risk classification. Even those studies that use more flexible models (e.g., discriminant analysis, CART) generally make a linearity assumption that limits the flexibility of the model to fit the subtle interactions of traits that predispose species to colonization. Given the relatively impressive success of linear models, one expects that careful application of newer, more flexible methods such as neural networks may well return decision trees with even greater accuracy (see below).

A particular advantage of trait-based risk assessment is that analyses can be conducted very rapidly using information available on the Internet or in the published literature. As an example, it takes approximately 2 days to screen a species using the Australian WRA (Pheloung 1995). The speed with which assessments can be performed is critical, particularly for species that are candidates for trade, meaning that the assessment procedure must respond to the demands of commerce. However, speedy assessments are valuable for governments, too, because they minimize the costs of performing analysis, thereby reducing the burden of approving or disapproving species for trade. Trait-based analyses also have a policy advantage. In the United States, there is a precedent in the Lacey Act of 1900 (amended 1988) and the Federal Noxious Weeds Act of 1974 for regulations to categorically prohibit the introduction of particular species. This "blacklist" approach provides reasonable standards for enforcement (inspectors need only to be able to identify the relatively small fraction of species that have been deemed an unacceptable risk) and minimizes the need for costly to collect postintroduction information such as average rates of spread, interactions with native species, and ability to predict species impacts.

Model Estimation and Validation

Although a large number of risk assessment tools are now available, validation of trait classification models has not received sufficient attention. Indeed, hypothesis tests notwithstanding, given the high economic and political costs of prohibiting species introduction, for any risk classification to be useful it is imperative to demonstrate that it improves the accuracy of classification compared with blind guessing.

Actually, this is a relatively low bar that should be readily met by any careful analysis. Ideally, the pool of species under consideration and the environment to which they are being introduced are constant, and introductions are independent, so that the set of relevant traits can be viewed as coming from an appropriately independent identically distributed distribution. In this case, validation would proceed by splitting the available data into two sets—for model estimation and model validation—prior to fitting a model. This was done in the studies by Rejmánek and Richardson 1996, Reichard and Hamilton (1997), and Goodwin et al. (1999) described above. Reichard and Hamilton (1997) used a random subset of 75% of the 235 woody plant species in their data set to train the model, with the remaining species used to determine how well the model performs on independent data. Kolar and Lodge (2002) used a similar approach for estimating the accuracy of their CART decision tree for fish in the Great Lakes (figure 3.1). They took a random subset of 10% of the 45 species and calculated a decision tree using the remaining 90% of records. The tree was then tested on the species that were put aside. This procedure was repeated 100 times, and the total accuracy (82%) is the percentage of times that species in the reserved group were correctly classified by the model. This can be compared to the classification tree constructed using all data, which correctly classifies 93% of species.

Where the total data set is not large enough to withhold a subset for testing (e.g., Keller et al. 2007b), the model can be estimated using a bootstrap procedure such as leave-one-out cross-validation or jackknife. Then, the error reported would be the average cross-validation error and not the error on a testing data set. The downside of this approach is that while splitting testing/training data sets might alert the analyst to inhomogeneity in the training data (i.e., if the data were split along chronological lines and the error on the test data set and the cross-validation error were very different), a jackknife would never reveal this inconsistency. By contrast, a virtue of the jackknife is that, if set up appropriately, it can be used to ensure that hierarchical phylogenetically relevant contrasts are treated appropriately.

ECONOMICS OF RISK ASSESSMENT

Although risk assessment can produce large environmental benefits by supporting programs that prevent the introduction of invasive species, the question of whether it can produce financial benefits for a nation has long been debated. This debate has arisen because introduced species generally become invasive at a low rate, and because accuracy of risk assessment predictions has been low. These can combine to give a high rate of false positives (i.e., noninvaders predicted to be invaders). Because it is costly to prohibit a species that poses a low risk of becoming invasive, it is possible for risk assessment to do more damage than good to an economy. Smith et al. (1999) demonstrated this with a theoretical model showing that a risk assessment would need extremely high accuracy, and invasive species would need to cause much more damage than noninvasive species caused benefits, to be financially worthwhile.

Keller et al. (2007a) have recently constructed a bioeconomic simulation model to determine when the application of risk assessment tools will produce net economic benefits for the importing country. Over the last several years, the data available regarding invasion costs and base rates have increased, and Keller et al. (2007a) were able to create a model for testing specific cases. The model includes ecological realism (e.g., lag times between establishment and invasion) and is flexible for a number of economic assumptions (e.g., discount rate, time horizon of simulation). Once these assumptions are set and the data gathered, the model is used to determine whether a particular risk assessment will produce economic benefits when applied to a particular suite of species (see appendix 3.1 for full model).

Keller et al. (2007a) applied the model to the Australian trade in nonindigenous ornamental plants and showed that risk assessment produces large net financial benefits for the Australian economy over the range of reasonable economic assumptions (figure 3.2). Indeed, for a number of reasons, the economic benefits from applying risk assessment to other taxa and trades are likely to be even greater. First, plants have a low base rate of invasion compared to other taxa, with only 5% of plant species introduced to Australia for the ornamental trade having become invasive (Virtue et al. 2004). In contrast, fish and mammals introduced to the United States from Europe become invasive at base rates of 25% and 62%, respectively (Jeschke and Strayer 2005). Second, the estimates of cost of invasions used by Keller et al. (2007a) were limited to market costs. Invasive species also have many impacts that

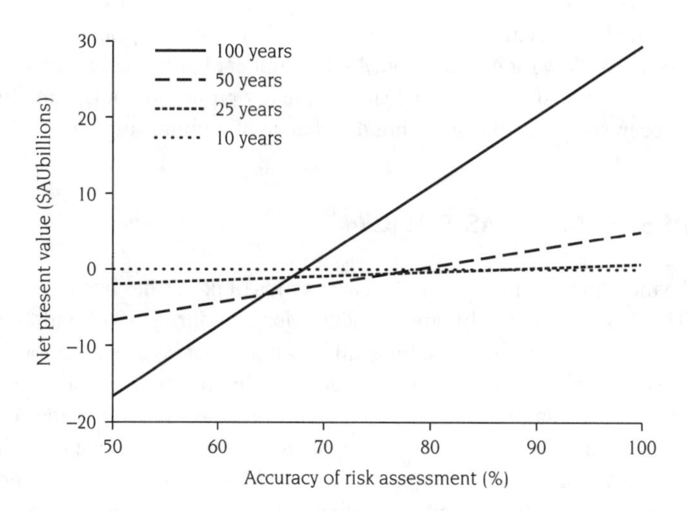

FIGURE 3.2.
Net present financial benefits from applying risk assessment to the Australian trade in ornamental plants. Accuracy of risk assessment is variable (x-axis) and leads to a value (y-axis) that depends on the time horizon of interest. Value is calculated as the total financial benefits from applying risk assessment minus the total financial benefits from allowing all species. Discount rate is hyperbolic (i.e., decreasing over time). If a policy maker is interested in using a risk assessment that has 90% accuracy and is interested in financial outcomes over 100 years, the value derived from this policy would be approximately $15 billion greater than the value of allowing all species. Reprinted from Keller et al. (2007a), with permission of The National Academy of Sciences of the USA.

are difficult to quantify in economic terms, such as loss of biodiversity and loss of recreation opportunities. If these could be included in economic models, the benefits of risk assessment would increase (see chapter 8 for examples of quantifying nonmarket costs). The available evidence from bioeconomic modeling suggests that risk assessment for invasive species is likely to produce large net financial benefits across a range of taxa and geographic regions.

Once a decision has been made to adopt a program of risk assessment, policy makers can further decide how cautious to be (Horan et al. 2002). The fate of any introduction remains uncertain, and the possibility for surprise—the introduction of a catastrophic invader that could not have been predicted as such—should be included in considerations, especially because many invasions are irreversible. For example, the introduction history of mollusks to the Great Lakes prior to 1987 indicated that invasions did occur but that impacts were generally low (Mills et al. 1993). An analysis assuming that the economic costs of any future invaders would be the same as the average cost of previous invaders would have indicated that it was not necessary to be particularly risk averse. Since this time, however, zebra and quagga mussels have invaded the Great Lakes, with annual economic impacts that are at least an order of magnitude greater than the sum of impacts from all other mollusk invaders (see chapter 12). In practical terms, an aversion to surprise outcomes from species introductions should encourage policy makers to use a lower risk threshold than would be considered optimal from the history of invasions in the region (Moffitt and Osteen 2006; Finnoff et al. 2007). This can be built into any program for determining which species to allow for introduction. As the number of invasions increases, the risk threshold may rationally increase because a greater proportion of potential impacts will have been sampled, thus reducing the chance of a future surprise.

RISK ASSESSMENT AS POLICY

By far the most successful risk assessment in terms of policy has been the Australian WRA. This was developed by government scientists during 1994 and adopted as national policy in 1997. Since that time, all new plant species proposed for introduction to Australia have been assessed and put on a whitelist or blacklist depending on the results. New Zealand has adopted a slightly modified version of the Australian WRA, and scientists in Hawaii (Daehler et al. 2004), Eastern Europe (Krivánek and Pysek 2006), and the United States (Jefferson et al. 2004) have tested how well it applies to different floras. The Australian WRA has been demonstrated to have consistently high accuracy across this range of geographies (Gordon et al. 2008). To the best of our knowledge, Australia and New Zealand are the only countries that use scientific risk assessment to screen all plant species for introduction.

The low international adoption of risk assessment stands in contrast to the relatively large number of tools that are available to discriminate between invasive and noninvasive species. For example, Kolar and Lodge's (2002) risk assessment for fish introductions to the Great Lakes has high accuracy, and although it has received much scientific and media attention, we are aware of no regulatory efforts during the 6 years since it was published to use it directly as a basis for policy (but see chapter 13 for a discussion of its indirect effects on policy). Most other scientific risk assessments have met the same fate.

One reason for the low uptake of risk assessments is that the trades that import live nonindigenous species, which stand to lose the most from such policies, have strongly argued that such policies would do more harm than good. In the United States, for example, although President Carter's Administrative Order 11987 of 1977 called for federal agencies to restrict the import of potentially harmful nonnative species, there is still no system in place for systematically screening species for invasiveness. Several trade groups have realized that such policies would reduce the number of species that they could sell and have formed a strong opposition. In the absence of a large lobby in support of risk assessment, this presidential order has been effectively ignored by federal agencies (Wade 1995).

At an international scale, however, the political will to implement risk assessment has increased over the last several years. Europe, South Africa and the United States are all considering and/or developing risk assessment programs for plant imports. New Zealand and Australia continue to have the strongest policies, with the Biosecurity New Zealand agency now responsible for, among other things, preventing the introduction of new invasive species in that country. Thus, over the next few years, it is likely that more countries will begin to use risk assessment as a tool for preventing introductions of invasive species.

Another reason for the low uptake of risk assessments may be the mismatch between political and ecological spatial scales. Kolar and Lodge's (2002) risk assessment applies to the Great Lakes basin and therefore includes part of two countries and portions of seven U.S. states and two Canadian provinces. Although this is an ecologically logical spatial unit, there is no single legislative body that could implement such a policy. Indeed, even if such a body existed, or even if the states and provinces agreed to multilaterally implement the risk assessment, there is no effective way to prevent the movement of species into the basin. The Internet trade in aquarium fish, for example, is largely unregulated in the United States, and species posing a high risk to the Great Lakes could easily be sent from other regions (Keller and Lodge 2007). Although this mismatch in scales may be partly responsible for slow uptake, we note that a risk assessment for woody plants in North America has been available since 1997 without being implemented (Reichard and Hamilton 1997), illustrating that other factors are also important.

LIMITATIONS OF RISK ASSESSMENT

Although risk assessment is already applied by some nations and offers great promise to others, a number of caveats need to be kept in mind. First, a disadvantage of trait-based risk assessment is that not all taxonomic groups have been sufficiently studied for trait data to be available. In general, the most data will be available for mammals and birds. Invertebrates and plants, in contrast, are less well studied, and the selection of characteristics may be limited as a result. Keller et al. (2007b) were not able to include any characteristics related to environmental tolerances in their risk assessment for mollusks in the Great Lakes because such data have not been gathered for all species. This situation is improving with the increasing number and accessibility of databases describing basic species biology. The availability of resources such as FishBase (www.fishbase.org) will thus make the construction of risk assessments faster, easier, and more accurate (Drake 2007).

Second, because trait data are usually easier to obtain for invasive than for benign nonindigenous species (more time is spent studying damaging species), risk assessments have often been created based on data from more invasive than noninvasive species. In turn, the resulting risk assessments are better at predicting invasive than noninvasive species. Caley et al. (2006) demonstrated this with a reevaluation of the Australian WRA, which was originally developed from a data set containing 77% invasive species. Given that only about 5% of the total number of plant species introduced to Australia have become invasive (Virtue et al. 2004), this assessment is biased toward successfully identifying invaders and against successfully identifying noninvaders. Indeed, Caley et al. (2006) showed that rates of false-positive predictions are probably much higher than estimated from the original data, meaning that a large number of noninvasive species are likely to be excluded from trade.

Third, risk assessment is not equally applicable to all vectors that introduce nonindigenous species. Risk assessments are conducted before species are introduced and are thus most straightforwardly applied to intentional introductions because the identities of species are known. Risk assessment tools can also be applied to accidental introductions, for example, to identify high-risk species from a suite of species that have the potential to be introduced. However, because the assessments are species specific, and because the identities of species in vectors of accidental introduction are rarely known, it is likely that more general attempts to limit the number of species in vectors of accidental introduction will be more effective (i.e., pathway-based risk management).

Fourth, current risk assessments assume that species and ecosystems are static over time. An increasing body of work, however, shows that invasive species can rapidly evolve in their new habitats and that native species can also adapt to the presence of invaders (Cox 2004). Additionally, the abiotic conditions of ecosystems can change naturally or as a result of anthropogenic forces such as climate change or improving water quality. Each of these factors could conceivably change a system to the extent that risk assessment based on patterns of previous invasions will not apply to future invasions. To make a risk assessment robust to these factors, it is important to use only those traits of species that are likely to be conserved. In reality, however, it is often difficult to use any other traits because sufficient data are generally unavailable for the types of traits that may rapidly change.

THE FUTURE FOR RISK ASSESSMENT

The two goals of risk assessment should be to accurately discriminate between species that will and will not become invasive and to generate economic benefits for the importing country. As noted above, it appears likely that risk assessment for invasive species will become increasingly adopted in the future. We hope that this spurs scientists to develop and adopt new and better methods, and we offer some suggestions to this end here.

Current risk assessments are only able to assign species to one of two categories—invasive or not. We know, however, that species have costs and benefits along a continuous scale, and a binary classification system is thus a large simplification. The development of tools that can predict not just whether a species will be invasive, but how much damage it will cause, would add significantly to the ability of policy makers to determine which species should be allowed for introduction. Likewise, better information about the benefits that a species proposed for introduction will bring is needed. For example, there are undoubtedly species that will have minor impacts as invaders, but that might have great benefits as ornamental or food species. In such cases, it may be permissible to allow the species and accept the damage it causes in return for the greater benefits. Conversely, a species may be predicted to be not invasive, but the predicted benefits may be so small that a cost–benefit analysis determines that it is optimal to prohibit import. Here, the potential for surprise should weigh in the decision.

The implementation of more powerful discrimination techniques, such as neural networks and a host of other machine learning tools, would likely both improve the accuracy of risk assessment and allow the incorporation of the more detailed data discussed in the preceding paragraph. Species traits interact in complex ways, but the currently available tools generally assume linearity. More complex analytical tools would be able to find the nonlinear patterns in data and apply them to species categorization.

Finally, we believe that greater collaboration among ecologists, economists, and policy makers will lead to risk assessments that are more likely to be adopted. Ecologists have developed techniques for risk assessment and have demonstrated that they can accurately discriminate between invasive and noninvasive species. The challenge now is to create tools that include economic realism and that meet the needs of policy makers. The greatest justification for the implementation of a risk assessment will occur when it can be demonstrated that the program will have both environmental and economic benefits.

APPENDIX 3.1. MODEL FOR DETERMINING ECONOMIC OUTCOMES FROM APPLYING TRAIT-BASED RISK ASSESSMENT AS POLICY

Keller et al. (2007a) produced a bioeconomic model for testing the conditions under which it is economically beneficial to implement a policy of risk assessment for invasive species. This model considers the accuracy of the risk assessment in question, the value of nonnative species to trade, the economic costs of invasive species, the base rate of invasion, the cost of implementing risk assessment, and the number of species introduced. The model development is as follows:

First, let the annual expected economic benefit (B_N) from allowing the importation of θ new species be

$$B_N = \theta V_T, \tag{1}$$

where V_T is the annual benefit generated by trade in a single species. The annual expected loss, C_N (i.e., that caused by species that become invasive), incurred by importing the same θ species is given by

$$C_N = \alpha\theta V_I, \tag{2}$$

where α is the base rate of invasion (= number of invaders introduced/total number of species introduced) and V_I is the annual economic loss caused by an invasive species. Combining equations 1 and 2 gives an annual expected net benefit (E_N) of allowing θ species for introduction as

$$E_N = \theta V_T - \alpha\theta V_I. \tag{3}$$

When a risk assessment with proportional accuracy A is used, the annual expected benefit (B_R) from introducing the θ species becomes

$$B_R = [(1-\alpha)\theta A + \alpha\theta(1-A)]V_T. \tag{4}$$

Equation 4 accounts for all correctly identified noninvaders and all incorrectly identified invaders having a positive value for trade. Although invasive species cause economic losses, this model also accounts for their benefits to trade, which is not affected by their invasiveness. Invasive species misidentified as harmless (i.e., false negatives) cause annual costs (C_R):

$$C_R = \alpha\theta\,(1-A)\,V_I \tag{5}$$

Including a fixed annual cost of administering the risk assessment (D), the annual expected net benefit from using risk assessment (E_R) is

$$E_R = [(1-\alpha)\theta A + \alpha\theta(1-A)]V_T - \alpha\theta\,(1-A)\,V_I - D. \tag{6}$$

Equation 6 can be used to project costs and benefits into the future to inform policy decisions. To do this, an appropriate discount rate must be chosen. Additionally, the lag time between a species being introduced and having benefits in trade (i.e., arriving at market), and the lag time between a species being introduced and becoming invasive (i.e., causing costs) must be considered and modeled. These lag times will depend on the specifics of the taxa and trade in question. Projections using equation 6 can be performed quite simply within standard spreadsheet software.

Acknowledgments We thank A. Silletti for assistance in preparing the literature review. Comments from Leif-Matthias Herborg, Kevin Drury, and one anonymous reviewer improved the chapter.

References

Baruch, Z., and G. Goldstein. 1999. Leaf construction cost, nutrient concentration, and net CO_2 assimilation of native and invasive species in Hawaii. Oecologia 121:183–192.

Caley, P., and P. M. Kuhnert. 2006. Application and evaluation of classification trees for screening unwanted plants. Austral Ecology 31:647–655.

Caley, P., W. M. Lonsdale, and P. C. Pheloung. 2006. Quantifying uncertainty in predictions of invasiveness, with emphasis on weed risk assessment. Biological Invasions 8:1595–1604.

Cox, G. W. 2004. Alien species and evolution: the evolutionary ecology of exotic plants, animals, microbes, and interacting native species. Island Press, Washington, DC.

Crawley, M. J., O. H. Harvey, and A. Purvis. 1996. Comparative ecology of the native and alien floras of the British Isles. Philosophical Transactions: Biological Sciences 351:1251–1259.

Daehler, C. C., J. S. Denslow, S. Ansari, and H. Kuo. 2004. A risk assessment system for screening out invasive pest plants from Hawai'i and other Pacific Islands. Conservation Biology 18:360–368.

Daehler, C. C., and D. R. Strong. 1993. Prediction and biological invasions. Trends in Ecology and Evolution 8:380.

De'ath, G., and K. E. Fabricius. 2000. Classification and regression trees: a powerful yet simple technique for ecological data analysis. Ecology 81:3178–3192.

Drake, J. M. 2007. Parental investment and fecundity, but not brain size, are associated with establishment success in introduced fishes. Functional Ecology doi:10.1111/j.0269-0269.2007.01318.x.

Drury, K. L. S., J. M. Drake, D. M. Lodge, and G. Dwyer. 2007. Immigration events in space and time: factors affecting invasion success. Ecological Modelling 206:63–78.

Finnoff, D., J. F. Shogren, B. Leung, and D. Lodge. 2007. Take a risk: preferring prevention over control of biological invaders. Ecological Economics 62:216–222.

Garcia-Berthou, E. 2007. The characteristics of invasive fishes: what has been learned so far? Journal of Fish Biology 71(supplement D):33–55.

Goodwin, B. J., A. J. McAllister, and L. Fahrig. 1999. Predicting invasiveness of plant species based on biological information. Conservation Biology 13:422–426.

Gordon D. R., D. A. Onderdonk, A. M. Fox, and R. K. Stocker. 2008. Consistent accuracy of the Australian weed risk assessment system across varied geographies. Diversity and Distributions 14:234–242.

Grotkopp, E., M. Rejmánek, and T. L. Rost. 2002. Toward a causal explanation of plant invasiveness: seedling growth and life-history strategies of 29 pine (*Pinus*) species. American Naturalist 159: 396–419.

Hamilton, M. A., B. R. Murray, M. W. Cadotte, G. C. Hose, A. C. Baker, C. J. Harris, and D. Licari. 2005. Life-history correlates of plant invasiveness at regional and continental scales. Ecology Letters 8:1066–1074.

Hayes, K. R., and S. C. Barry. 2008. Are there any consistent predictors of invasion success? Biological Invasions 10:483–506.

Hennessey, M. K. 2004. Quarantine pathway pest risk analysis at the APHIS plant epidemiology and risk analysis laboratory. Weed Technology 18:1484–1485.

Horan, R. D., C. Perrings, F. Lupi, and E. H. Bulte. 2002. Biological pollution prevention strategies under ignorance: the case of invasive species. American Journal of Agricultural Economics 84:1303–1310.

Jefferson, L., K. Havens, and J. Ault. 2004. Implementing invasive screening procedures: the Chicago Botanic Garden model. Weed Technology 18:1434–1440.

Jeschke, J. M., and D. L. Strayer. 2005. Invasion success of vertebrates in Europe and North America. Proceedings of the National Academy of Sciences of the United States of America 102:7198–7202.

Keller, R. P., J. M. Drake, and D. M. Lodge. 2007b. Fecundity as a basis for risk assessment of nonindigenous freshwater molluscs. Conservation Biology 21:191–200.

Keller, R. P., and D. M. Lodge. 2007. Species invasions from commerce in live organisms: problems and possible solutions. BioScience 57:428–436.

Keller, R. P., D. M. Lodge, and D. C. Finnoff. 2007a. Risk assessment for invasive species produces net bioeconomic benefits. Proceedings of the National Academy of Sciences of the United States of America 104:203–207.

Kolar, C. S., and D. M. Lodge. 2001. Progress in invasion biology: predicting invaders. Trends in Ecology and Evolution 16:199–204.

Kolar, C. S., and D. M. Lodge. 2002. Ecological predictions and risk assessment for alien fishes in North America. Science 298:1233–1236.

Krivánek, M., and P. Pysek. 2006. Prediction invasions in woody species in a temperate zone: a test of three risk assessment schemes in the Czech Republic (Central Europe). Diversity and Distributions 12:319–327.

Lodge, D. M. 1993. Biological invasions: lessons for ecology. Trends in Ecology and Evolution 8:133–137.

Lodge, D. M., S. Williams, H. MacIsaac, K. Hayes, B. Leung, L. Loope, S. Reichard, R. N. Mack, P. B. Moyle, M. Smith, D. A. Andow, J. T. Carlton, and A. McMichael. 2006. Biological invasions: recommendations for policy and management. Ecological Applications 16:2035–2054.

Mack, R. N. 1996. Predicting the identity and fate of plant invaders: emergent and emerging approaches. Biological Conservation 78:107–121.

Mills, E. L., J. H. Leach, J. T. Carlton, and C. L. Secor. 1993. Exotic species in the Great Lakes: a history of biotic crises and anthropogenic introductions. Journal of Great Lakes Research 19:1–54.

Moffitt, L. J., and C. D. Osteen. Prioritizing invasive species threats under uncertainty. Agricultural and Resource Economics Review 35:41–51.

Orr, R. 2003. Generic nonindigenous aquatic organisms risk analysis review process. Pages 415–438 *in* G. M. Ruiz and J. T. Carlton, editors. Invasive species: vectors and management strategies. Island Press, Washington, DC.

Pheloung, P. C. 1995. Determining the weed potential of new plant introductions to Australia. Agriculture Protection Board, Perth, Australia.

Pyšek, P., and D. M. Richardson. 2007. Traits associated with invasiveness in alien plants: where do we stand? Pages 97–125 *in* W. Nentwig, editor. Biological invasions. Springer-Verlag, Berlin.

Reichard, S. H., and C. W. Hamilton. 1997. Predicting invasions of woody plants introduced into North America. Conservation Biology 11:193–203.

Rejmánek, M., and D. M. Richardson. 1996. What attributes make some plant species more invasive? Ecology 77:1655–1661.

Rejmánek, M., D. M. Richardson, S. I. Higgins, M. J. Pitcairn, and E. Grotkopp. 2005. Ecology of invasive plants: state of the art. Pages 104–161 *in* H. A. Mooney, R. N. Mack, J. A. McNeely, L. E. Neville, P. J. Schei, and J. K. Waage, editors. Invasive alien species: a new systhesis. Island Press, Washington DC.

Sachs, J. D. 2005. The end of poverty: economic possibilities for our time. Penguin, New York.

Smith, C. S., W. M. Lonsdale, and J. Fortune, 1999. When to ignore advice: invasion predictions and decision theory. Biological Invasions 1:89–96.

Thompson, K., J. G. Hodgson, and T. C. G. Rich. 1995. Native and alien invasive plants: more of the same? Ecography 18:390–402.

Virtue, J. G., S. J. Bennett, and R. P. Randall. 2004. Plant introductions in Australia: how can we resolve "weedy" conflicts of interest? Pages 42–48 *in* B. M. Sindel and S. B. Johnson, editors. Proceedings of the 14th Australian Weeds Conference. Weed Society of New South Wales, Sydney.

Wade, S. A. 1995. Stemming the tide: a plea for new exotic species legislation. Journal of Land Use and Environmental Law 10:343–370.

Williamson, M. 1999. Invasions. Ecography 22:5–12.

Williamson, M. H., and A. Fitter. 1996. The characters of successful invaders. Biological Conservation 8:163–170.

World Trade Organization. 2005. Sanitary and phytosanitary agreement. World Trade Organization, Geneva.

4

Identifying Suitable Habitat for Invasive Species Using Ecological Niche Models and the Policy Implications of Range Forecasts

Leif-Matthias Herborg, John M. Drake, John D. Rothlisberger, and Jonathan M. Bossenbroek

In a Clamshell

Ecological niche modeling is a recently developed tool for predicting the distribution of organisms based on their environmental limitations. Its value in invasive species research is the ability to forecast the potential geographic distribution of a species in the introduced range based on occurrence points (and absence points for some methods) and appropriate environmental data layers. Such predictions are important for understanding the invasion process and for developing effective intervention strategies to reduce the ecological and economic impacts of invasions. Due to the variety of little tested techniques for ecological niche modeling, we provide a review of the currently available techniques and their strengths and weaknesses. One focus of this chapter is the set of unique challenges of environmental niche models for freshwater species. We provide an in-depth description of one of the most commonly used methods (genetic algorithm for rule-set prediction, or GARP), including studies that evaluate the approach. We also list potential data sources for researchers.

Ecological niche modeling seeks to address a fundamental question in ecology: What is the potential geographic distribution of an organism? The answer is central to a wide range of environmental questions (Guisan and Thuiller 2005). It can identify potential conservation zones or guide selection of areas for the reintroduction of threatened and endangered species (Araujo et al. 2004) based on their potential range. Similarly, the modeling of suitable habitat can reveal at which unsurveyed sites rare species are most likely to occur (Elith and Burgman 2002; Raxworthy et al. 2003). Based on scenarios of future climate change, land use, and habitat availability, ecological niche modeling can predict potential future shifts in species distributions (Thuiller et al. 2004). It can also provide background information to sharpen the focus of management efforts and risk analyses (Leung et al. 2002; Drake and Bossenbroek 2004).

Ecological niche modeling can also be used to predict suitable ranges for invasive species in an introduced habitat (Peterson 2003; Peterson et al. 2003; Drake and Bossenbroek 2004; Iguchi et al. 2004; Herborg et al. 2007c). With the increasing availability of high-resolution spatial environmental data sets, more accessible species distribution data, and faster computers, an increasing number of scientists, conservationists, and managers are using ecological niche models to predict invasions. One controversial, but potentially advantageous, feature of ecological niche modeling is its ability to predict distributions of little-studied species, based solely on georeferenced presence points (and absence points for some approaches) and spatially explicit environmental data for a range of parameters deemed important in determining the species range.

A good example of the scope of ecological niche models is a recent study of the invasive Chinese mitten crab, which has a long history of invasion in European rivers. Besides its catadromous life cycle (it grows and matures in freshwater but has to return to the sea to reproduce), little else is known of its environmental tolerances and ecological requirements. Since the recent colonization of San Francisco Bay by the species, there is growing concern for its spread throughout North America. The initial steps for developing an ecological niche model included the collection of presence data for Chinese mitten crab in its native Asian range and its introduced European range, as well as climate data (mean, minimum, and maximum air temperature, frost frequency, precipitation, wet day index), hydrological data (river discharge and river temperature), and oceanographic data (spring ocean temperature) (see table 4.1 for details). Since the available data for the native range was limited and the invaded range in Europe is still expanding, a widely tested ecological niche modeling approach (genetic algorithm for rule-set prediction, or GARP) using presence-only data was chosen. After identifying which environmental variables contributed significantly to model performance, the model was validated by successfully predicting the invaded European range based on the native range (Herborg et al. 2007c). To identify ports in the United States at the highest risk of mitten crab establishment, two separate predictions, one based on the native Asian range and one on the invaded European range, were combined with the amount of ballast water released into each major U.S. port (see figure 4.1; Herborg et al. 2007a). The models identified Chesapeake Bay as the location with the highest risk in the United States, a prediction that has recently been substantiated by the discovery of several mitten crabs in that area (Ruiz et al. 2006). This example highlights how combining biological factors (species potential distribution) and economic activities (maritime trade) can cause particular locations (ports and their associated rivers) to be at particular risk of introduction. Once high-risk locations are identified, economic analysis of the effect particular invaders can have on valuable industries can be initiated.

Our objective here is to highlight the importance of ecological niche modeling as a tool for understanding the invasion process and for developing effective intervention strategies to reduce the impact of invasions. Because niche modeling is a

TABLE 4.1. List of environmental niche model methods and their data requirements (modified from Elith et al. 2006).

Model	Type of model	Data
GLMs	Generalized linear models (logistic regression, autologistic regression, others)	p, a
GAMs	Generalized additive models	p, a
NPMR	Nonparametric multiplicative regression	p, a
ENFA	Ecological niche factor analysis	p
GARP	Genetic algorithm for rule-set prediction	p, a
OM-GARP	Open modeler-GARP	p, a
ANNs	Artificial neural networks	p, a
CLIMEX	Environmental-threshold-based software	p
MAXENT	Maximum entropy	p, a
SVM	Support vector machines	p, a
BIOCLIM	Envelope model	p

p, presence data; a, some form of absence data, which might include pseudo-absence data.

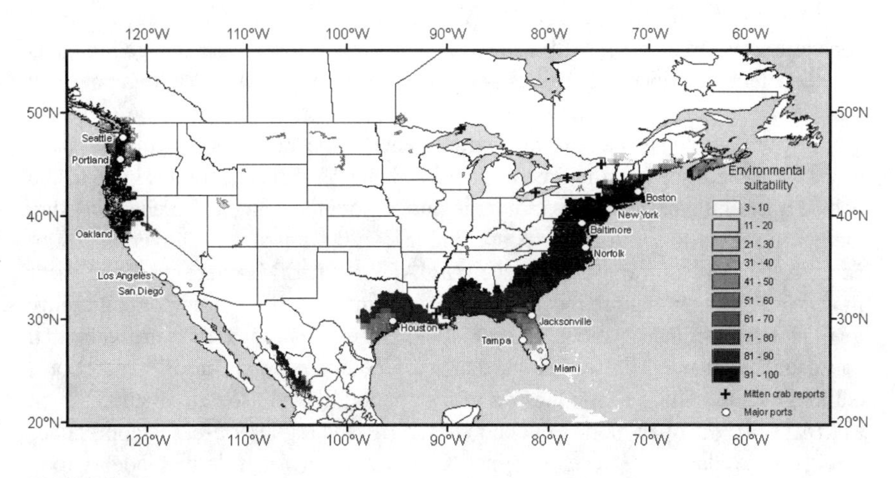

FIGURE 4.1.
Environmental suitability, on a scale from 0 to 100, combined with inland migration distances for the Chinese mitten crab. Reprinted from Herborg et al. (2007a), with permission of the Ecological Society of America.

rapidly developing methodology, there is ongoing discussion about the most appropriate and accurate methodology (Elith et al. 2006). We provide a mini-review of the available techniques and their strengths and weaknesses. The methods discussed here can be applied to terrestrial or aquatic environments, but most of our examples are based on aquatic organisms due to the expertise of the authors and the particular challenges researchers face when developing an ecological niche model for an aquatic species.

SUMMARY OF AVAILABLE TECHNIQUES FOR ENVIRONMENTAL NICHE MODELS

Environmental niche modeling is a rapidly developing area of research, and the variety of available methods is overwhelming. While there is no recipe for selecting the most appropriate method for a particular problem, there are several considerations that can provide guidance to a potential user. An important first step is to determine if the model will include only presence data, presence and absence data, or biological community information in addition to presence/absence data (see table 4.1). Clearly, presence data is prerequisite for any prediction and either can be obtained from observations or, in the absence of observation points, can be randomly subsampled from the range of a species using geographic information systems (Herborg et al. 2007b). While predictions based on this second approach will likely lead to less accurate results, in some cases it may be the only option given the available information. Reliable absence data are more difficult to obtain, and their availability depends on the type of species, its habitat, and the amount of research conducted. Additionally, species can be absent from areas with suitable environments because of local extirpation, historical events such as glaciation or plate tectonics, and dispersal constraints (Peterson et al. 1999). Absence data for species in their invaded range is very problematic, since the species is unlikely to have spread through its potential range.

The second step is selecting an environmental niche model method that has been tested successfully given the type of data available. Two recent studies have tested and compared the performance of a large number of environmental niche modeling methods (Seguardo and Auraujo 2004; Elith et al. 2006) in detail. Elith et al. (2006) compared widely used approaches, such as GARP, BIOCLIM, and generalized linear models (GLMs), with more recent or rarely used models with respect to their ability to model current distributions of species accurately. The models were validated based on independent observations that were compared with model predictions. Although all modeling techniques performed generally well, several methods, such as MAXENT, the open modeler version of GARP, and regression-based approaches (multivariate adaptive regression splines, GLMs, general dissimilarity models) stood out from the rest. Another study looked at the performance of eight ecological niche modeling approaches (Seguardo and Auraujo 2004), but only three were in common with Elith et al. (2006). In that study, neural networks performed the best, but not significantly better than more traditional methods like GLMs and generalized additive models (GAMs). The worst performers were DOMAIN and BIOMAP.

Considering the wide variety of methods available, we discuss only a subset here, beginning with statistical approaches to environmental niche models, including GLMs, GAMs, and nonparametric multiplicative regression (NPMR). GLMs are a class of regression models in which the mean response is a nonlinear "link" function of the predictors, and the data may belong to one of several distributions, that is, need not be normally distributed. Logistic regression, a model for binary data in which covariates linearly affect the logit-transformed data, is a type of GLM that is based on the logit link function and binomial sampling distribution (McCullagh and Nelder 1989), resulting in a sigmoidal model. In practice, logistic regression is the main GLM used for niche modeling (Guisan et al. 2002), but others are available, for example, log-linear models. Since the simple logistic regression assumes independence, a version called the autologistic model is available to account for spatial autocorrelation (Augustin et al. 1996). GAMs are a generalization of GLMs in which the link function is replaced with a nonparametric estimator (Hastie and Tibshirani 1999; Wood 2006), making the underlying model considerably more flexible. As with GLMs, the logistic regression version is most commonly used for niche modeling (Wood and Augustin 2002). Spatial autocorrelation of the dependent variables is readily incorporated into GAMs, in which case they are sometimes called geoadditive models (Wood and Augustin 2002; Kammann and Wand 2003). NPMR takes this approach a step further by modeling the interaction of environmental covariates multiplicatively (McCune 2006). Finally, a new nonparametric approach, MAXENT, has been introduced based on the maximum entropy principle (Phillips et al. 2006). Like statistical density estimation, MAXENT aims to identify a probability density of species occurrence.

Nonstatistical methods use computation to find rule-based or model-based criteria for discriminating suitable from unsuitable habitat (e.g., GARP, CLIMEX, artificial neural networks [ANNs]). GARP requires input of binary presence data and pseudo-absence data. Pseudo-absence data are generated by GARP from a random selection of locations without confirmed species presence (Stockwell 1999), so they are commonly simulated, and much of the discussion of GARP has focused on this feature. GARP represents a species niche according to a collection of logical rules that "evolve" over the course of model iterations to provide an optimal fit. In contrast with GARP's rule-based approach, CLIMEX is a model-based approach (Sutherst and Maywald 1985). The model is a biologically motivated ecological index that supposes climate factors are the most important determinants of species fitness and distribution. The aim of the interactive CLIMEX software is to identify the parameters of submodels (e.g., thresholds, slopes) that best match observed and predicted species distributions. CLIMEX is much less automated and depends more heavily on user-supplied guidance (expert knowledge) than other methods. Like GARP, the classification mode uses a genetic algorithm to estimate the parameters of the stress index component of the CLIMEX model. Estimation is accomplished by iterative model tuning and evaluation directed entirely by the user.

ANNs are a class of algorithms for pattern recognition in data. ANNs may be applied to classification or boundary estimation problems. For ecological niche modeling applications, classification has been the more common approach. ANNs are extremely flexible and may be applied to data structures that would be awkward to address with other methods. The downside is that ANNs can be very sensitive to model specification and user-supplied optimization parameters, thus requiring considerable expertise to be used effectively.

BIOCLIM is another nonstatistical approach, which is based on a multidimensional bounding box defined by the empirical quantiles of a data set. A problem with the BIOCLIM approach is that for most species there will be physiological trade-offs such that the corners of the box are generally uninhabitable; that is, an organism might tolerate extreme conditions in one variable if all the other variables are ideal, but the organism cannot tolerate extremes in all environments simultaneously. This problem is exacerbated when the dimensionality of the data is increased (i.e., as niche axes are added), but this can be addressed in a number of ways. One method introduced by Guo et al. (2005) to address this problem is the support vector machine (SVM). SVMs were developed for automated classification and belong to a class of machine learning algorithms called kernel methods. The basic idea is that the raw data are presumed to exhibit some property that is very complicated in the natural dimensions in which the data were collected. However, by transforming the data with a functional relation called a kernel into a new, higher dimensional space, these properties appear much simpler.

DATA SOURCES AND APPROACHES TO GATHERING DATA TO CONSTRUCT ENVIRONMENTAL NICHE MODELS FOR AQUATIC SPECIES

Independent of the environmental niche model selected, environmental data relevant to the species in question is essential for modeling ecological niches. While an increasing amount of environmental data are available at broad extents or even globally, the freshwater environment probably has the least data available. Global coverages for watersheds of various resolutions, flow directions, slope, aspect, and other elevation-derived parameters have been generated (HYDRO1k database; table 4.2); only limited data on water temperature, flow rates and volumes, water chemistry, benthic substrate, and so on, are available on a broader scale (see table 4.2). Many of the data sources only have limited value in smaller scale analysis and/or require lengthy processing into data layers. Small-scale hydrological data might be more readily available through government agencies, in particular, in the United States.

As with environmental data, species distribution data can be gathered from a range of sources. Clearly, museum records and scientific publications are a primary source of distribution data, but native distributions in particular are rarely published in the recent primary literature due to the perceived limited scientific value of such reports. Several national and international databases on the current distribution of invasive species are available online (table 4.3). In the case of recent invasions, absence data are of limited value because in most cases it will be impossible to determine if the species is absent due to environmental conditions or lack of colonization opportunities.

EXAMPLES OF ECOLOGICAL NICHE MODELING FOR AQUATIC INVADERS

Ecological niche modeling has been applied to several aquatic invasive species. Zebra mussels (*Dreissena polymorpha*) have caused large economical (Leung et al.

TABLE 4.2. Sources of environmental layers relevant for aquatic ecological niche modeling.

Environmental variables	Source	Geographic area
Ground frost frequency; maximum, minimum, mean, and diurnal range of air temperature; precipitation; vapor pressure; cloud cover; solar radiation; wind speed; wet day index; elevation	The Intergovernmental Panel on Climate Change's Climate Research Unit global climate data set at www.ipcc-data.org/obs/get_30yr_means.html	Global
Topographic index, slope, aspect, flow accumulation, drainage basins, flow directions, streams	The U.S. Geological Survey HYDRO1k elevation-derived database at edc.usgs.gov/products/elevation/gtopo30/hydro/index.html	Global
Mean river temperature (°C)	Generated from the Global Environment Monitoring System at www.gemswater.org/publications/index-e.html	Global but few data points
Global runoff data for selected sites	Runoff data in table form from the Global Runoff Data Centre at grdc.bafg.de/servlet/is/2781/	Global but few data points
Global runoff data for selected sites	Runoff data points can be selected from a global map, and the data exported into table formats, from the Global River Discharge Database at www.rivdis.sr.unh.edu/maps/	Global but few data points
Air temperature and precipitation, and wide range of derived bioclimatic variables (e.g., warmest temperature in driest quarter, precipitation in coldest quarter)	Data can be downloaded from the WorldClim database at www.worldclim.org/	Global
Water chemistry, temperature, nutrients, pollution measures as point data per watershed	U.S. Environmental Protection Agency's STORET database at www.epa.gov/storet/index.html	United States
Land cover with a focus on different tree types, percentage tree cover, terrestrial primary production	University of Maryland's Global Land Cover Facility at glcf.umiacs.umd.edu/data/	Global
Hydrological data sources	World Meteorological Association's Hyperlinks in Hydrology at www.nwl.ac.uk/ih/devel/wmo/hhcdbs.html	Global and European

TABLE 4.3. Sources of species distribution data.

Species distributions	Source	Geographic area
Sea Grant Nonindigenous Species site containing a large number of publications on invasive species	sgnis.org/[a]	Global aquatic
U.S. Geological Survey Nonindigenous Aquatic Species site, species distribution by watershed	nas.er.usgs.gov/	U.S. freshwater
FishBase, very large database on distribution data of fish from museums, collections, and other sources	filaman.ifm-geomar.de/search.php	Global freshwater and marine
Invasive Species Specialist Group's Invasive Species Database contains references identifying the native and invasive range of a wide range of invasive species	www.issg.org/database/welcome/	Global terrestrial, freshwater, marine
Regional Biological Invasions Centre provides occurrence data on some aquatic invaders in Europe	www.zin.ru/rbic/	Baltic Sea states, brackish-marine
Northern Europe and Baltic Network on Invasive Alien Species, general information on the distribution of a wide range of species	www.nobanis.org	Baltic Sea states, northern Europe, terrestrial, freshwater, marine

[a] Not updated since 2008.

2002) and ecological (Ricciardi et al. 1998) impacts since their introduction into the Laurentian Great Lakes in North America. A GARP model was applied to the invaded distribution in the United States to predict the potential distribution of the species in North America (Drake and Bossenbroek 2004). The contribution of an initial set of geological, hydrological, and terrestrial climate layers (bedrock geology, elevation, flow accumulation, frost frequency, maximum temperature, precipitation, slope, solar radiation, and surface geology; see table 4.2) on the numbers of false positives (omission error) and false negatives (commission error) was utilized to select a subset of layers with higher predictive accuracy. Further combinations of the subset of environmental layers highlighted a strong effect of elevation in some predictions. The model (figure 4.2) concluded that while large parts of the Colorado, Columbia, and Missouri rivers are unsuitable environments, coastal areas along

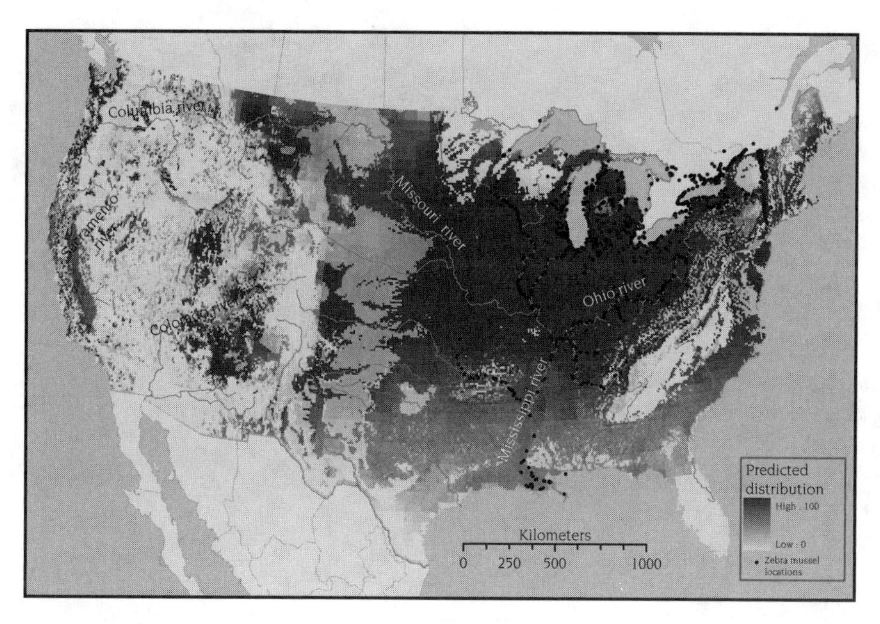

FIGURE 4.2.
Prediction of suitable environment for the zebra mussel in United States based on its current invaded range. Reprinted from Drake and Bossenbroek (2004), with permission of the American Institute of Biological Sciences.

the West Coast, including the San Joaquin and Sacramento rivers and the lower reaches of the Columbia, are suitable for establishment (Drake and Bossenbroek 2004).

Largemouth bass (*Micropterus salmoides*) and smallmouth bass (*M. dolomieu*) are two North American predatory fishes with a high impact on the native fauna of Japan. Topographical, hydrological, land-use, land-cover, and terrestrial climate data (elevation, slope, aspect, topographic index, land use, land cover, percent tree cover, flow accumulation, and flow direction; table 4.2) were utilized to predict the potential range of both species using GARP (Iguchi et al. 2004). Due to the widespread invasion of these two game fish species, a validation of the predictions was possible (see the following section). A different approach (artificial neural network) was taken to predict the vulnerability of lakes in Ontario to smallmouth bass colonization, based on a wide range of climate, habitat, and biological factors (Vander Zanden et al. 2004).

Eurasian ruffe (*Gymnocephalus cernuus*) and rainbow smelt (*Osmerus mordax*) are both recent invaders in North America, and ecological niche modeling based on ecological and terrestrial climate data (annual precipitation, elevation, ground frost frequency, slope, wet day frequency, and maximum, mean, and minimum air temperature; table 4.2) was used to predict their potential range (Drake and Lodge 2006). A subset of layers was selected for each species according to Drake and Bossenbroek (2004), and final predictions indicated large areas of suitable environments in North

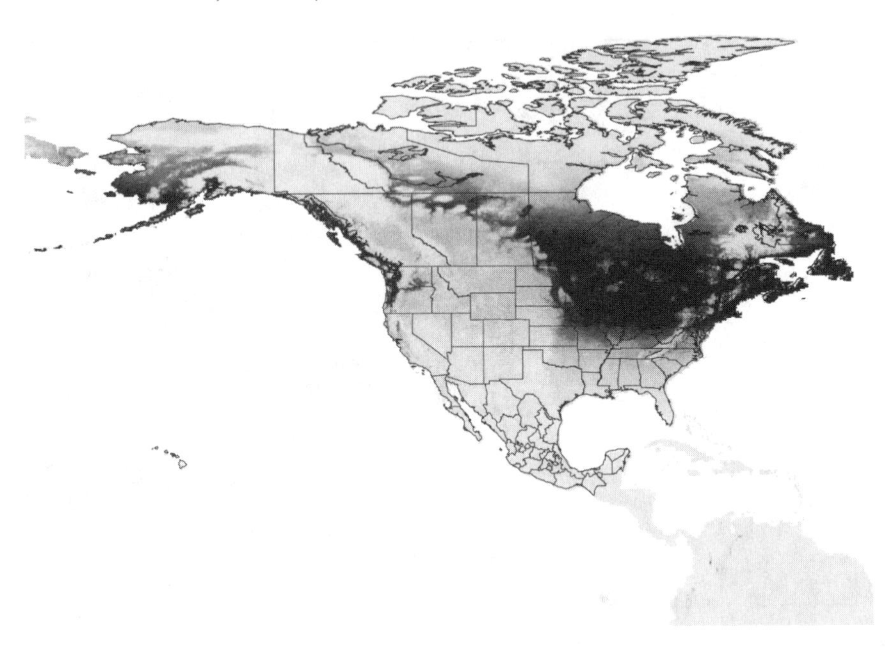

FIGURE 4.3.
The potential distribution of the Eurasian ruffe in North America. Reprinted from Drake and Lodge (2006), with permission of the American Fisheries Society.

America (figure 4.3), underlying the importance of rapid management actions (Drake and Lodge 2006).

The combination of ecological niche modeling and estimates of propagule pressure can provide information on locations where a species can survive and where it is likely to arrive. Predictions of suitable environments (figure 4.1) and the volume of ballast water released were used to quantify the relative risk of establishment of the Chinese mitten crab (*Eriocheir sinensis*) in U.S. ports (Herborg et al. 2007a).

EVALUATION OF ECOLOGICAL NICHE MODELS

One challenge in using ecological niche modeling to forecast potential geographic range of invasive species is that experimental validations of model predictions are not ethically acceptable. For example, releasing zebra mussels into a lake where they are absent, but where models predict they can persist, for the sake of model validation is not permissible. More creative approaches are required to surmount the problem of model validation. Several validation approaches have been employed, including (1) generating models based on a few known point localities of a cryptic species and then searching for the species at sites predicted suitable by the model, (2) dividing highly resolved and detailed data sets into training and validation subsets to test the ability of the model to accurately predict what is already known but has been reserved from model training, and (3) the generation of ecological niche modeling based on simulated point occurrence data according to investigator specified rule-sets and then determining how accurately the model recovers the rule-sets.

While several studies have compared different modeling approaches (see "Data Sources and Approaches," above), validating predictions of invasive species ranges is inherently difficult, and only limited studies are available. Iguchi et al. (2004) validated their GARP-generated ecological niche modeling of smallmouth and largemouth bass in their native North American range using independent point occurrence data (i.e., occurrence points not used during model development). Their analyses showed that the predictive ability of the models is statistically significant ($p < 0.001$) (Iguchi et al. 2004). This model, validated for the native range of the bass, is then projected on the environmental conditions in Japan to predict the geographic range of both bass species there, where bass are not native. While limited data were available for the validation of model projections, the models predicted Japanese occurrences accurately.

Herborg et al. (2007c) used native occurrences of the Chinese mitten crab (*Eriocheir sinensis*) to validate a GARP prediction of environmental suitability based on the well-documented invasion in Europe. The comparison of the model prediction and the observed occurrences indicated high accuracy in the prediction of occurrence points: 84% of all reported occurrences were in areas predicted suitable by >80% of the models. Also, watersheds with established populations of mitten crabs had significantly higher environmental suitability in the model than did uninvaded watersheds.

LIMITATIONS IN THE APPLICATION OF ENVIRONMENTAL NICHE MODELS FOR AQUATIC SPECIES

Species distribution modeling is conceptually connected to Hutchinson's (1957) ideas on fundamental and realized niches, where the fundamental (potential) niche represents all environments under which the species can persist, whereas the realized (occupied) niche is the area where the species is actually present. The discrepancy between the occupied and the potential niches can occur because of biogeographic dispersal barriers, glacial refugias (e.g., Nekola 1999), and biotic interactions (e.g., competition, predation) that result in range limitation. These differences can be a source of error associated with ecological niche modeling because the current range of a species, whose environmental characteristics are used to calibrate the model of its potential range, may not represent the potential niche.

Freshwater species are particularly likely to be restricted by dispersal limitations between watersheds and hydrological barriers. Thus, for many aquatic species, numerous unoccupied sites may exist that are environmentally different from presently inhabited sites but that would provide habitat sufficient to maintain a positive net population growth rate. When this is the case, any estimates of potential species distribution based on point occurrence observations will be overly conservative. Human-aided long-range dispersal, which often accompanies invasion events, can overcome this type of range limitation and could lead to establishments of species in locations not predicted as suitable by ecological niche models.

Another challenge associated with ecological niche modeling is that models are generally constructed using only presence data. True absence data are extremely difficult to collect and are almost never available. These data limitations may obscure the signal of the ecological requirements of a species and lead to inaccuracies in model predictions (Brotons et al. 2004). Also of concern is that an ecological niche model would ideally be based on data for each cell of a hypothetical grid overlaid on the current range of the species. It would also be desirable to have the frequency of points in the data set be representative of the frequency of species occurrence in the various regions and types of environments being used to fit the model. These idealistic requirements pose a particular challenge for aquatic species because such species are difficult, relative to many terrestrial species, to survey. Having a stratified random sample of species occurrence is critical to producing accurate ecological niche models because most modeling techniques are sensitive to sample size and how evenly the data are distributed (Stockwell and Peterson 2002). Even invasive aquatic species, which typically attain large population sizes as they reach nuisance levels at a particular site, may persist as small populations, unnoticed for many years, at other sites. Thus, initially cryptic populations can limit the quality of ecological niche models for aquatic invasive species because a subset of suitable sites will not be included in the point occurrence data set that is used to train and validate the model. In cases where substantial and relatively representative point occurrence data are available, this is a minor concern, but this is not so when the known sites of established populations are few, as is often the case early in invasions.

Training and projecting environmental niche models for aquatic species are also limited by the paucity of aquatic environmental data of suitable extent or resolution. Broad, continental-scale, digitized data are available for climatic and terrestrial parameters. While freshwater habitat data might be available for smaller scale studies, forecasts of potential distribution for aquatic species on a larger scale have relied on climatic and terrestrial parameters as surrogates for aquatic conditions (Drake and Bossenbroek 2004; Iguchi et al. 2004). However, a comparison of predictions of fish distributions based on environmental factors at broad spatial scales and small-scale aquatic habitat data revealed that broad-scale variables can predict fish distribution successfully (Marsh-Matthews and Matthews 2000).

Another concern is the selection of environmental parameters included at the outset of the ecological niche modeling process. The inclusion of some parameters can have a nontrivial influence on the model's final predictions. For example, in predicting the potential distribution of zebra mussels in the United States, Drake and Bossenbroek (2004) included elevation as a predictor variable. In the original model results, the pattern of potential habitat was clearly related to elevation, even though there is no reason to believe that zebra mussels actually respond to elevation. In other words, these models are only as good as the data with which they are developed. Removal or addition of a single parameter as a potential predictor of species distribution at the outset of a modeling exercise can result in substantial discrepancies in modeled predictions of potential distribution (Guisan and Zimmerman 2000).

Some ecological niche modeling approaches are computationally intensive, which, despite ever-increasing processor speeds and ever-growing RAM banks, can limit their usefulness. Managing and interpreting the copious output of some environmental niche modeling methods can be burdensome. Finding a balance between model complexity and interpretability will require continued effort by ecological niche modelers.

BIOECONOMIC FORECASTS/MANAGEMENT GUIDANCE

Ecological niche modeling has the potential to be a valuable tool for the management of invasive species, due to its ability to predict the potential geographic distribution of a species before an introduction. Such predictions allow a prioritization of management effort for a range of invaders based on their potential distribution in an introduced range. For example, the predictions of potential distributions of a number of species of snakeheads and Asian carps helped identify those fish species that pose the highest risk for Canadian freshwater systems (Herborg et al. 2007b). Another important application of environmental niche models for managers and bioeconomic analysis is the possibility to incorporate future scenarios for climate change, altered land use, or management, species, or habitat modification into predictions of potential distributions of invaders. A lake-based environmental niche model incorporating a range of different climate change scenarios predicted the range expansion of smallmouth bass in Canada, highlighting its potential impact on northern fish communities in Canada (Sharma et al. 2007). Other scenarios that could be incorporated in future environmental niche models could include a range of potential management strategies that would alter habitat suitability and availability. This would allow a quantitative test of different management strategies, crucial for a cost-benefit analysis. As is noted often throughout these chapters, the spread of invasive species has and will continue to have a large economic impact on society. The potential spread of invasive species, such as zebra mussels and the emerald ash borer, have resulted in large, multistate (if not international) education programs to prevent the further spread of these organisms (e.g., the 100th Meridian Initiative [www.100thmeridian.org/] and the Emerald Ash Borer website [www.emeraldashborer.info/]). Having accurate predictions of the potential habitat of these species would allow focused outreach and prevention efforts to areas considered most at risk—there is no need to spend money to educate people in a pine forest about the risks of the emerald ash borer.

Predicting the potential economic benefits of slowing the spread of an invasive species requires the integration of forecasts of potential range, consideration of species-specific dispersal capabilities, and evaluation of potential economic impacts and the costs and benefits of different control and prevention strategies. The integration of these issues is a general goal of this book and has been specifically addressed with regard to the spread of zebra mussels to the western United States (see chapter 12) and the emerald ash borer in Michigan and Ohio (see chapter 6). The general objectives of these research projects are twofold: (1) to provide estimates of the regional economic impact that invasive species will inflict upon the region of concern and (2) to provide policy makers with quantitative guidance for cost-effective alternative strategies to control, prevent, or slow the spread of these species.

Achieving these objectives requires that we (1) determine the potential geographic range of the invasive species, (2) predict its spread via natural and human-mediated dispersal, (3) estimate the economic impact of the invasion, and (4) determine its regional economic consequences. Determining the potential geographic range involves generating an environmental niche model. In the case of the emerald ash borer, it is currently assumed that the potential habitat of the borer is equivalent to the distribution of ash trees, which can be determined by U.S. Forest Service data (Iverson et al. 2006). As the emerald ash borer spreads, or as more information is discovered about its native range, environmental niche models can be used to make refined predictions of potential habitat. Estimating the economic impact of the invasive species involves estimating the regional labor and capital employment in a spatially explicit manner. For the emerald ash borer, one assessment suggests that their damage to the urban communities of Ohio, based on losses in landscape value and the costs of tree removal and replacement, could reach $2 billion to $8 billion (Sydnor et al. 2007). Determining the regional economic consequences as the invasive species spreads can be done through the development of a regional computable general equilibrium model (see chapter 12 for details). The emerald ash borer and zebra mussel projects thus provide clear examples of the need for accurate assessments of the ecological niche of an invasive species for management purposes.

The ecological niche modeling techniques discussed above have not yet been directly linked to economic analyses, but they have been used to make recommendations about the management of invasive species. An ecological niche model of the Asian longhorn beetle suggested that most areas of the Pacific Coast of the United States are unsuitable habitat, which would therefore allow a focus of all management efforts on the Atlantic Coast (Peterson and Vieglais 2001). A model combining ecological niche modeling and ballast water discharge calculated the relative invasion risk of Chinese mitten crab for major ports in the United States (figure 4.1). Herborg et al. (2007a) determined Norfolk, Baltimore, and Portland as having the highest risk, while other major ports such as Los Angeles–Long Beach have a very low risk, thus identifying key areas for monitoring and risk management. An ecological niche model of potential zebra mussel habitat in the United States found less suitable habitat west of the 100th meridian than previously predicted (figure 4.3; but see Whittier et al. 2008) but concluded that the Columbia, San Joaquin, and Sacramento river drainages were at risk. These results had a direct effect on the management strategy of the 100th Meridian program, which aimed to stop the spread of zebra mussels in the United States (Drake and Bossenbroek 2004). Ecological niche modeling was used to produce a potential habitat map as a template for a vector model of the spread of West Nile virus by birds and mosquitoes. This approach advanced understanding of how the pathogen is spreading throughout the United States (Peterson et al. 2003). Arriaga et al. (2004) identified the risk posed by the buffelgrass invasions of desert scrub in the arid and semiarid regions of the northern Mexican Sonoran Desert. These examples demonstrate the usefulness of ecological niche modeling for informing management strategies for specific species.

Ecological niche models have also been used to make management recommendations for multiple species or species groups. Peterson et al. (2003) demonstrated that native ranges of four plant species could be used to predict their potential ranges in North America. Each of the four species analyzed by Peterson et al. (2003) were currently in North America, and the niche model predictions contained the current distribution. Two of the species, garlic mustard and Russian olive, were predicted to have a potential distribution much larger than currently exists in North America, suggesting that continued spread is likely. The high predictive ability of the Peterson et al. (2003) results suggest that niche modeling can be successful in estimating the potential habitat of species that have not yet been introduced. Likewise, Thuiller et al. (2005) found that for 96 plant taxa (species and subspecies) the distribution data in the native country enabled accurate predictions of the climate in regions where these taxa are invasive. The ability to accurately model plant niches to predict the invasive range of these taxa highlights the potential of ecological niche modeling as an important management tool for other taxonomic groups, including fishes (Kolar 2004).

INTEGRATION OF ECOLOGY AND ECONOMICS

While there are currently no formal applications linking environmental niche models and economic analysis, there is clearly opportunity for future research that would greatly benefit policy makers, managers, and the public. First, the economic impact of most invasive species is correlated with population abundance. Thus, if ecological niche modeling could be enhanced to estimate population abundance (i.e., the potential severity of impacts), economic predictions would have a range of potential impacts as opposed to merely presence or absence. For example, damage functions linking predicted population size with financial losses could be generated and incorporated in economic analyses. Second, risk assessments are often (if not inherently) subjective, and such subjectivity could affect the application of ecological niche models. Further involvement of economists in the interpretation of environmental niche models and resulting policy suggestions may help us to better understand the consequences of such subjectivity. Predictions of environmental niche models can vary substantially based on the variables chosen for the model (for an example, see Drake and Bossenbroek 2004). Therefore, if different analysts include different parameters when modeling the potential distribution of the same species, disagreements in range forecasts are possible. Moreover, even if potential range maps are identical, different analysts may draw different conclusions because their interpretations are based on their own inherent biases, of which they may be unaware. For example, property owners or natural resource managers with a vested interest in the consequences of environmental impacts of invasive species may overestimate the future risk posed by such threats (Burgman 2005). On the contrary, an industry that engages in activities that may introduce nonindigenous species or that may be affected in the future by an invasive species may be more conservative in their predictions, not wanting to invest in prevention or control earlier than is necessary. Economists are equipped to analyze the relative benefits of such diverse approaches to risk management (Shogren 2000).

Generally speaking, certain types of economic models can provide guidance on the timing and magnitude of many kinds of investments, from the allocation of capital in the stock market to public spending on HIV intervention programs (Zaric and Brandeau 2002). In the context of invasion biology, economic models could be used to address similar questions such as when preventative measures should be put in place, how much should be spent on outreach and education versus quarantine efforts or population control, and whether management activities produce a net economic benefit.

The ability of a model to accurately describe or predict nature can be limited by the quality of empirical observations used to generate the model. This is the case for ecological niche models. High-quality data sets of species presence, not to mention absence, for large geographic regions are desirable in producing and validating environmental niche models. While the majority of available occurrence data, particularly from the native range of a species, is maintained by museums, acquiring such data can be prohibitively expensive. Economists could help ecologists and the agencies that fund them determine how much to invest in obtaining accurate point locality data. Analyses of this sort could be based on the projected financial losses associated with a potential invader, the cost of data collection, and the prospects for model improvement with each additional data point available.

ROLES FOR RESEARCHERS

Improving the accuracy and reliability of ecological niche modeling will require continued effort from researchers in the field—both those that develop the theory of ecological niche modeling and those that test the validity of its predictions. This is a field of research with ample opportunities for collaboration among biologists, ecological modelers, and statisticians. Opportunities to collaborate with economists would also be valuable because further considerations of spatial predictions of the potential range of a nonnative species may enhance forecasts of the economic impacts of invasive species. In other words, if the abiotic conditions that predict the potential for species presence can be delimited on maps, then current risk and future change in risk should be predictable. This would allow economists to generate dynamic models of financial impacts. Furthermore, understanding the relationship between species presence and anthropogenic habitat change could lead to dynamic models of the locations of the suitable habitat conditions themselves, which may change in response to social trends. One example of this is irrigation in the Desert Southwest of the United States that has allowed the establishment of the red imported fire ant

(*Solenopsis invicta* Buren) in that region, which would otherwise be too dry for this species (Morrison et al. 2004).

Additionally, several procedural measures can help to improve the accuracy and robustness of ecological niche modeling predictions. Given the effects of discrepancies between the observed and potential niche on model predictions, it is essential to be aware of the quality of data incorporated and its potential limitation. In the case of invasive species, models should be regenerated as the geographic extent of the invasion grows and more presence data become available.

Other areas in need of further investigation include the importance of variable selection and the effects of the relative size of geographic area used for model training. Also, the question of whether to include invaded range point localities in training data sets, as opposed to only native range data, remains open. The answer to this interesting question may depend on whether there has been local adaptation in the invaded range.

The areas needing the most attention in species distribution modeling, unsurprisingly, reflect the limitations of ecological niche modeling already discussed in this chapter. Araujo and Guisan (2006) discusses five areas needing attention, each corresponding to one or more of the current limitations of ecological niche modeling: (1) niche concept clarification, (2) sampling designs for the data used in model construction, (3) parameterization improvements, (4) techniques for understanding parameter contribution and selecting the best models, and (5) methods for evaluating models (Araujo and Guisan 2006).

Ecological niche modeling is an area of active research as well as a useful tool for understanding and dealing with biological invasions. The ecological modeling community continues to develop new techniques that will enhance the accuracy of ecological niche modeling (Elith et al. 2006), but it is already valuable for predicting the regions most at risk for invasion by particular species. Ecological niche modeling helps ecologists understand what factors determine the distribution of species, native or otherwise, and, among other things, test hypotheses about species evolution (Peterson et al. 1999). Managers and policy makers can also benefit from the use of environmental niche models. Management actions targeting different stages in the invasion process (see chapter 1, figure 1.1) can be informed by potential range forecasts. Identifying potential habitat for nonindigenous species can focus prevention and control efforts and can inform early detection/rapid response activities. Moreover, environmental niche models can guide these efforts at a variety of spatial scales, from global to regional. The formal incorporation of potential geographic ranges of invasive species in economic analyses is an exciting prospect. Some early efforts in this direction have been described here, but many important advances await further collaboration between ecologists and economists. Such advances will surely increase our capacity to efficiently reduce the negative effects of invasive species on society.

Acknowledgments This chapter was substantially improved by the editors of this book and reviews by Joanna McNulty, Jim Muirhead, and an anonymous reviewer. This material is based on work supported by the Integrated Systems for Invasive Species project (D. M. Lodge, principal investigator) funded by the National Science Foundation (DEB 02-13698). This is publication no. 2009-01 from the University of Toledo Lake Erie Center.

References

Araujo, M. B., P. J. Densham, and P. H. Williams. 2004. Representing species in reserves from patterns of assemblage diversity. Journal of Biogeography 31:1037–1050.

Araujo, M. B., and A. Guisan. 2006. Five (or so) challenges for species distribution modelling. Journal of Biogeography 33:1677–1688.

Arriaga, L., A. E. Castellanos, E. Moreno, and J. Alarcon. 2004. Potential ecological distribution of alien invasive species and risk assessment: a case study of buffel grass in arid regions of Mexico. Conservation Biology 18:1504–1514.

Augustin, N. H., M. A. Mugglestone, and S. T. Buckland. 1996. An autologistic model for the spatial distribution of wildlife. Journal of Applied Ecology 33:339–347.

Brotons, L., W. Thuiller, M. B. Araujo, and A. H. Hirzel. 2004. Presence-absence versus presence-only modelling methods for predicting bird habitat suitability. Ecography 27:437–448.

Burgman, M. A. 2005. Risks and decisions for conservation and environmental management. Cambridge University Press, Cambridge, UK.

Drake, J. M., and J. M. Bossenbroek. 2004. The potential distribution of zebra mussels in the United States. BioScience 54:931–941.

Drake, J. M., and D. M. Lodge. 2006. Forecasting potential distributions of nonindigenous species with a genetic algorithm. Fisheries 31:9–16.

Elith, J., and M. A. Burgman. 2002. Predictions and their validation: rare plants in the Central Highlands, Victoria, Australia. Pages 303–314 *in* J. M. Scott, P. J. Heglund, M. L. Morrisonet, J. B. Haufler, M. G. Raphael, W. A. Wall, and F. B. Samson, editors. Predicting species occurrences: issues of accuracy and scale. Island Press, Covelo, CA.

Elith, J., C. H. Graham, R. P. Anderson, M. Dudik, S. Ferrier, A. Guisan, R. J. Hijmans, F. Huettmann, J. R. Leathwick, A. Lehmann, J. Li, L. G. Lohman, B. A. Loiselle, G. Manion, C. Moritz, M. Nakamura, Y. Nakazawa, J. M. Overton, A. T. Peterson, S. J. Phillips, K. Richardson, R. Scachetti-Pereira, R. E. Schapire, J. Soberon, S. Williams, M. S. Wisz, and N. E. Zimmermann. 2006. Novel methods improve prediction of species' distributions from occurrence data. Ecography 29:129–151.

Guisan, A., T. C. Edwards, and T. Hastie. 2002. Generalized linear and generalized additive models in studies of species distributions: setting the scene. Ecological Modelling 157:89–100.

Guisan, A., and W. Thuiller. 2005. Predicting species distribution: offering more than simple habitat models. Ecology Letters 8:993–1009.

Guisan, A., and N. E. Zimmermann. 2000. Predictive habitat distribution models in ecology. Ecological Modelling 135:147–186.

Guo, Q., M. Kelly, and C. H. Graham. 2005. Support vector machines for predicting distribution of Sudden Oak Death in California. Ecological Modelling 182:75–90.

Hastie, T. J., and R. J. Tibshirani, R. J. 1999. Generalized additive models. Chapman and Hall/CRC, Boca Raton, FL.

Herborg, L. M., C. J. Jerde, D. M. Lodge, G. M. Ruiz, and H. J. MacIsaac. 2007a. Predicting invasion risk using measures of introduction effort and environmental niche model. Ecological Applications 17:663–674.

Herborg, L. M., N. E. Mandrak, B. Cudmore, and H. J. MacIsaac. 2007b. Comparative distribution and invasion risk of snakehead and Asian carp species in North America. Canadian Journal of Fisheries and Aquatic Sciences 64:1723–1735.

Herborg, L. M., D. A. Rudnick, Y. Siliang, D. M. Lodge, and H. J. MacIsaac. 2007c. Predicting the range of Chinese mitten crabs (*Eriocheir sinensis*) in Europe. Conservation Biology 21:1316–1323.

Hutchinson, G. E. 1957. Concluding remarks. Population studies: animal ecology and demography. Cold Spring Harbor Symposia on Quantitative Biology 22:415–427.

Iguchi, K., K. Matsuura, K. M. McNyset, A. T. Peterson, R. Scachetti-Pereira, K. A. Powers, D. A. Vieglais, E. O. Wiley, and T. Yodo. 2004. Predicting invasions of North American basses in Japan using native range data and a genetic algorithm. Transactions of the North American Fisheries Society 133:845–854.

Iverson, L. R., A. M. Prasad, D. Sydnor, J. M. Bossenbroek, and M. W. Schwartz. 2006. Modeling potential Emerald Ash Borer spread through GIS/cell-based/gravity models with data bolstered by web-based inputs. Pages 12–13 *in* V. Mastro, R. Reardon, and G. Para, editors. Emerald ash borer research and technology development meeting, Pittsburgh, PA. U.S. Department of Agriculture, Forest Service, Animal and Plant Health Inspection Service, Otis, MA.

Kammann, E. E., and M. P. Wand. 2003. Geoadditive models. Journal of the Royal Statistical Society Series C 52:1–18.

Kolar, C. 2004. Risk assessment and screening for potentially invasive fishes. New Zealand Journal of Marine and Freshwater Research 38:391–397.

Leung, B., D. M. Lodge, D. Finnoff, J. F. Shogren, M. A. Lewis, and G. Lamberti, G. 2002. An ounce of prevention or a pound of cure: bioeconomic risk analysis of invasive species. Proceedings of the Royal Society of London Series B 269:2407–2413.

Marsh-Matthews, E., and W. J. Matthews. 2000. Geographic, terrestrial and aquatic factors: which most influence the structure of stream fish assemblages in the midwestern United States? Ecology of Freshwater Fish 9:9–21.

McCullagh, P., and J. A. Nelder. 1989. Generalized linear models. Chapman and Hall/CRC, Boca Raton, FL.

McCune, B. 2006. Non-parametric habitat models with automatic interactions. Journal of Vegetation Science 17:819–830.

Morrison, L. W., S. D. Porter, E. Daniels, and M. D. Korzukhin. 2004. Potential global range expansion of the invasive fire ant, *Solenopsis invicta*. Biological Invasions 6:183–191.

Nekola, J. C. 1999. Paleorefugia and neorefugia: the influence of colonization history on community pattern and process. Ecology 80:2459–2473.

Peterson, A. T. 2003. Predicting the geography of species' invasions via ecological niche modeling. Quarterly Review of Biology 78:419–433.

Peterson, A. T., M. Papes, and D. A. Kluza, D. A. 2003. Predicting the potential invasive distributions of four alien plant species in North America. Weed Science 51:863–868.

Peterson, A. T., J. Soberon, and V. Sanchez-Cordero. 1999. Conservatism of ecological niches in evolutionary time. Science 285:1265–1267.

Peterson, A. T., and D. A. Vieglais. 2001. Predicting species invasions using ecological niche modeling: new approaches from bioinformatics attack a pressing problem. BioScience 515:363–371.

Phillips, S. J., R. P. Anderson, and R. E. Schapire. 2006. Maximum entropy models of species geographic distribution. Ecological Modelling 190:231–259.

Raxworthy, C. J., E. Martinez-Meyer, N. Horning, R. A. Nussbaum, G. E. Schneider, M. A. Ortega-Huerta, and A. T. Peterson, A. T. 2003. Predicting distributions of known and unknown reptile species in Madagascar. Nature 426:837–841.

Ricciardi, A., R. J. Neves, and J. B. Rasmussen. 1998. Impending extinctions of North American freshwater mussels (Unionida) following the zebra mussel (*Dreissena polymorpha*) invasion. Journal of Animal Ecology 67:613–619.

Ruiz, G. M., L. Fegley, P. Fofonoff, Y. Cheng, and R. Lemaitre. 2006. First records of *Eriocheir sinensis* H. Milne Edwards, 1853 (Crustacea: brachyura: varunidae) for Chesapeake Bay and the mid-Atlantic coast of North America. Aquatic Invasions 1:137–142.

Seguardo, P., and M. G. Auraujo. 2004. An evaluation of methods for modelling species distributions. Journal of Biogeography 31:1555–1568.

Sharma, S., D. A. Jackson, C. K. Minns, and B. J. Shuter. 2007. Will northern fish populations be in hot water because of climate change? Global Change Biology 13:2052–2064.

Shogren, J. F. 2000. Risk reduction strategies against the "explosive invader." Pages 56–69 in C. Perrings, M. H. Williamson, and S. Dalmazzone, editors. The economics of biological invasions. Edward Elgar, Cheltenham, UK.

Stockwell, D. 1999. The GARP modelling system: problems and solutions to automated spatial prediction. International Journal of Geographic Information Science 13:143–158.

Stockwell, D. R. B., and A. T. Peterson. 2002. Effects of sample size on accuracy of species distribution models. Ecological Modelling 148:1–13.

Sutherst, R. W., and G. F. A. Maywald. 1985. Computerized system for matching climates in ecology. Agriculture, Ecosystems and Environment 13:281–299.

Sydnor, T. D., M. Bumgardner, and A. Todd. 2007. The potential economic impacts of emerald ash borer (*Agrilus planipennis*) on Ohio, U.S., communities. Arboriculture and Urban Forestry 33:48–54.

Thuiller, W., L. Brotons, M. B. Araujo, and S. Lavorel. 2004. Effects of restricting environmental range of data to project current and future species distributions. Ecography 27:165–172.

Thuiller, W., D. M. Richardson, P. Pysek, G. F. Midgley, G. O. Hughes, and M. Rouget. 2005. Niche-based modelling as a tool for predicting the risk of alien plant invasions at a global scale. Global Change Biology 11:2234–2250.

Vander Zanden, M. J., J. D. Olden, J. H. Thorne, and N. E. Mandrak. 2004. Predicting occurrences and impacts of smallmouth bass introductions in north temperate lakes. Ecological Applications 14:132–148.

Whittier, T. R., P. L. Ringold, A. T. Herlihy, and S. M. Pierson. 2008. A calcium-based invasion risk assessment zebra and quagga mussels (*Dreissena* spp). Frontiers in Ecology and the Environment 6:180–184.

Wood, S. N. 2006. Generalized linear models: an introduction with R. Chapman and Hall/CRC, Boca Raton, FL.

Wood, S. N., and N. H. Augustin. 2002. GAMs with integrated model selection using penalized regression splines and applications to environmental monitoring. Ecological Modelling 157: 157–177.

Zaric, G. S., and M. L. Brandeau. 2002. Dynamic resource allocation for epidemic control in multiple populations. IMA Journal of Mathematics Applied in Medicine and Biology 19:235–255.

5

Stochastic Models of Propagule Pressure and Establishment

John M. Drake and Christopher L. Jerde

In a Clamshell

Invasion of a nonindigenous species is initiated by the immigration or introduction of propagating organisms. Empirical studies show that rates of propagule introduction into an ecosystem are correlated with rates of species invasion overall and, for a particular species, the chance that it has invaded. In the literature on species invasions, this is called ***propagule pressure***. This correlation suggests a basis for risk analysis of invasive species establishment. Establishment is an ambiguous concept, however, which we aim to make precise by describing different ***establishment events*** that are states of the nascent population corresponding to thresholds important in detection, growth, and spread of the colonizing species. Depending on the establishment event of interest, different mechanisms of the introduction pathway and/or the population dynamics might be considered. Here we discuss establishment events of interest to researchers and provide a framework linking introduction pathways and population dynamics underlying establishment. We emphasize stochastic processes to motivate a biologically based theory for risk analysis of species invasions.

Most research on risk analysis for species invasions has focused on predicting species that are capable of invading on the basis of biological characteristics (Ricciardi and Rasmussen 1998; Kolar and Lodge 2001; Keller et al. 2007; see also chapter 3) and/or a match between species biology and environments (Herborg et al. 2007). In what follows, we presuppose that a focal species or set of species has already been identified. We focus on the next step in the risk analysis and management process—identifying conditions under which establishment is likely to occur and when (Carlton 1996; Sakai et al. 2001). In particular, we focus on theories that assign probabilities to establishment events. Such estimated probabilities could be useful for evaluating the economic costs and benefits of preventing invasions (Leung et al. 2002) and optimizing management plans (see chapter 7).

Although predicting particular ecological events is difficult (Gilpin 1990), the statistical relationship between propagule pressure and establishment rates is clear (Bossenbroek et al. 2001; Drake and Lodge 2004; Leung et al. 2004; Lockwood et al. 2005). Estimates of these relationships may be obtained from models of species redistribution, for example, gravity models (Bossenbroek et al. 2001; Leung et al. 2004); from relative measures and surrogates, for example, the volume of ballast water entering a shipping port (Herborg et al. 2007); or from the underlying biological processes leading to population growth and decline (Drake 2004; Drake and Lodge 2006). Given the scarcity of high-quality data on both propagule pressure and establishment, we recommend that risk analysis proceed on as many of these fronts as is feasible. In each of these cases, arrival is linked to establishment by including propagule pressure as a parameter or "subprocess" in a model of establishment (Jerde and Lewis 2007). For many species, other factors such as Allee effects (inverse density dependence or depensation) and demographic stochasticity (Lande 1993) will interact with propagule pressure. These factors can be identified and estimated from observational data (Leung et al. 2004) or obtained from laboratory or field experiments (Jerde and Lewis 2007). In either case, controlling for confounding factors is crucial for obtaining an accurate picture of invasion risk.

In what follows, we highlight new ideas about establishment, joining the arrival and establishment stages of the invasion process. In our view, the arrival–establishment link is crucial to an economic perspective and any perspective that aims at understanding human intervention and control, since the arrival process itself is an expression of human choices and behavior and therefore is indirectly a response to market forces and trade (Thuiller et al. 2005; see also chapter 2). Several recent studies have reviewed the many and conflated concepts collected under the heading of "propagule pressure" (Lockwood et al. 2005). Because this chapter is about the relationship between propagule pressure and establishment, we visit the complementary question: What do we mean by establishment? We then consider the conceptual basis of a dose–response curve to relate establishment events and propagule pressure. We emphasize biological models implying a causal relationship between propagule pressure and establishment. These models complement statistical approaches in two ways. First, mechanistic formulations help to generalize by identifying external conditions under which species establish, as compared with the characteristics of the invasive species more commonly sought. Second, mechanistic models of establishment can be used to inform management about processes that may be targeted to reduce the probability of establishment (e.g., Wonham et al. 2005). In contrast, phenomenological models are basically limited to targeting only the input, propagule pressure. We explore the relationship between dose–response curves and waiting times, illustrating with an empirical example drawn from our research on scentless chamomile, *Matricaria perforata*. Lastly, we discuss our approach in the context of decision making with the goal of identifying potentially fruitful connections to be made between the economics and biological theory of species invasions.

ESTABLISHMENT: A CONCEPTUAL ANALYSIS

Establishment as Persistence, or Nonextinction

As a concept, "establishment" is vague and has been used in multiple ways by researchers (table 5.1). One view is that establishment is persistence or the "opposite" of extinction, that is, populations that do not go extinct are said to establish. But all (real) populations go extinct in the long run, so although there is ultimate extinction, there is no such thing as ultimate establishment. Naively, establishment could be the present condition of not being extinct, but then even declining populations and populations soon-to-be-extinct would qualify as established. More carefully, we might consider establishment to be persistence over some definite interval. Indeed, our intuitive idea about "being established" is neither about ultimate behavior of the population nor about past and present behavior, but has more to do with what a population will do in the near future. Thus, intuitively, an established population is one that is "unlikely" to go extinct in the near term. One approach to risk analysis therefore is to relate the chance of extinction in the near term to characteristics of the population, for example, population size, expected population dynamics, and (most important, from our perspective) propagule pressure.

TABLE 5.1. Concepts of establishment from the literature on biological invasions.

Definition	Reference (review [R] or experiment [E])
The presence of at least one individual in a location for one discrete time step	Jerde and Lewis 2007 (R + E) Levine et al. 2004 (R + E)
A measured population having more than 10 individuals present	Drake and Lodge 2004 (E)
The detected presence of the species	Bossenbroek et al. 2001 (E) Leung et al. 2004 (E) MacIsaac et al. 2004 (E)
A viable self-sustaining population	Sakai et al. 2001 (R)
A species with a self-sustaining population outside of its native range	Kolar and Lodge 2001 (R) Puth and Post 2005 (R) Pyšek et al. 2004 (R)
A new population that can sustain itself through local reproduction and recruitment, which thus augments or replaces dispersal from the donor region as a means for the population's persistence	Vermeij 1996 (R)
Survival to form a reproducing population	Wonham et al. 2000 (R + E) Carlton 1985 (R + E)
A fixed period of time over which the invader is present	Taylor and Hastings 2005 (R)
Survival and growth of at least one individual	Von Holle 2005 (E)

There is a trend in that empirical applications (E) must define a population threshold to define what it means to establish. Alternatively, conception reviews (R) of invasion biology and theoretical studies emphasize population persistence.

Establishment as a State of Abundance

Alternatively, we might think about establishment as corresponding to a particular population size; that is, a population is said to have established when its size first exceeds a particular value. Of course, this threshold is a choice that must be stipulated, whether by the researcher or by policy, and might differ depending on context and application (e.g., the time horizon over which probable extinction is relevant to management decisions, or the kind of data that are available for monitoring establishment). Thus, when Bossenbroek et al. (2001) developed a gravity model for the spread of zebra mussel (*Dreissena polymorpha*), the parameters for which were estimated from observed spread, they were forced to use the simple binary presence or absence of zebra mussel as a measure of establishment, because that is what was available. Alternatively, in a microcosm experiment to study dynamics of zooplankton populations at small size, Drake and Lodge (2004) arbitrarily considered any population composed of 10 or more individuals to be established.

In general, empirical studies have typically used thresholds as indicators of establishment ("E" in table 5.1) while conceptual reviews and theoretical studies consider persistence ("R" in table 5.1). This ambiguity is not unwarranted because conventional stochastic population models generally find a functional relation between extinction probability and population size, so the two concepts of establishment are convertible. Further, it is important that theoretical terms are operationally useful. In our view, it is therefore neither likely nor desirable that a consensus be made on what exactly it means to establish. Although there are technical approaches to accommodating such pluralism—Regan et al. (2002) suggest models based on supervaluation and fuzzy logic—we suggest the simplest solution is to provide clear definitions (chapter 7). Thus, in this chapter we discuss models that refer to both kinds of establishment events (nonextinction intervals and abundance thresholds). When we refer to the chance of establishment, we simply mean the probability of the defined establishment event within or at the stipulated interval of time.

REPRESENTING RISK: DOSE–RESPONSE CURVES

We turn now to the concept of risk. Risk is conventionally defined to include components of both probability and severity. Thus, in the context of invasive species, one might consider hazards of increasing severity, for example, (1) introduction, (2) establishment, (3) spread, (4) dominance, and (5) widespread environmental damage. Focusing here on a single "level" of severity, establishment, it remains to relate this to a stressor, propagule pressure. For this we use the concept of a dose–response curve. From ecotoxicology, a dose is a concentration of a chemical released into the environment (Suter 1993). In place of chemicals we consider propagules, in which case "concentration" refers to a density that may be expressed with respect to space (Drake et al. 2005), time (Drake and Lodge 2006), or both (Drury et al. 2007). For instance, we may consider the rate at which zebra mussels are introduced to isolated lakes by recreational anglers (Leung et al. 2004), rates of introduction between connected lakes through streams (Bobeldyk et al. 2005), or introductions of multiple species together in ballast water (Drake and Lodge 2007a) or on vessel hulls (Drake and Lodge 2007b). Given a dose of a potential invader, the problem for risk analysis is to estimate the probability of establishment (Drake and Lodge 2006). These are related by the dose–response curve. As argued above, establishment carries multiple definitions, which results in various dose–response curves for the same event data (figure 5.1). For example, establishment may be tied to the actual number of individuals establishing (figure 5.1A), the expected number of individuals establishing (figure 5.1B), or the probability of a specified number of individuals establishing (figure 5.1C,D).

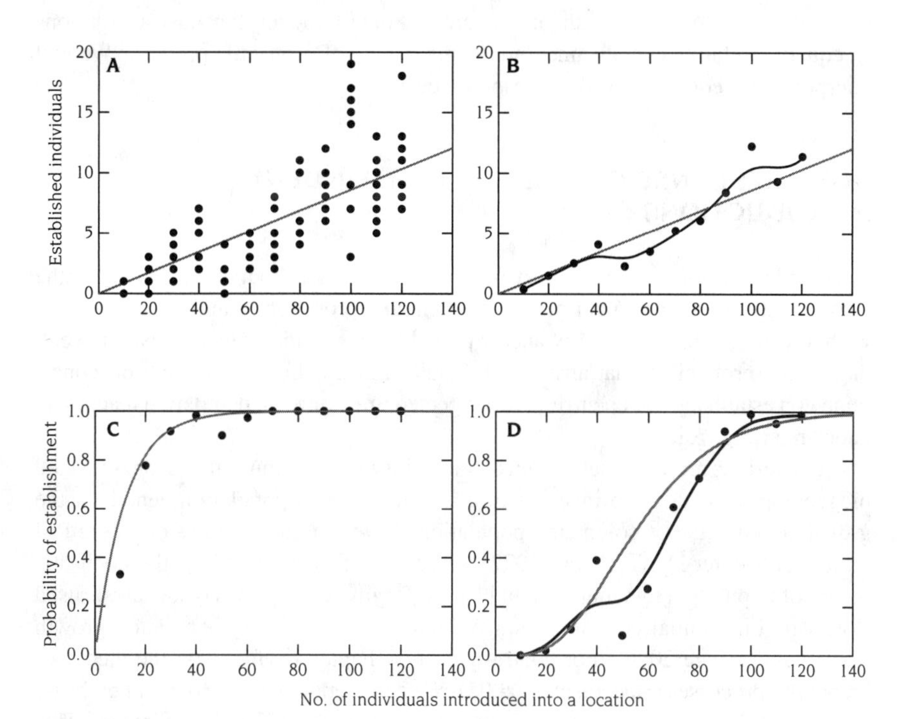

FIGURE 5.1.
Four different dose–response curves for the establishment of the invader M. *perforata*. (A) The raw data of the number of established individuals given the propagule pressure. For each dose, there are 10 observations where the observed number of established individuals are repeated. The individual probability of survival across all doses is 0.086. The response curve (gray line) is the expectation given the propagule pressure and the individual probability of survival. (B) The expectation of each dose has the same 0.086(ϕ) linear trend in A. (C) Alternatively, if a Poisson distribution is fit to each dose, then the probability of at least one individual establishing can be estimated. The model fit comes from using the individual probability of survival for different doses of individuals. (D) The probability of five or more individuals establishing produces a sigmoid dose–response curve of the establishment process. B and D have localized regression smooth curves (lowess curves) to identify the phenomenological trend in the data (black lines).

Often, dose–response curves are phenomenological descriptions of data. Figure 5.1, B and C, shows smooth fitted curves (splines) applied to data on the establishment of *M. perforata*. Such empirical dose–response curves are summary descriptions of the observed data, typically without a mechanistic biological or physical basis. They do not provide a causal relationship between propagule pressure and establishment. Phenomenological dose–response curves may be useful for risk assessment of invasive species when details of the population dynamics are unknown but observational data are available. Other fitting methods, such as probit analysis and logistic regression, may provide useful descriptions of observed patterns in the data for testing the effect of control treatments, for instance, when fitted to experimental data, again, when the underlying establishment dynamics are unknown and unrepresented by the model. A goal for invasion theory is to buttress these curves by biological explanation. As we turn to biological theory, we are looking for models that have as solutions an equation relating establishment to the magnitude of propagule pressure. We will interpret these equations as dose–response curves.

DOSE–RESPONSE CURVES FOR INDEPENDENT INTRODUCTIONS

In considering dose–response curves for independent introductions, imagine the arrival of a particular number of viable potentially propagating individuals, X_0. We wish to estimate the probability and/or probable magnitude of the undesired event, that is, the probability that arrival will result in an establishment event or, conditioned on establishment occurring, the expected time until the population reaches an undesired (large) size.

For various reasons (mathematical tractability, application to the theory of island biogeography, analogy to invasion of a mutant allele in population genetics), the growth and decline of colonizing populations have long been topics of theoretical research (Richter-Dyn and Goel 1972). A previous discussion of how the stochastic theory of population dynamics might be used for invasive species risk analysis used stochastic differential equations as a framework for modeling population growth (Drake and Lodge 2006). For contrast, in this chapter we focus on the theory of branching processes (Haccou et al. 2005). We denote population size at time t by X_t for $t = 0, 1, 2, \ldots$. To be biologically realistic, we require X_t to be an integer. This is important for understanding the population biology of establishment because it is this requirement that gives rise to *demographic stochasticity*, a main determinant of extinction of colonizing populations (Lande 1993). Variables and parameters are defined in table 5.2.

Now, if the exact trajectory of future population growth or decline were known at the time of introduction, the time of extinction (for any population that would in fact go extinct) could be known with certainty. This is rarely if ever the case (Gilpin 1990), and it is the purpose of risk analysis to probabilistically represent limitations to knowledge that might be used for prediction. Here, we cover two important sources of uncertainty. First, individual organisms will differ in their life trajectories.

TABLE 5.2. Parameter and variable definitions used in the modeling approaches of propagule pressure and establishment.

Parameter or variable	Definition
X_t	Number of individuals at time t
$g(x)$	Distribution of offspring
$f(z)$	Probability-generating function of $g(x)$
ξ	Nonnegative root of $f(x)$; probability of extinction
k	Threshold of establishment
Q_k	Transition probability matrix
p_{ij}	Transition probability from i to j
ρ	Smallest nonnegative root of $f(z)$
n	Fecundity
p_s	Survival probability
ϕ	Number of arriving individuals
Y	Random variable of the number of establishing individuals
T	Random variable of time

Some will reproduce once, and some twice. Some individuals will escape predation, disease, and misfortune and live long lives; others die young. We cannot say for each individual what its particular life trajectory will be, so we draw it randomly from a statistical distribution that we intend to represent the collection of all life trajectories in proportion to their realized prevalence in the real world. In practice, we summarize these distributions with theoretical quantities, some of which may be interpreted as the vital rates of the population. This is the idea of demographic stochasticity. Unless we wish to break down our population into individuals of different types (i.e., those that reproduce a lot and live a long time vs. those that reproduce only a little and live short lives), this type of uncertainty is irreducible. Fortunately, stochastic models such as the Galton-Watson process handle population structure when it is biologically relevant (Haccou et al. 2005).

The second source of uncertainty results from the fact that we have to choose a probability distribution to represent individual life trajectories. Particularly, two choices must be made. First, we must choose what equation or set of equations will represent the probability distribution; second, we must specify values for the parameters that enter into the equations. We will never choose these exactly right for at least three reasons: (1) the real world is too complex to be so simply represented (i.e., our model is an approximation); (2) even if the world were simple enough, we would never have enough information to know we had chosen the right simple model; (3) even if we had chosen the right model, we could not estimate the parameters of the model with arbitrary precision using finite data. Therefore, we have both structural uncertainty and parameter uncertainty. Note that both structural uncertainty and parameter uncertainty can be reduced by collecting additional information, although they cannot be reduced to zero. We mention structural uncertainty here because we think it is an important problem (chapter 7). It is also a difficult problem, and we do not treat it further here, but rather refer the reader to the literature on model selection as a start (Burnham and Anderson 2002). Neither do we further address parameter uncertainty, but this is a relatively familiar topic to empiricists. Thus, with the principle source of uncertainty in mind—demographic stochasticity—we turn to the theory of branching processes and how population dynamics represented in terms of stochastic processes can be used to develop dose–response curves for invasion risk.

To begin, we represent the initial population size by X_0 and wish to consider the dynamics of this population over the time interval immediately after arrival. Depending on the biology of the species under consideration, we could choose to study population dynamics in either continuous time or discrete time, in constant or fluctuating environments, and populations consisting of one class of individuals or of many classes of individuals. For simplicity, consistency, and generality, in this chapter we use discrete-time models of single-type populations where simple analytical solutions are most readily available. Where such models are not appropriate, numerical solutions may be computed.

To proceed, we allow that each individual gives rise to a random number of offspring in the next time step (including itself, if we are interested in species with overlapping generations) given by the probability mass function $g(x)$ called the offspring distribution. We make the biologically realistic assumptions that $g(0) > 0$ (an individual might not reproduce any viable offspring) and $0 < g(0) + g(1) < 1$ (it is possible that an individual would produce more than one offspring; otherwise, extinction would be guaranteed). If we assume that individuals do not interact (e.g., no density dependence), then if there are X_t individuals at time t, at time $t + 1$ the number of individuals is the sum of X_t random draws from $g(x)$, that is, a random number that is drawn from the X_tth convolution of $g(x)$. Such a process $\{X_t\ , t \geq 0\}$ is a discrete-time Markov chain known as the Galton-Watson process (Haccou et al. 2005). There are two possible eventual behaviors of this process. First, if the mean of the offspring distribution is ≤ 1, the population is guaranteed to go extinct. However, if the mean is > 1 (i.e., average individual fitness is > 1), then the population either goes extinct or explodes to infinity (Haccou et al. 2005). For initial population size $X_0 = 1$, the chance of extinction is given by the smallest nonnegative root $\xi = q$ of the equation (Haccou et al. 2005, theorem 5.2),

$$f(z) = \sum_{x=0}^{\infty} g(x)z^x = z, \tag{1}$$

where $f(z)$ is known as the probability-generating function and $g(x)$ is called the offspring distribution. Equation 1 is known as the probability-generating function of the distribution $g(x)$. For initial population size $X_0 > 1$, the chance of extinction is given by $\xi = q^{X_0}$ (Allen 2003, corollary 4.1). Accordingly, the chance of population explosion, $1 - \xi = 1 - q^{X_0}$, is seen to be an increasing function of propagule pressure X_0 that approaches 1 as X_0 gets large (figure 5.2).

Now that we have a way of calculating the extinction probability, we must ask if this is at all useful for risk analysis. After all, our notion of "population explosion" is unbounded exponential growth, a biologically unrealistic scenario.

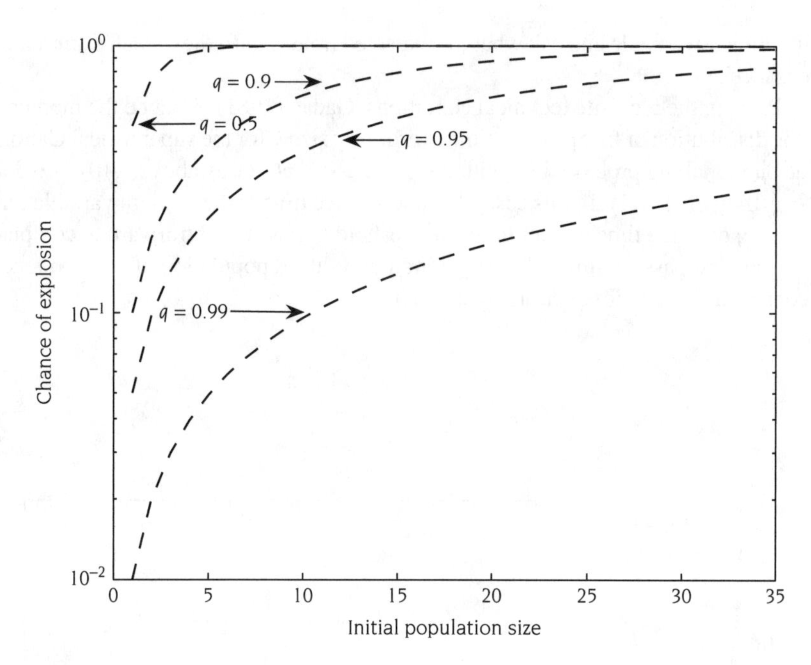

FIGURE 5.2.
Chance of explosion as a function of initial population size for different values of q (the smallest nonnegative root of equation 1) and the extinction probability for a population with initial size $x_0 = 1$.

Has our ignoring density dependence cost too much in terms of biological realism? We do not think so, because we really are interested in the early "growth" phase of an invasion. During this phase, we can think of the Galton-Watson branching process as a (stochastic) linear approximation to a density-dependent model. In a population in which regulation is primarily compensatory, and for the growth phase in which we are interested, the density-independent model will generally track the density-dependent model quite closely until the population approaches regulating population sizes such as carrying capacity. In situations where this is not the case (e.g., highly nonlinear dynamics with strong negative feedbacks, Allee effects), the model can be simulated rather than solved directly. But feedbacks are not known to be common in invasions, and the tractability of the simpler model gives some insight into the invasion process. For these reasons, the approximation of density-dependent growth by a density-independent model may well be justified, certainly for our purposes here. A greater problem lurks, however: dose–response as eventual nonextinction is insensitive to the amount of time it may take for population extinction or explosion to occur. (Consider the well-known phenomenon of lag times in invasions [Kowarik 1995].) In particular, where the mean of the offspring distribution is near 1, populations that will eventually explode may take a long time in getting there. A better approach might be to focus only on those populations that will ultimately invade and ask what the expected time is until they reach a stipulated "nuisance" size.

Proving the requisite technical conditions, Gadag (1981) obtained the moments of the distribution of first passage time to arbitrary size k for the supercritical Galton-Watson branching process (i.e., with the same assumptions as above: $g(0) > 0$ and $0 < g(0) + g(1) < 1$). In this case, the first passage time is the random variable that corresponds to the time it takes to reach threshold k. It is straightforward to compute the mean first passage time, $E[T_k(i)]$, for an introduced population of size i to reach k conditioned on nonextinction. Specifically,

$$E[T_k(i)] = (I - Q_k)^{-1}\,\eta, \tag{2}$$

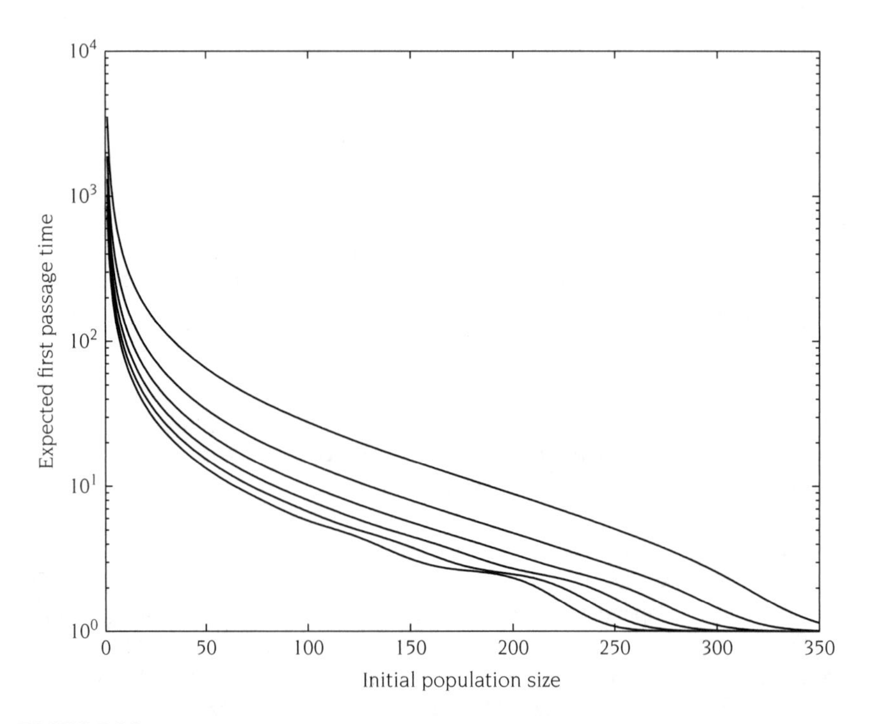

FIGURE 5.3.
Expected conditional first passage time as a function of initial population size for a representative Galton-Watson branching processes. The y-axis shows the expected first passage time to population size $k = 350$ as a function of initial population size. The offspring distribution is geometric with one-parameter distribution function $g(x) = p(1-p)^x$. The mean (μ) of the offspring distribution is determined by p through the relation $p = \mu/(1+\mu)$. By rearranging, we obtain the p associated with $\mu = \{1.1, 1.2, 1.3, 1.4, 1.5, 1.6\}$. First passage time is inversely correlated with μ so that the highest curve in the plot is $\mu = 1.1$ and the others are ordered accordingly. The individual transition probabilities p_{ij} are negative binomial with parameters p and j, and the extinction probability is $1/\mu g$. The expected first passage time is computed using equation 2.

where I is the identity matrix, η is the $k \times 1$ vector with all elements equal to 1, $Q_k = (q_{ij})$ is the matrix of one-step transition probabilities of the conditional process,

$$q_{ij} = \left[(1 - \rho^j)/(1 - \rho^i)\right] p_{ij}, \quad (3)$$

and p_{ij} are the one-step transition probabilities for the unconditional process. (Here, to avoid confusion and be consistent with Gadag's notation, we use ρ instead of q to represent the smallest nonnegative root of $f(z) = z$.) Figure 5.3 shows how this first passage time changes with initial population size for some representative models. We now turn to an example.

AN EXAMPLE: MATRICARIA PERFORATA

Recalling the different senses of propagule pressure, we now turn to an example to demonstrate some of the ways that quantitative reasoning can inform considerations of establishment risk. *M. perforata* (scentless chamomile) is a widespread, short-lived perennial that was first introduced to Canada in the early twentieth century either through contaminated livestock feed or for commercial horticulture. *M. perforata* has a highly variable fecundity. A plant can produce a single flowering head with less than a hundred seeds or multiple flowering heads resulting in thousands of seeds. The seeds generally do not naturally disperse far from the adult plant. De Camino Beck (2006) showed that 99% of seeds fall within 5 m of the adult. However, *M. perforata* is often mixed with agricultural crops, such as wheat (Douglas et al. 1991) or hay harvested for livestock. Thus, invasions typically result from transport by humans rather than natural dispersal. Variation in survival and seed production is strongly affected by spring precipitation, soil disturbance, and interspecific competition. Above-average spring precipitation leads to increased survival, and disturbed soil where competitors are removed allows for increased densities of plants (Bowes et al. 1994). Under common conditions, *M. perforata* can increase to densities of more than 25 plants per square meter, leading to wheat yield losses greater than 55% (Douglas et al. 1991). Taken in total, the life history of *M. perforata* in the Canadian prairies results in a lifetime reproductive value $\gg 1$, suggesting that establishment of just one individual is sufficient to allow a population to grow and invade a location (de Camino Beck 2006). Our example analysis below confirms this impression.

Jerde and Lewis (2007) estimated the probability that an individual seed would successfully germinate, develop, and flower to be 0.085 (90% confidence interval: 0.081, 0.091) for seeds sown in experimental agricultural plots in Vegreville, Alberta, Canada. As noted above, and like many invasive species, *M. perforata* is extremely prolific. For our example, we assume that introduction has been to a marginal habitat, that individuals produce an average of only a couple flowers, and that these each have 250 viable seeds. Production of a plant producing two flowers would be 500 seeds; 15 flowers would yield 3,750 seeds. We use these values as benchmarks to examine how establishment risk changes with propagule pressure. Now, denoting

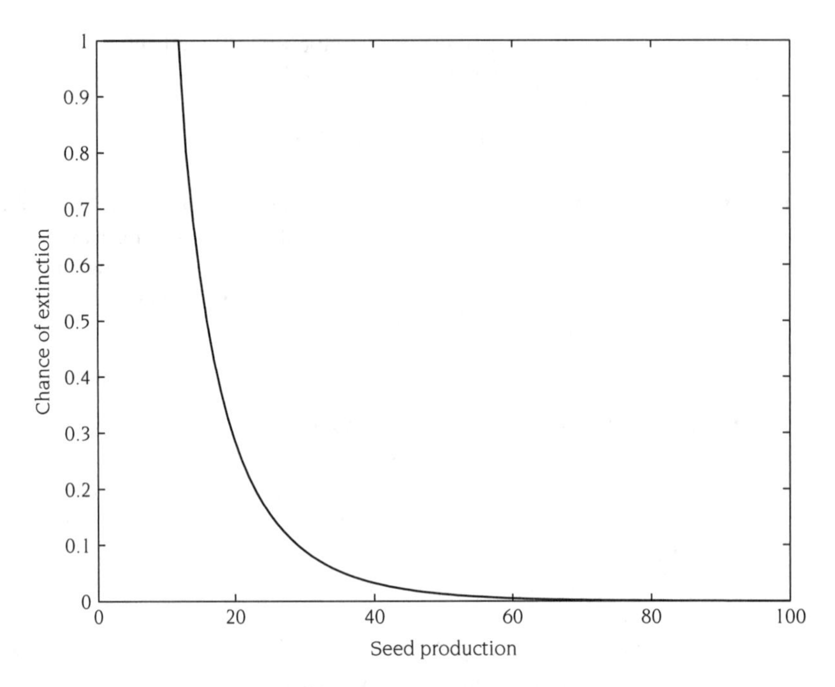

FIGURE 5.4.
Chance of extinction for a population of M. *perforata* modeled as a binomial survival process. The model assumes that every seed individually survives with probability $p = 0.085$. Conditioned on surviving, each plant deterministically produces a number of seeds given by the x-axis. Chance of extinction is given by the smallest positive root of equation 4.

production (500 seeds, 3,750 seeds, etc.) by n and assuming that survival probability of each seed is independent with probability $p_s = 0.085$, the number of plants in the second generation is a binomial random variate with probability-generating function

$$f(z) = (1 - p + pz)^n. \tag{4}$$

Numerically solving for the roots of this equation, we study how the chance of extinction/explosion increases with n, our measure of fecundity (figure 5.4). As expected, the chance of nonextinction is virtually nil for even modest seed set, and indeed, the time it takes to invade is small, too (figure 5.5). This is not surprising. For a species with seed production $n = 500$ (the extreme low end for *M. perforata*) and individual seed survival $p_s = 0.085$, the average population size of the second generation is 42.5 plants. Given the assumption that only demographic stochasticity matters, the chance (from the binomial distribution) that zero seeds would survive to the second generation is approximately 5.1×10^{-20}.

Note that the ability to overcome demographic stochasticity is a feature of the extraordinary potential reproduction rates of this species. It is for this reason that

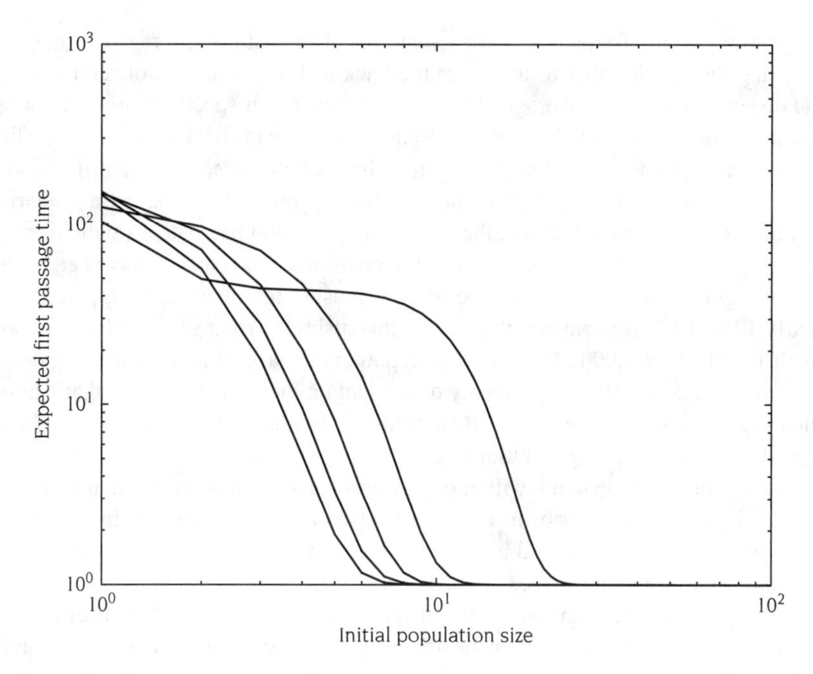

FIGURE 5.5.
Expected conditional first passage time as a function of initial population size for a Galton-Watson branching processes model for M. *perforata*. The y-axis shows the expected first passage time to population size $k = 500$ as a function of the number of seeds introduced. As described, we assume the offspring distribution is geometric. We stipulate that mean reproduction is the product of fecundity, $n = \{500, 1500, 2500, 3500, 4500\}$, and survivorship ($p_S = 0.085$), giving a mean offspring production range of $np_S = 43 - 383$.

high reproduction has traditionally been associated with weediness (de Camino Beck 2006). However, other species, including some that are capable of becoming a nuisance, may be much more affected by demographic stochasticity. Perhaps, for such weedy species, spatial heterogeneity or year-to-year variation (i.e., environmental stochasticity) is a more important determinant of fitness.

SPATIAL HETEROGENEITY AND ENVIRONMENTAL STOCHASTICITY

Although we do not explore effects of spatial heterogeneity and environmental stochasticity in detail here, we can anticipate their effects through some thought experiments. First, consider spatial heterogeneity. Suppose the environment is made of small patches that may be classified according to an arbitrary number of categories (e.g., good/bad), for each of which there is a unique offspring distribution. Assume the grain of heterogeneity is smaller than the characteristic dispersal distance of seeds, so each seed will independently arrive in a patch with a probability proportional to the frequency of the categories. The spatially averaged compound offspring distribution that results from the back-to-back processes of reproduction and dispersal will be the average of the means of the patch-specific unique offspring distributions weighted by their relative frequencies. The upshot is that if low-quality patches are not common enough to reduce the spatially averaged offspring distribution to be quite close to 1, the chance of explosion will still be extraordinarily high. Suppose, for example, that there are only good and bad patches with average offspring of 1,000 and 0, respectively. For a plant in a good patch to have negligible probability of establishment, the bad patches must outnumber the good patches by nearly 1,000 to 1. By contrast, the chance that a single errant seed lands on a good patch is only 1 in 1,000. Thus, in this hypothetical case of a paradigmatic weed, the chance of establishment is hugely dependent on the accidental arrival at a good patch—the outcome is almost wholly determined by the events between arrival and reproduction in the next generation.

If spatial heterogeneity will rarely prevent establishment, what remains to explain is why there are not in fact more invasions than there are. In particular, we wonder if the reason could be environmental stochasticity. We can incorporate fluctuating environments in the Galton-Watson process by conditioning the offspring distribution on another, randomly varying quantity (Haccou et al. 2005). In effect, we replace the parameters that govern the offspring distribution (and that we have sought to interpret biologically above) with random variables. The resulting model that includes demographic and environmental stochasticity is therefore doubly stochastic in the sense that both the population size and some parameter that determines the distribution of possible population sizes in a time step are random. At this point we have a model that is sufficiently complex that making headway analytically will be difficult—though it is a classical problem and has been treated at length (Haccou et al. 2005). In short, environmental stochasticity may well diminish establishment probability as population size asymptotically grows as the geometric mean of annually realized reproduction. One relevant finding is that there are subcritical processes in random environments (i.e., processes guaranteed to go extinct) for which the average of realized offspring distributions is in fact > 1. This classical result (Lewontin and Cohen 1969) has led to more recent studies that have shown, among other things, that sequential introduction of propagules is more likely to result in establishment in a variable environment than is a single introduction of size equal to the sum of the sequential introductions (Haccou and Iwasa 1996; Haccou and Vatutin 2003).

DOSE–RESPONSE CURVES FOR RECURRENT INTRODUCTIONS

We turn now, briefly, to consider situations when introductions are repeated. This is an important area for further study, and we refer the reader to theoretical papers (Haccou and Iwasa 1996; Haccou and Vatutin 2003) and to work by the authors in the context of invasions (Drake and Lodge 2006; Drury et al. 2007; Jerde and Lewis 2007).

Repeated introductions present a new source of uncertainty—when and how many additional propagules arrive. As an example, we compare what we can learn statistically from data with the additional insight we obtain by considering the invasion process dynamically. The gray lines in figure 5.1 represent simple mechanistic models using the experimentally determined probability of survival for a single seed of *M. perforata* ($p_s = 0.085$; for details, see Jerde and Lewis 2007). Figure 5.1, A and B, models the expectation of the number of establishing individuals (Y) as a function of the number introduced (ϕ) when we assume the seeds are independent and identically distributed. Given the assumptions of independence and identically distributed realizations (Y), the number to reach flowering is Poisson distributed with average 0.085ϕ. (Recall from above that, for *M. perforata*, for a plant to reach flowering is practically to ensure establishment.) Figure 5.1, C and D, shows the probability of at least one individual reaching flowering, $\Pr(Y \geq 1) = 1 - \Pr(Y = 0) = 1 - e^{-0.085\phi}$, and the probability of at least five individuals reaching flowering is $\Pr(Y \geq 5) = 1 - [\Pr(Y = 0) + \cdots + \Pr(Y = 4)]$, respectively.

As before, we can extend our analysis to consider the waiting time until establishment, that is, a temporal extension of the dose–response curve. In general, for a given dose (i.e., constant initial population size, repeatedly introduced), there is an associated probability of an establishment event occurring. If the expected response is constant for discrete time steps, then the time at which an establishment event will occur follows a geometric distribution. Identifying the dose–response curve with the waiting time for establishment (i.e., flowering) of at least y individuals, we have

$$\Pr(T = t) = [\Pr(Y \geq y)][1 - \Pr(Y \geq y)]^{t-1}, \qquad (5)$$

where the left-hand term is the probability of establishment at time T, and the right-hand term is the geometric probability mass function; $y = 1, 2, 3, \ldots$ is the predetermined threshold. The random variable T is the time at which establishment occurs and can take on values, $t = 1, 2, 3, \ldots$; as above, our interest is in the expectation of T, that is, the average waiting time.

It is known that establishment of one *M. perforata* plant ($y = 1$) is sufficient to invade a new location (de Camino Beck 2006). The waiting time distributions for $\phi = 5, 10$, and 30 introduced *M. perforata* seeds, corresponding to establishment probabilities, $\Pr(Y \geq 1) = 0.35$, 0.57, and 0.922 (from figure 5.1C), are plotted in figure 5.6. The waiting time distribution for *M. perforata* with an establishment threshold of one individual is therefore

$$\Pr[(T = t)|p_e = 0.085] = \left[\left(1 - e^{-0.085\phi}\right)\right]\left[\left(1 - \left(1 - e^{-0.085\phi}\right)\right)^{t-1}\right]. \qquad (6)$$

With only five seeds being introduced, there is $> 33\%$ chance that establishment will occur at the next time step. When the propagule pressure (ϕ) is raised to 30 individuals, establishment of *M. perforata* becomes almost certain. Reducing ϕ from 30 to 5 individuals does reduce the probability of an establishment event from

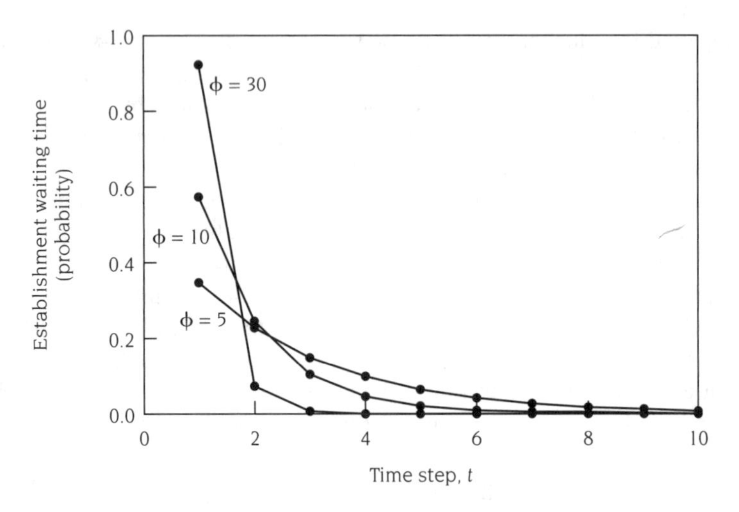

FIGURE 5.6.
The waiting time distributions for scentless chamomile for three different levels of arriving individuals, $\phi = 5$, 10, and 30. As ϕ becomes smaller, the waiting time distribution becomes flat and the probability of an establishment event at the next time step becomes small.

approximately 1 to 0.33, for one time step. This is the same interpretation that could be inferred from the dose–response curves (figure 5.1C). However, waiting times allow for extrapolation of the invasion process to determine the probability of invasion for a location under constant propagule pressure and individual probability of survival by some future time t, by using the cumulative mass function of the waiting time distribution,

$$\Pr(T \leq t) = \left[1 - (1 - \Pr(Y = y))^t\right], \tag{7}$$

where the left-hand term is the probability of establishment by time T, and the right-hand term is the geometric cumulative mass function. For *M. perforata*, $\Pr(T \leq t|\phi) = 1 - [1 - (1 - e^{-0.085\phi})]$, and after $t = 10$ time steps, irrespective of the propagule pressure (5, 10, or 30 individuals), the probability of an establishment event is approximately 1.

SYNTHESIS

To conclude, we first summarize the ecology of invasions as conceptualized by the simple models presented here. Next, we reflect on how these models need to be advanced both biologically and to integrate with the real-time bioeconomic decisions that must be made by land and resource managers. Finally, we point to some outstanding issues that will require additional consideration.

First, what have we learned about the process of invasion? Reflecting on both the general arguments advanced early in the chapter and the example of *M. perforata*, one, perhaps surprising, commonality emerges: when we are precise about what we think are the factors that determine establishment (e.g., demographic stochasticity), we find that it is actually rather difficult *not* to establish. That establishment seems far more common than failure suggests our current understanding of the very early stages of population growth remains limited. We expect that explaining nonestablishment might be straightforward if additional information about environmental heterogeneity and stochasticity were incorporated into models. Unfortunately, it is precisely these factors that are most likely to vary greatly from place to place, precluding the kind of generalization that is the goal of risk assessment in the first place. Then, perhaps the best that may be hoped is to bring the uncertainty associated with species introductions to within reasonable bounds. When the problem has been understood as precisely as possible at the level of generality needed and some uncertainty remains, how to best manage invasion risk can be determined only by economic arguments and instruments.

What should these economic instruments look like? In an ideal world, we would have both a reliable, accurate stochastic model for the introduction process and for the postintroduction growth process, and a perfect understanding of the future costs of control and establishment. In such a case, we might represent the problem in terms of a Markov decision process (e.g., Mangel 1985). That such reliable information is rarely forthcoming notwithstanding, this approach is probably the most common among ecologists and conservation biologists, at least in idea, if not also in practice. By contrast, having reflected on the difficulty of obtaining such information (and the low quality of decisions made pretending that bad information is good), we suggest that more creative approaches are required. What will these approaches look like? That is hard to say. The problem would be hard enough with good information. How to make robust decisions with poor or corrupt information is correspondingly daunting. As a start, we suggest that paying close attention to the difference between information already in hand and the amount and cost (and therefore feasibility) of obtaining whatever information would be required to warrant changing course from intuition (see Ben-Haim 2006).

In this light, we recommend carefully considering Simberloff's (2003) provocative question: How much population biology is needed to manage introduced species? The World Trade Organization's Agreement on Application of Sanitary and Phytosanitary Measures requires a quantitative risk assessment that demonstrates an exotic species poses a substantial threat in order to ban its import. Due to the WTO's World Trade Organization's requirements, population parameters are almost certainly necessary in order to perform such a risk assessment. Perhaps such investment in biological information would be better spent on control. Simberloff (2003) provides examples where physical or chemical removal of exotic species successfully eradicated or contained invasions when population data were not available. Therefore, one progressive approach to future management of invasive species would be to build quantitative risk assessment protocols applicable to different kinds of data and to variation in data quantity and quality. For example, invasion waiting times provide a probability of establishment by a time, T, but require estimates of survivorship. For new locations, survivorship will generally be unknown. However, if it can be assumed these locations have the same but unknown survivorship, then relative waiting times can be formulated (Jerde and Lewis 2007; see also chapter 7).

We conclude by highlighting some outstanding theoretical issues. First, perhaps the most useful theoretical advance would be to better understand the transient and long-run behavior of density-dependent branching processes. For our purposes here, we have assumed that a linear approximation is appropriate over the range of densities that are relevant. This assumption is patently inappropriate for species exhibiting Allee effects (probably not a minority of species), and perhaps even for species regulated by compensatory dynamics, particularly for populations highly regulated by local density dependence introduced to highly heterogeneous environments. To know on what time scales to expect extinction for discrete time/density-dependent branching processes would be particularly useful. Second, the model for recurrent introductions that we use here does not allow any carryover of established plants from year to year. For our *M. perforata* example here this is fine, because we were mostly interested in the chance that a single plant germinates and matures to flowering. However, for species with overlapping generations and for which establishment thresholds are set somewhat higher, the contribution of already present individuals must be tracked over time. The classical birth-death-immigration process (Matis and Kiffe 2000) represents one version of this "rescue effect" (Brown and Kodrick-Brown 1977), and for a spatial context this problem has been studied by Drury et al. (2007). However, to our knowledge the effect of immigration on supercritical and density-dependent discrete time branching processes remains to be addressed.

Acknowledgments We thank D. Lodge, R. Keller, T. Purucker, J. Bossenbroek, J. Muirhead, and three reviewers for helpful suggestions on earlier versions of this chapter. We gratefully acknowledge editorial assistance by Andrea Silletti and the support of the U.S. National Science Foundation for the Integrated Systems for Invasive Species project (DEB 02-13698). The Alberta Research Council provided the plots for the *M. perforata* experiments. C.L.J. was supported by a National Sciences and Engineering Research Council of Canada Collaborative Research Opportunity grant and a Discovery grant to M. A. Lewis and by the Canadian Aquatic Invasive Species Network. J.M.D. was supported by a grant from the Great Lakes Protection Fund.

References

Allen, L. J. S. 2003. An introduction to stochastic processes with applications to biology. Pearson Education, Inc., Upper Saddle River, NJ.

Ben-Haim, Y. 2006. Info-gap decision theory: decisions under severe uncertainty, 2nd edition. Academic Press, San Diego, CA.

Bobeldyk, A. M., J. M. Bossenbroek, M. A. Evans-White, D. M. Lodge, and G. A. Lamberti. 2005. Secondary spread of zebra mussels (*Dreissena polymorpha*) in coupled lake-stream systems. EcoScience 12:339–346.

Bossenbroek, J. M., J. C. Nekola, and C. E. Kraft. 2001. Prediction of long-distance dispersal using gravity models: zebra mussel invasion of inland lakes. Ecological Applications 11:1778–1788.

Bowes, G. G., D. T. Spurr, A. G. Thomas, D. P. Peschken, and D. W. Douglas. 1994. Habitats occupied by scentless chamomile (*Matricaria perforata* Merat) in Saskatchewan. Canadian Journal of Plant Science 74:383–386.

Brown, J. H., and A. Kodrick-Brown. 1977. Turnover rates in insular biogeography: effect of immigration on extinction. Ecology 58:445–449.

Burnham, K. P., and D. R. Anderson. 2002. Model selection and multimodel inference. Springer, New York.

Carlton, J. T. 1985. Transoceanic and interoceanic dispersal of coastal marine organisms: the biology of ballast water. Oceanography and Marine Biology: Annual Review 23:313–371.

Carlton, J. T. 1996. Pattern, process, and prediction in marine invasion ecology. Biological Conservation 78:97–106.

de Camino Beck, T. 2006. Theoretical considerations for biological control: a case study with *M. perforata*. Ph.D. thesis, University of Alberta.

Douglas, D. W., A. G. Thomas, D. P. Peschken, G. G. Bowes, and D. A. Dersken. 1991. Effects of summer and winter annual scentless chamomile (*Matricaria perforata* Merat) interference on spring wheat yield. Canadian Journal of Plant Science 71:841–850.

Drake, J. M. 2004. Allee effects and the risk of biological invasion. Risk Analysis 24:795–802.

Drake, J. M., and D. M. Lodge. 2004. Effects of environmental variation on extinction and establishment. Ecology Letters 7:26–30.

Drake, J. M., and D. M. Lodge. 2006. Allee effects, propagule pressure and the probability of establishment: risk analysis for biological invasions. Biological Invasions 8:365–375.

Drake, J. M., and D. M. Lodge. 2007a. Rate of species introductions in the Great Lakes via ships' ballast water and sediments. Canadian Journal of Fisheries and Aquatic Sciences 64: 530–538.

Drake, J. M., and D. M. Lodge. 2007b. Hull fouling is a risk factor for intercontinental species exchange in aquatic ecosystems. Aquatic Invasions 2:121–131.

Drake, J. M., D. M. Lodge, and M. Lewis. 2005. Theory and preliminary analysis of species invasions from ballast water: controlling discharge volume and location. American Midland Naturalist 54:459–470.

Drury, K. L. S., J. M. Drake, D. M. Lodge, and G. Dwyer. 2007. Immigration events dispersed in space and time: factors affecting invasion success. Ecological Modelling 206:63–78.

Gadag, V. G. 1981. First passage time for a supercritical Galton-Watson process restricted to the non-extinction set. Journal of Theoretical Biology 93:585–589.

Gilpin, M. 1990. Ecological prediction. Science 248:88–89.

Haccou, P., and Y. Iwasa. 1996. Establishment probability in fluctuation environments: a branching process model. Theoretical Population Biology 50:254–280.

Haccou, P., P. Jagers, and V. A. Vatutin. 2005. Branching processes: variation, growth, and extinction of populations. Cambridge University Press, Cambridge, UK.

Haccou, P., and V. Vatutin. 2003. Establishment success and extinction risk in autocorrelated environments. Theoretical Population Biology 64:303–314.

Herborg, L-M., C. Jerde, D. Lodge, G. Ruiz, and H. MacIsaac. 2007. Predicting the North American distribution of Chinese mitten crabs (*Erioncheir sinensis*) using measures of propagule pressure and environmental niche models. Ecological Applications 17:663–674.

Jerde, C. L., and M. A. Lewis. 2007. Waiting for invasion: a framework for the arrival of non-indigenous species. American Naturalist 170:1–9.

Keller, R. P., J. M. Drake, and D. M. Lodge. 2007. Fecundity as a basis of risk assessment of nonindigenous freshwater molluscs. Conservation Biology 21:191–200.

Kolar, C. S., and D. M. Lodge. 2001. Progress in invasion biology: predicting invaders. Trends in Ecology and Evolution 16:199–204.

Kowarik, I. 1995. Time lags in biological invasions with regard to the success and failure of alien species. Pages 15–38 *in* P. Pysek, K. Prach, M. Rejmanek, and M. Wade, editors. Plant invasions: general aspects and special problems. SPB Academic Publishing, Amsterdam.

Lande, R. 1993. Risks of population extinction from demographic and environmental stochasticity and random catastrophes. American Naturalist 142:911–927.

Leung, B., J. Drake, and D. Lodge. 2004. Predicting invasions: propagule pressure and the gravity of Allee effects. Ecology 85:1651–1660.

Leung, B., D. Lodge, D. Finnoff, J. Shogren, M. Lewis, and G. Lamberti. 2002. An ounce of prevention or a pound of cure: bioeconomic risk analysis of invasive species. Proceedings of the Royal Society Series B Biological Sciences 269:2407–2413.

Levine, J., P. Adler, and S. Yelenik. 2004. A meta-analysis of biotic resistance to exotic plant invasions. Ecology Letters 7:975–989.

Lewontin, R. C., and D. Cohen. 1969. On population growth in a randomly varying environment. Proceedings of the National Academy of Sciences of the United States of America 62:1056.

Lockwood, J., P. Cassey, and T. Blackburn. 2005. The role of propagule pressure in explaining species invasions. Trends in Ecology and Evolution 20:223–228.

MacIsaac, H., Borbely, J., Muirhead, J., and Graniero, P. 2004. Backcasting and forecasting biological invasions of inland lakes. Ecological Applications 14:773–783.

Mangel, M. 1985. Decision and control in uncertain resource systems. Academic Press, New York.

Matis, J.H., and T. Kiffe. Stochastic Population Models. Lecture Notes in Statistics 145. Springer, New York.

Puth, L., and D. Post. 2005. Studying invasion: have we missed the boat? Ecology Letters 8:715–721.

Pyšek, P., D. M. Richardson, M. Rejmánek, G. L. Webster, M. Williamson, and J. Kirschner. 2004. Alien plants in checklists and floras: towards better communication between taxonomists and ecologists. Taxon 53:131–143.

Regan, H. M., M. Colyvan, and M. A. Burgman. 2002. A taxonomy and treatment of uncertainty for ecology and conservation biology. Ecological Applications 12:618–628.

Ricciardi, A., and J. B. Rasmussen. 1998. Predicting the identity and impact of future biological invaders: a priority for aquatic resource management. Canadian Journal of Fisheries and Aquatic Sciences 55:1759–1765.

Richter-Dyn, N., and N.S. Goel. 1972. On the extinction of a colonizing species. Theoretical Population Biology 3:406–433.

Sakai, A. K., F. W. Allendorf, J. S. Holt, D. M. Lodge, J. Molofsky, K. A. With, S. Baughman, R. J. Cabin, J. E. Cohen, N. C. Ellstrand, D. E. McCauley, P. O'Neil, I. M. Parker, J. N. Thompson, and S. G. Weller. 2001. The population biology of invasive species. Annual Reviews of Ecology and Systematics 32:305–332.

Simberloff, D. 2003. How much population biology is needed to manage introduced species? Conservation Biology 17:83–92.

Suter, G. 1993. Ecological risk assessment. Lewis Publishers, Chelsea, MI.

Taylor, C. M., and A. Hastings. 2005. Allee effects in biological invasions. Ecology Letters 8:895–908.

Thuiller, W., D. M. Richardson, P. Pyšek, G. F. Midgley, G. Hughes, and M. Rouget. 2005. Niche-based modelling as a tool for predicting the risk of alien plant invasions at a global scale. Global Change Biology 11:2234–2250,

Vermeij, G. 1996. An agenda for invasion biology. Biological Conservation 78:3–9.

Von Holle, B. 2005. Biotic resistance to invader establishment of a southern Appalachian plant community is determined by environmental conditions. Journal of Ecology 93:16–26.

Wonham, M., J. Carlton, G. Ruiz, and L. Smith. 2000. Fish and ships: relating dispersal frequency to success in biological invasions. Marine Biology 136:1111–1121.

Wonham, M., M. Lewis, and H. MacIsaac. 2005. Minimizing invasion risk by reducing propagule pressure: a model for ballast-water exchange. Frontiers in Ecology and the Environment 3:473–478.

6

Estimating Dispersal and Predicting Spread of Nonindigenous Species

Jim R. Muirhead, Angela M. Bobeldyk, Jonathan M. Bossenbroek, Kevin J. Egan, and Christopher L. Jerde

In a Clamshell

The ability to estimate dispersal and the subsequent spread of nonindigenous species is important because preventative methods most readily apply to these stages of the invasion sequence. We address several methods used to calculate dispersal kernels for nonindigenous species, including natural movement, hydrological, aerial, and anthropogenic dispersal. In particular, we focus on gravity and random-utility models because they are suitable for modeling anthropogenic dispersal among spatially discrete habitat such as freshwater systems. We then address mechanisms of population spread and link dispersal kernels to models forecasting spread. We compare the merits and data requirements for different models of dispersal and population spread, and illustrate which models are best suited to deal with long-distance dispersal and heterogeneous habitat according to model assumptions. We also highlight the role of Allee effects on management strategies for invasive population spread where controlling population spread may be more effective when individuals in an invading population experience difficulty in finding mates, for example. Finally, we show how some of these models may be used as tools in making bioeconomic and management decisions in terms of risk assessment and risk management.

Invasions of aquatic ecosystems by nonindigenous species (NIS) are occurring at increasing rates in terms of both the number of novel species entering a system (Ricciardi 2006) and the cumulative number of systems or geographic areas that have been invaded (e.g., Veit and Lewis 1996; MacIsaac et al. 2004; Gilbert et al. 2005). NIS are also becoming increasingly important agents in altering ecosystem processes and contributing to the global homogenization of biodiversity (Crooks 2002; Rahel 2002; Olden et al. 2004). The best way to reduce the impacts of many species is to prevent their movement into uninvaded areas. Thus, the ability to accurately forecast spread enables prediction of high-risk areas and may lead to management strategies that contain or reduce spread.

NIS movement occurs during two stages of the invasion sequence: the initial transport of propagules from the native region, and secondary spread once populations become established in novel habitats (Kolar and Lodge 2001). Management efforts focusing on the initial transport of propagules (see chapter 1) are potentially easier and less costly than mitigation after the invasion has occurred because curbing species dispersal will likely have the greatest influence in determining establishment of the species (Leung et al. 2002; Jeschke and Strayer 2005). Through the identification of mechanisms, pathways, and spatially explicit invasion "hotspots," management efforts become more feasible and cost-effective. For example, by distinguishing high-risk sites, forecasts can guide prevention or rapid response efforts that do not waste resources on low-risk sites. In addition, focusing on dispersal vectors allows for a management strategy applicable to multiple species transported by similar mechanisms as opposed to a strategy specific to each new NIS.

This chapter is divided into three sections: dispersal, spread, and risk management. Here, *dispersal* is defined as the process that describes the movement or redistribution of individuals from one place to another (e.g., Wiens 2001), and *spread* as the change in population density as a function of space and time. It is important to make this distinction between dispersal and population spread, because models of population spread implicitly address the additional invasion "filters" of habitat suitability and biotic integration into the recipient community (chapters 4 and 5). That is, at each step of the invasion sequence, only a subset of propagules are able to make the transition from the previous stage. First, we address methods used to collect data and to calculate dispersal kernels for different mechanisms such as animal movement or hydrological or anthropogenic dispersal. We define the dispersal kernel as the probability of an individual moving a specified distance from its last location per unit time, or the proportion of the population dispersed a specified distance within a time interval. In particular, we focus on two methods to describe the dispersal kernel for the anthropogenic transport of NIS, gravity and random-utility models, because they are suitable for modeling dispersal among spatially discrete habitat. Second, we address mechanisms of population spread and link dispersal kernels to models forecasting spread. For each section, we also compare the merits of different modeling approaches and data requirements for dispersal models and models of population spread. Finally, we highlight applications of these models for bioeconomic and management decisions in terms of risk assessment and risk management.

DISPERSAL

Dispersal is an essential component in organismal biology, affecting individual survival and reproduction as well as population- and ecosystem-level dynamics (Turchin 1998). Dispersal continues to be one of the most studied aspects of ecology and has been fundamental to works on island biogeography, metapopulations, and more recently, metacommunities (e.g., MacArthur and Wilson 1963; Hanski 1999; Leibold et al. 2004). Depending on the transport mechanism, dispersal patterns are often characterized by two overlapping modes: local and long-distance dispersal. Patterns of local dispersal are frequently observed when the organism disperses under its own power (e.g., walking). Here, range expansion proceeds continuously from the periphery of the population (Shigesada and Kawasaki 2002). Long-distance dispersal, on the other hand, is more of a passive process where the organism is transported by vectors such as water currents or wind or through interaction with human movement. In recent years, studies have recognized the importance of long-distance dispersal and have focused on its role in determining rates of spread and persistence of populations, the genetic structure of populations, and natural versus human-mediated vectors (e.g., Okubo 1980; Avise 1994; Suarez et al. 2001; Trakhtenbrot et al. 2005). Understanding long-distance dispersal of NIS is important because in both theoretical and empirical examples, long-distance dispersal events result in a change from linear to exponential rates of population spread (Lewis 1997). As a consequence, characterizing long-distance dispersal and predicting the location of future "satellite" colonies become increasingly difficult (Nathan et al. 2003).

Mechanisms involved in the introduction and subsequent dispersal of NIS can be divided into two major categories: natural and anthropogenic. An understanding of this distinction is necessary when addressing preventative measures of further spread or applying appropriate educational or legislative management efforts. However, several dispersal mechanisms may be operating simultaneously and interactively for many NIS, thus increasing the number of vectors that need to be assessed and the level of uncertainty involved when forecasting spread. In addition, dispersal mechanisms may have indirect effects that alter habitat invasibility. For example, as a byproduct of anthropogenic activity, the quality of habitat may also become stressed resulting in increased invasibility.

Natural Dispersal

Without human intervention, species may spread under their own power and by the aid of abiotic factors. Life history characteristics of NIS, such as individual mobility, pelagic larval stages, or the production of resistant resting eggs, lend themselves to natural dispersal. Chance abiotic events, such as wind storms or floods, also contribute to the natural dispersal of species across distances far greater than would be possible under individual mobility. Finally, the loss of a restrictive barrier (e.g., glacial ice sheet, waterfall) may allow species to invade a new environment.

Hydrologic Dispersal

Hydrologic pathways, such as rivers and streams (i.e., lotic systems), provide a natural mechanism for aquatic species to disperse into connected systems. For example, the veliger stage (planktonic larval stage) of zebra mussels (*Dreissena polymorpha*) has been widely recognized as a key factor in the invasion of entire river systems. After primary introductions, zebra mussel veligers drift through outflowing streams, colonizing downstream reaches and lakes (Horvath et al. 1996; Stoeckel et al. 1997; Bobeldyk et al. 2005). Similar dispersal has been documented for the waterflea (*Daphnia lumholtzi*) drifting downstream from invaded reservoirs (Shurin and Havel 2002). Even species without planktonic larval stages, such as rusty crayfish (*Orconectes rusticus*), can spread naturally through lakes and stream corridors to invade other systems (Puth and Allen 2005).

In these systems, dispersal data can be obtained through mark-recapture studies or radiotelemetry tracking of individuals (table 6.1). Under the assumption that populations in lotic systems form along a one-dimensional strip of habitat, the dispersal kernel can then be fitted from the dispersal data of the tracked individuals (Lewis et al. 2006). Movements of the invasive signal crayfish (*Pacifastacus leniusculus*), for example, were tracked by radiotelemetry every 2 days (Bubb et al. 2004). The frequency distribution of movements upstream and downstream from the release location, as well as the total distance dispersed, was described by inverse power laws:

$$m = cx^{-n}, \tag{1}$$

where m is the probability of a movement, x is the displacement from the release point, and c and n are fitted constants.

However, if the underlying shape of the dispersal kernel is unknown, the moment-generating function for the kernel is estimated from one-dimensional dispersal data:

$$M^E(s) = \frac{1}{N}\sum_{i=1}^{N} \exp(sz_i), \tag{2}$$

where s is the unknown slope, z_i is the observed individual dispersal, N is the number of individuals, and the superscript E refers to the empirical estimate of the moment-generating function (Lewis et al. 2006). A moment-generating function is unique to each dispersal kernel and is used to describe characteristics (i.e., moments) of the dispersal kernel such as the mean and variance in dispersal distance.

Alternatively, data on the movement of organisms may be collected such that the density of organisms (number per unit area) is recorded at distance intervals from the putative source. Horvath et al. (1996) recorded the density of zebra mussels in two streams in the St. Joseph River basin (Indiana and Michigan, USA) and found that density within the streams decreased exponentially with distance from invaded upstream lakes. In order to avoid bias due to the width of the stream or river when fitting the dispersal kernel, the density of organisms should be multiplied by the width of the stream and the dispersal kernel constrained so that the integrand is equal to 1 (Lewis et al. 2006).

Occasionally, data are in the form of frequency distributions, and the dispersal kernels are fit to histograms. Care should be taken with this approach because the

TABLE 6.1. Examples of natural and human-mediated dispersal, and methods to collect data and estimate dispersal kernels.

Dispersal mechanism	Types of organisms	Method of data collection	Method to estimate dispersal kernel	Reference
Aerial	Insects, plant spores, seeds	Pheromone, seed traps	Moment-generating functions fit to density data	Turchin and Thoeny (1993)
Hydrologic	Plankton, fish, macroinvertebrates	Mark-recapture, radiotelemetry	Moment-generating functions fit to density or dispersal distances	Bubb et al. (2004)
Terrestrial (animal movement)	Reptiles, mammals	Mark-recapture, radiotelemetry	Moment-generating functions fit to dispersal distances	Murray et al. (1986)
Anthropogenic		Gravity and random-utility models: surveys, government databases of recreationalist/commercial movement	Moment-generating functions fit to dispersal distances	Bossenbroek et al. (2001)

width of the distance intervals may bias the shape of the dispersal kernel. More formally, the associated moment-generating function $M^H(s)$ is represented as

$$M^H(s) = \frac{1}{s}\sum_{i=1}^{L} f_i\,[\sinh(sb_i) - \sinh(sb_{i-1})], \tag{3}$$

where f_i is the bin height of the histogram for bin i, b_i and b_{i-1} are the end points of bin i, L is the maximum number of bins in the histogram, and s as the unknown slope (Lewis et al. 2006).

Within a lake, diffusion of species through water currents will increase colonization rates of the entire lake. Hydrologic pathways can also assist in spreading aquatic plants, especially those with free-floating life forms, such as water hyacinth (*Eichhornia crassipes*) and aquarium water moss (*Salvinia molesta*). Even for submerged macrophytes such as water milfoil (*Myriophyllum spicatum*), dispersal occurs via advective transport of vegetative fragments (Madsen et al. 1988).

Animal-mediated Dispersal

Animal-mediated dispersal has been shown to spread several taxonomic groups over long distances. Plant seeds, for example, have the potential to be transported in the digestive tract of animals such as white-tailed deer and waterfowl (DeVlaming and Proctor 1968; Myers et al. 2004). Waterfowl have also been shown to carry species such as snails and zooplankton on their feathers and feet (Boag 1986; Green and Figuerola 2005). Fish may also act as dispersal agents by moving to a new environment after consuming species that can withstand gut passage. For example, the diapausing eggs of the invasive spiny waterflea (*Bythotrephes longimanus*) have been shown to survive passage through fish guts (Jarnagin et al. 2000). As with hydrological dispersal, models of animal-mediated dispersal may be parameterized by collecting data through mark-recapture or radiotelemetry (table 6.1).

Aerial Dispersal

Advective transport by air currents or gusts of wind is another dispersal mechanism for some NIS, even aquatic ones. Unlike dispersal via water currents, species that disperse aerially can move outside of closed systems and into new environments and are thus not limited to downstream movement. For example, small species such as zooplankton that have desiccation-tolerant resting stages can be carried by the wind outside of isolated aquatic environments. Although ample anecdotal evidence exists for wind-mediated species dispersal, manipulative studies necessary to calculate dispersal rates are rare (but see Cáceres and Soluk 2002; Skarpaas et al. 2005). Most studies have shown that zooplankton, especially rotifers, are capable of aerial transport but that dispersal events are infrequent and limited to a few species and that dispersal and colonization potential differ among different zooplankton species (Cáceres and Soluk 2002). As an example, in all samples of a year-long study, only four species of rotifers from regionally available zooplankton species were collected from rain and wind samples, and all were within 1 km of the source body of water (Jenkins and Underwood 1998). For aquatic species, one consequence of wind dispersal is that lower densities of species disperse and colonize new systems than if dispersed through hydrologic pathways. Seeds from aquatic macrophytes carried by the wind, for example, have a lower chance of successfully colonizing new habitats than if dispersed through streams (Soons 2006).

Aerial dispersal of organisms is frequently estimated by collecting data on the density of individuals or on measured dispersal of propagules. For insects, mark-recapture methods based on pheromone traps provide an empirical measure of the mean number of recaptures as a function of distance from release (table 6.1). For example, Turchin and Thoeny (1993) fit a two-parameter exponential distance decay function to the number of insects as a function of distance from the release source:

$$C(r) = Ar^{-1/2} \exp(-r/B), \tag{4}$$

where $C(r)$ is the number of recaptures as a function of the radial distance from the point of release, r, and A and B are the fitted parameters.

Greene and Johnson (1989) were able to describe the dispersal of winged seeds based on a ballistic formulation. The resultant dispersal from a point source using the geometric mean and variance of wind speed in the downwind direction, u_g and σ_x, release height H, and falling speed W_s is described by

$$K(x) = \frac{n}{\sqrt{2\pi}x\sigma_x} \exp\left[\frac{-(\ln(W_s x/Hu_g))^2}{2\sigma_x^2}\right], \tag{5}$$

where n is total seed production. Jung and Croft (2001) used this model to parameterize mite dispersal with measurements of falling speeds in a greenhouse.

Anthropogenic Dispersal

Anthropogenic dispersal of NIS can result in the transport of propagules at a faster rate and greater distance beyond their native range than they could otherwise achieve (Wonham et al. 2000; Mack and Lonsdale 2001; Hebert and Cristescu 2002). As a result of this rapid transport, propagules often have a greater chance of surviving the trip. For example, both transoceanic ships and ships that remain in the Laurentian Great Lakes make multiple stops for unloading and loading cargo at several ports within the Great Lakes. Many of these ships unload cargo in Lake Erie and Lake Ontario and take on ballast water at these locations for stability. These ships then proceed to Lake Superior ports, where ballast water is pumped out and cargo is picked up for the return trip. As a result, biotic exchange of NIS among the Great Lakes occurs on the order of days, orders of magnitude faster than would be possible under the intrinsic dispersal abilities of the species.

The spread of the pathogen *Phytophthora ramorum*, which causes sudden oak death, is a well-documented example of human-mediated dispersal. This pathogen was first reported in central California in 1995 (Garbelotto et al. 2001) and has since

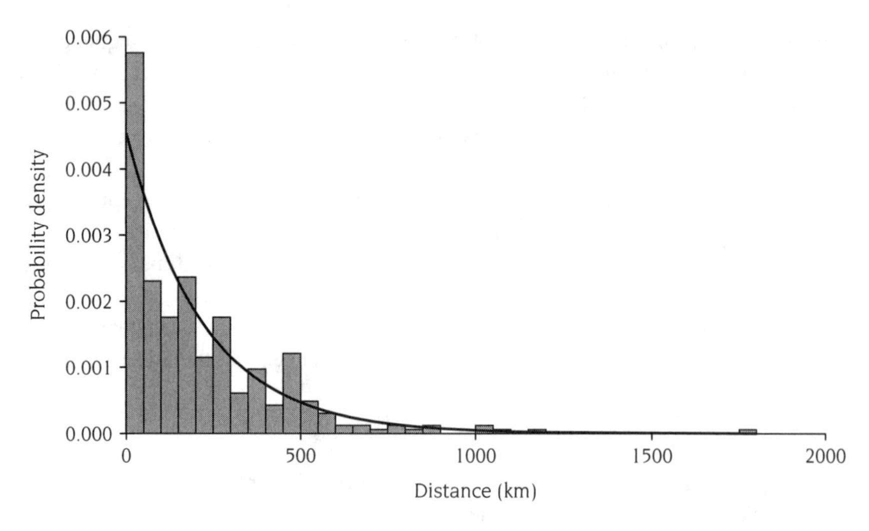

FIGURE 6.1.
Exponential dispersal kernel fitted to distances recreationalists traveled in Ontario (z) after visiting a lake invaded by the spiny water flea (*Bythotrephes longimanus*). The histogram and fitted dispersal kernel, $k(z) = 0.005\exp(-0.005z)$, have been scaled so that $\int k(z)dz = 1$.

been transported long distances from infested nurseries via commercially important hosts susceptible to disease (e.g., oak saplings, rhododendrons), in addition to local dispersal.

Two methods well suited for modeling human-mediated dispersal of propagules are gravity models and random-utility models. In both cases, the dispersal kernel is modeled on overland human-mediated transport and describes the distribution of trip distances across a landscape or a road-based network (figure 6.1). Both methods are applicable for modeling dispersal of aquatic NIS, because lakes may be considered to be discrete patches in a heterogeneous landscape. Like advective flow through rivers or streams, overland human-mediated dispersal is constrained to a network composed of one-dimensional segments.

Gravity Models

A common tool for modeling spatial interaction is a gravity model, which is used to describe how the influences of distances and the "mass" or attraction of origins and/or destinations affect the flow of people (Thomas and Huggett 1980; Roy and Thill 2004). The attractiveness of a location can be described as the property that creates an incentive for trips to be made to that location. The simplest formulation of a gravity model is

$$T_{ij} = kA_iB_jc_{ij}^{-\alpha}, \tag{6}$$

where T_{ij} is the interaction between locations i and j, k is a constant, A_i and B_j define the propulsion from the origin and the attraction to the destination, respectively,

c_{ij} is the distance between i and j, and α is a distance coefficient, or distance-decay parameter, which defines how much of a deterrent distance is to interaction. In a transportation context, A_i and B_j can be described as the number of individuals leaving and arriving at particular locations.

Gravity models can be used in heterogeneous landscapes, can incorporate various dispersal kernels, and are based on known transport mechanisms. The cognitive process of humans in making decision about where they travel is incorporated into gravity models. These models are also spatially explicit because they model flow from specific origins to destinations and thus enable the incorporation of large GIS databases as data sources for model parameterization and prediction. In addition, gravity models are created with an assumption that the specific perception of what is attractive to humans is correct and is the dominant driving force in destination choice.

Although gravity models were initially developed for use in economics (Reilly 1931; Linneman 1966) and other social sciences (Zipf 1946), they have more recently been used to predict the spread of diseases such as influenza (Viboud et al. 2006), plant pathogens (Ferrari et al. 2006), and NIS. The first published use of a gravity model to successfully assess the risk of human-mediated transport of an NIS was by Schneider et al. (1998), who used a doubly constrained gravity model to predict the rank order in which inland lakes and reservoirs of Illinois were expected to become invaded. Their concern was to predict the threat posed by zebra mussels to the native mussel communities of these water bodies. A doubly-constrained gravity model is used when information is known about the number of individuals leaving and arriving at each destination. In general, Schneider et al. (1998) predicted that those water bodies with high boat use and close to sources of zebra mussels were the most likely to become invaded with zebra mussels. An assessment of the predictions from this study shows that only 2 of the 55 lakes that were assessed are now invaded by zebra mussels. These lakes were predicted to be the first and the 52nd lakes to become invaded, suggesting a re-evaluation of the risk to these lakes is necessary. Since this publication, several additional lakes have become infested in Illinois, but most are in the greater Chicago area and close to Lake Michigan—a major source of zebra mussels.

Similarly, Bossenbroek et al. (2001) forecast the distribution of lakes invaded by zebra mussels in Michigan, Ohio, and northern Illinois and Indiana using a production-constrained gravity model. A production-constrained gravity model is used when the number of individuals leaving an origin is known, but not the number of individuals arriving at a particular destination. Bossenbroek et al. (2001) showed that a gravity model could be used to mimic the existing pattern of the zebra mussel invasion. On a national scale, Bossenbroek et al. (2007) predicted the relative probability that zebra mussels would be transported to different watersheds throughout the United States. This analysis suggested that the watersheds most likely to experience new introduction events are those that already contain lakes invaded by zebra mussels. Bossenbroek et al. (2007) also identified particular reservoirs in areas not currently invaded, such as the H. S. Truman Reservoir in Missouri and Lake Mead, which have a higher probability of becoming invaded (see chapter 12).

In another example, MacIsaac et al. (2004) used a doubly-constrained gravity model to forecast the spread of the aquatic spiny water flea (*Bythotrephes longimanus*) throughout Ontario Canada. The authors used recreational boater surveys to identify the strength of vector flows between different lakes and thus the risk *Bythotrephes* spread to noninvaded lakes.

Gravity models have been successfully used to model rare, long distance disperal events. For example, the discovery of the zebra mussel congener, the quagga mussel (*Dreissena bugensis* (= *D. rostriformis bugensis* [Andrusov (1897)])), in Lake Mead identifies the relevance of this type of modeling and the likelihood of such events since it is almost certain that human-mediated transport was responsible for its introduction and not transport through connected waterways.

Gravity models have also been used to assess human-mediated movement of terrestrial NIS such as the emerald ash borer (*Agrilus planipennis*) (Muirhead et al. 2006; Iverson et al. in press). The emerald ash borer is an invasive beetle from China that rapidly kills the native ash tree species of North America. It was first discovered in North America in the summer of 2002 and since has spread from its point of introduction, Detroit, Michigan, to much of the lower peninsula of Michigan, and to southwestern Ontario and northern Ohio. This species can disperse by flight, but most movement is less than 1 km and only 1% travel farther than 4 km in a 24-h period (Taylor et al. 2004). Thus, the rapid spread of the emerald ash borer suggests that human-mediated dispersal, such as the movement of firewood by campers, or the movement of ash products for use in landscaping, has been important (figure 6.2). Both Muirhead et al. (2006) and Iverson et al. (in press) demonstrate the importance of considering long-distance spread when predicting the overall dispersal rate of the emerald ash borer. Additionally, Iverson et al. (in press) specifically predict the relative risk of campers moving wood from the core area of emerald ash borer infestation to campgrounds across Ohio (figure 6.2).

The initial uses of gravity models were either untested in their predictions (e.g., Schneider et al. 1998) or parameterized based on their ability to recreate known patterns or distributions of an invasion (e.g., Bossenbroek et al. 2001). These analyses, however, did not assess whether the model predictions were accurately simulating the mechanism behind the patterns. Leung et al. (2006) compared the results of a production-constrained gravity model to four different metrics of human movement behavior based on angler activity records and a mailed survey developed specifically for that project. Leung et al. (2006) found that gravity models were able to simulate the mechanism of long-distance zebra mussel transport, i.e. the movement of recreational boaters.

Gravity models used in invasion biology have mostly been used to assess the likelihood that an NIS will be transported to a particular location. These models have not included biological attributes of the NIS, such as colonization or reproduction potential. Leung et al. (2004), however, were able to use the gravity model framework to demonstrate that zebra mussels are subject to Allee effects within their invaded range in North America. Likewise, Leung and Delaney (2006) demonstrated that spread can be estimated with limited data sets, particularly when propagule pressure is accounted for. Leung et al. (2004) and Leung and Delaney (2006) show the

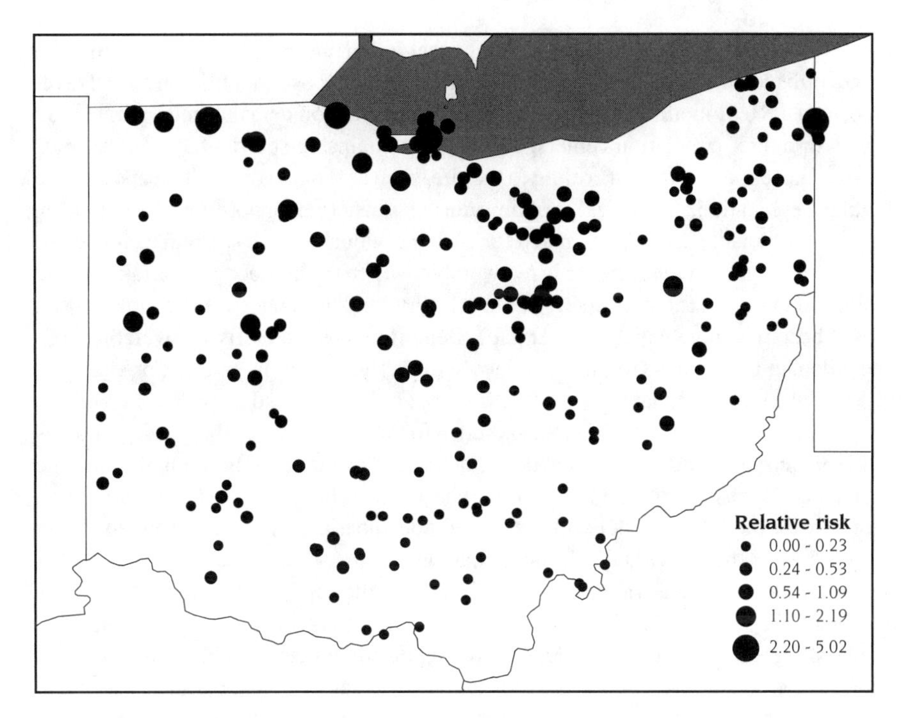

FIGURE 6.2.
Relative risk of an introduction of emerald ash borer to Ohio campgrounds due to higher attractiveness and/or travel from the core area in southern Michigan infested with emerald ash borer, based on a gravity model.

importance of accounting for population dynamics at early stages of an invasion when modeling patterns of dispersal.

The next steps in the development of gravity models for use in risk assessment require the inclusion of NIS population dynamics such as growth rate and mortality and more rigorous assessments of uncertainty (see chapter 7). This would enable gravity models to serve as a bridge from describing dispersal of individual propagules to describing population spread of NIS at a landscape level (Jerde and Lewis 2007). For example, models describing population growth can be coupled with gravity models describing immigration and emigration among systems, in a manner analogous to metapopulation models.

Random-Utility Models

An alternative to the gravity model for estimating the human-mediated movement of organisms is the random-utility maximization (RUM) model. This is the most widely used modeling framework for economists estimating the recreation demand of various sites. An advantage of RUM models over gravity models is that they explain more fully the individual economic behavior driving decisions by humans of where to recreate. The primary reason for moving away from gravity models is the necessity to include travel costs so that a demand function can be estimated and utilized in the calculation of benefit values for cost-benefit analysis. Travel costs in a RUM model (and all other models in recreation demand) incorporate road infrastructure, population centers, and varying attainable speeds. Included are out-of-pocket expenses from traveling to the site, such as the cost of gasoline and vehicle value deprecation, and the value of time during transit (i.e., opportunity cost of time).

Once travel costs are estimated for each individual to each site in the choice set, a demand function for each site is recoverable, where the number of trips taken to the site and travel costs are inversely related (i.e., individuals choose fewer trips to a site as it becomes more expensive per trip). Benefit values are easily recoverable from the demand functions (for an overview, see Corrigan et al. in press). For example, if one wanted to estimate the benefits from the reduced spread of an NIS by human-mediated movement, a RUM model is suited to this task, whereas the gravity model is a statistical allocation model that does not include estimates of individual economic benefits (Bockstael et al. 1986). Moreover, even if the goal is only the prediction of spread, the RUM model will most likely be superior, due to the more complete modeling of the individual's decision making.

One particular advantage of RUMs is that they model individual-level trips instead of aggregate zonal trips. With the current state-of-the-art RUM, the repeated mixed logit model, the individual makes repeated choices of which sites to visit. In economics, utility is a measure of the relative satisfaction or desiredness from the consumption of goods. Given this measure, one may speak meaningfully of increasing or decreasing utility and thereby explain economic behavior in terms of attempts to increase one's utility. The assumption is that individuals visit the sites that give them the most utility, constrained by their income and time to recreate. The RUM model assumes the utility of individual i choosing site j on choice occasion t is of the form

$$U_{ijt} = V\left(X_{ij};\beta_i\right) + \varepsilon_{ijt}, \quad i = 1,\ldots,N; \; j = 0,\ldots,J; \; t = 1,\ldots,T, \tag{7}$$

where V represents the observed portion of utility, and from the perspective of the researcher, ε_{ijt} represents the *random*, unobserved portion of utility (hence the name "random-utility model"). The observed portion of utility is a function of explanatory variables, X_{ij}, such as travel costs, site characteristics (e.g., lake size), and household characteristics; and β_i is the estimated parameters on the explanatory variables, where these parameters are routinely allowed to vary across the population of individuals, allowing for substantial heterogeneous trip-taking behavior (hence the random effect leading to the mixed model).

Conditional on knowing β_i, the probability that individual i chooses alternative j on choice occasion t is defined as

$$\Pr_{ijt}\left[\beta_i\right] = \frac{\exp\left[V_{ijt}(\beta_i)\right]}{\sum_{k=0}^{J} \exp\left[V_{ikt}\left(\beta_i\right)\right]}, \tag{8}$$

and the unconditional probability can be obtained by integrating the conditional probability over all possible values of β_i and then using maximum simulated likelihood to estimate the parameters (Train 2003). For the modeling of the human-mediated movement of organisms, it is convenient that RUM models estimate a trip probability (i.e., dispersal kernel) for each individual to each site. These estimated trip probabilities can easily be augmented with additional data on the biological attributes of the destination site concerning their candidacy for the accidentally transported NIS (Macpherson et al. 2006), as well as including biological attributes of NIS that may facilitate dispersal, such as the production of resistant resting stages by zooplankton or vegetative reproduction in plants.

While Macpherson et al. (2006) discuss a simulation exercise with a dynamic RUM model, no papers to date have applied the RUM model to the spread of invasive species with actual data, not even a simpler static model. This emphasizes the downside of utilizing RUM models—the more intensive data requirements, because individuals must be surveyed about which sites they visit and how often, along with other information such as their income. However, in the case of zebra mussels, Leung et al.'s (2002) high estimates of the partial economic costs from the spread of zebra mussels indicates the extra data collection costs would be worth undertaking to better predict NIS spread based on a more accurate modeling of boaters' recreation activity.

POPULATION SPREAD

The development of spatial models for the spread of NIS has a long history starting with diffusion theory and gas kinetics in physics and chemistry. In the biological literature, dispersal models have been developed for a wide range of applications, including gene flow (Fisher 1937), spread of and susceptibility to infectious diseases (Kendall 1957; Noble 1974), predator–prey dynamics (Okubo 1980), and aerial dispersal of fungal spores (Aylor 2003; see also reviews in Higgins and Richardson 1996; Hastings et al. 2005).

The application of these modeling approaches to forecasting NIS spread is a logical extension, and the choice of a particular model depends on the type of information available (Shigesada and Kawasaki 1997). For many invasions, species presence/absence data may be the only information available from historical records. With this scenario, pattern-based, or spatial-phenomenological, models (using the terminology of Higgins and Richardson 1996) can be used to forecast range expansion (or, in economic terms, expanding the extensive margin). These types of models assume that the location of the invasion wavefront corresponds to the location of first recording, and forecast rates of spread are generally equivalent to past spread rates. These models are more applicable to terrestrial systems at regional spatial scales than to the discrete habitat of aquatic systems, and do not incorporate the ecology of the species or recipient habitat in predictions. As with many dispersal models, there is also the assumption that sampling effort is sufficient to detect the species if present beyond the invasion wavefront—otherwise, rates of spread would be underestimated. An example of a phenomenological invasion model is the spread of *Mimosa pigra* in Australia (Lonsdale 1993). *Mimosa pigra* is an invasive shrub that has been identified as one of the 100 worst invaders in the world by the Global Invasive Species Programme. The area invaded by *M. pigra* was positively related to the time of occurrence extracted from historical records and aerial photographs. Future spread was then predicted assuming that the rate of spread remains constant and expansion occurs similarly in all directions.

In contrast to the spatial-phenomenological class of models, patterns of spread from process-based or mechanistic models can be forecast using information on the ecology of NIS or properties of the transport mechanism. One of the first forms of a process-based approach is a reaction-diffusion (RD) model that predicts population density as a function of space and time. Reaction-diffusion models describe exponential or logistic population growth that diffuses randomly across homogeneous space (Okubo 1980). Population spread assuming exponential growth in a two-dimensional environment is represented as

$$\frac{\partial n}{\partial t} = D\left(\frac{\partial^2 n}{\partial x^2} + \frac{\partial^2 n}{\partial y^2}\right) + rn, \tag{9}$$

where n is the population size, r is the intrinsic per capita population growth rate, and D is the diffusion coefficient (distance^2/time) for propagules dispersing in x and y directions. Here, distances that individuals disperse from a source are assumed normally distributed, with corresponding directions uniformly distributed from 0 to 360° (i.e., isotropic). The resulting pattern of population spread forms a series of concentric circles spreading away from the source through time. The diffusion coefficient D is often measured using mark-recapture data, for example, and can be estimated by

$$D = \frac{\left(\sum_{n=1}^{m} x/m\right)^2}{\pi t}, \tag{10}$$

where x is the distance from the marking site, m is the number of marked individuals released, and t is the time since marking (Lockwood et al. 2007).

For marine or aerial dispersers, the basic diffusion model may be modified to include advective flow due to water or air currents. Two-dimensional spread in these systems is described by

$$\frac{\partial n}{\partial t} = D\left(\frac{\partial^2 n}{\partial x^2} + \frac{\partial^2 n}{\partial y^2}\right) + rn - w_x\frac{\partial n}{\partial x} - w_y\frac{\partial n}{\partial y}, \tag{11}$$

where w_x and w_y describe velocity down x- and y-axes, and other terms in the model are as in equation 9 (Holmes et al. 1994). Here, dispersal distances are also assumed to follow a normal distribution, but spread is not isotropic.

In one-dimensional RD models, the location of the invasion wavefront is expected to travel away from the epicenter at a constant rate of $\sqrt{4rD}$, where r is the intrinsic rate of population increase and D is the diffusivity coefficient, providing population density is sufficient for detection. This, however, holds true only when the dispersal kernels reach an asymptote at large distances (e.g., exponential decay). For fat-tailed dispersal kernels, the velocity of the wavefront increases through time.

In addition to assumptions about the homogeneity of space and random dispersal, this class of models assumes continuous reproduction and dispersal through time and that there is no finite border limiting geographic spread.

Reaction-diffusion models have been successfully applied to describing secondary spread for a number of species. Skellam's (1951) assertion that the European range expansion of muskrats followed a linear rate of increase as predicted from an RD model is generally supported by the observed spread of this species. Predicted rates of range expansion of the small cabbage white butterfly, *Pieris rapae*, based on estimates of intrinsic rates of increase and survivorship fit with observed expansion rates in North America (Andow et al. 1990). However, in the same study, predicted rates of spread of the cereal leaf beetle, *Oulema melanopus*, from diffusion models were underestimated by at least two orders of magnitude. The authors suggested that other mechanisms, such as human-mediated dispersal or advection in air currents, were responsible for long-distance transport. Current patterns of *O. melanopus* spread in the United States are consistent with "stratified diffusion" (Hengeveld 1989), in which spread proceeds by a combination of local and long-distance dispersal events (Shigesada et al. 1995). These long-distance dispersal events appear to be important in determining both the speed of the invasion and the variability in population density at the invasion wavefront.

Although RD models that assume a continuous landscape have been successful for predicting geographic range extensions (e.g., Hengeveld 1989), they are limited in their applicability to heterogeneous systems. Consequently, spatially discrete RD models have been developed that convert a continuous landscape into a lattice of grid cells representing suitable and unsuitable habitat (e.g., Flather and Bevers 2002). In the Flather and Bevers (2002) model, individuals of a hypothetical species dispersed from one habitat cell to another with distances according to a Weibull distribution and with uniformly distributed directions. Habitat amount relative to habitat arrangement was key to the overall regional population size. No studies to date, however, have applied this method to predict NIS spread, although it seems like a promising approach for terrestrial species.

The limitations of RD models for predicting the spread of aquatic invasive species were first highlighted with zebra mussels. Buchan and Padilla (1999) attempted to fit an RD model to the spread of zebra mussels in the State of Wisconsin. They determined that due to the heterogeneous nature of the landscape and the need to understand the mechanism of dispersal (in this case, recreational boaters), RD models were not sufficient for this system. Due to the limitations of RD models, researchers have developed more sophisticated statistical methods for dealing with long-distance dispersal (see below) and have adapted techniques from geography and economics to model different mechanisms of dispersal (see "Dispersal," above).

Several modeling approaches have been developed to forecast range expansion without the restrictive assumptions underlying RD models. Integrodifference (ID) models (e.g., Kot et al. 1996; Veit and Lewis 1996; Krkošek et al. 2007) have two primary advantages over RD models: (1) they estimate dispersal of individuals according to their life stage at discrete time intervals, and (2) they are flexible due to their use of non-Gaussian dispersal kernels. The shape of the dispersal kernel can be fitted to data from mark-recapture experiments and include rare, long-distance dispersal events observed with invasive spread (i.e., leptokurtic, or "fat-tailed" dispersal kernels; Kot et al. 1996). With flowering plants, for example, seeds are the primary dispersing stage transported by vectors such as wind, water, and animals. Sensitivity analysis on the contribution that each life stage provides to the overall rate of spread may then guide management efforts (Neubert and Caswell 2000). Like their RD counterparts, basic ID models assume a homogeneous landscape and that population growth and dispersal are the same at each point in space. Consequently, they are better suited for aerial or terrestrial invasions than for aquatic invasions, which occur across a matrix of habitat suitability. In ID models, population density at time $t + 1$ and location x is represented as

$$n_{t+1}(x) = \int k(x - y)f[n_t(y)]\,dy, \tag{12}$$

where n is population size, t is time interval, k is the dispersal kernel for displacement from y to x, and $f(n_t)$ describes population growth as a function of population size in the previous time step at location y.

Population-based models such as the RD or ID equations have been adapted for scenarios when propagules of the NIS are first introduced into a novel habitat and mate location is difficult due to low population size. As a result, the population experiences reduced or negative growth rates at low density, called Allee effects, which then translate to fewer individuals able to undergo secondary spread. In theory, Allee effects may serve as an alternate mechanism to explain increasing rates of spread as a contrast to long-distance dispersal events, because Allee effects may cause a lag in the initial stages of secondary dispersal (Kot et al. 1996). Allee effects have been shown to be present for the invasive weed smooth cordgrass (*Spartina alterniflora*; Davis et al. 2004) and zebra mussels (Leung et al. 2004) and have been modeled for several organisms, including the plant pathogen *Tilletia indica* (Garrett and Bowden 2002), the house finch (*Carpodacus mexicanus*; Veit and Lewis 1996), and the gypsy moth (*Lymantria dispar*; Liebhold and Bascompte 2003). In the latter study, the authors suggest that 100% eradication of the species may not be necessary as long as the population density is reduced to a threshold below which Allee effects will cause the population to crash.

MANAGEMENT GUIDANCE

Historically, dispersal models have been used to describe the patterns and processes by which organisms colonize novel habitat (Shigesada and Kawasaki 1997), but these models are increasingly being used to predict locations likely to become invaded (e.g., Leung et al. 2004; Bossenbroek et al. 2007) so that monitoring and control actions can be implemented. However, success in predicting biological invasions has been limited, and Gilpin (1990) advocates a probabilistic approach for predicting biological invasions. That is, the probability of a prescribed undesired effect (i.e., biological invasions) is by definition risk (Suter 1993). Risk analysis involves estimating probabilities of invasion, while management activities involve deciding how to respond to each risk. Together, risk analysis and risk management form a risk assessment approach to biological invasions (Suter 1993). The development of spread and dispersal models provides a useful framework in estimating the risk of invasion and evaluating management decisions (Leung et al. 2002).

Risk analysis has a variety of modeling approaches (Suter 1993). Each of the models discussed in this chapter represent a different approach that is more or less suited for predicting the dispersal of an NIS, depending on the mechanisms of dispersal, the environment the organism is being introduced into, the life history of the organism, the risk assessor's understanding of these processes, and the data available to parameterize the models (Stohlgren and Schnase 2006). In general, the goal of risk analysis for a single species is to predict when and where the species is likely to invade (MacIsaac et al. 2004; Bossenbroek et al. 2007) and report the probability associated with an invasion event (Jerde and Lewis 2007).

Management efforts will change depending on whether the goal is to stop the initial dispersal of propagules into a new habitat or to contain a spreading population. In the first scenario, risk management involves a preventative approach and relies on some knowledge of the dispersal mechanisms. In this chapter, we have shown that it is possible to assess the risk of one destination being invaded relative to another. The models discussed can thus be used to prioritize the location of investments in NIS prevention. For containing a spreading population, transport mechanisms also should be considered, but management efforts may be focused on restricting the outbound flow of propagules through quarantine or other methods. Modeling the patterns of population spread also may indicate that novel management decisions are optimal. Sharov and Liebhold's (1998) model for spread reduction of the gypsy moth (*Lymantria dispar*), for example, suggests that the optimal method for reducing the rate of spread is to eliminate satellite colonies outside the expanding population front.

Leung et al. (2002) provide a general risk management framework that links risk analysis to an economic model and uses stochastic dynamic programming to evaluate the costs and benefits of management actions (chapter 9). This framework requires estimating the probability of invasion, evaluating the costs of an invasion and management actions, and measuring the reduction in the probability of invasion from the different management actions (Stohlgren and Schnase 2006). An overview of the process of valuing market and nonmarket costs of invasion is discussed in chapter 8. From the management framework, connecting models of dispersal and spread to risk management rests on the risk analysis and the potential changes in the probability of invasion due to management actions (Leung et al. 2002).

Although predicting invasions through modeling is gaining popularity, more attention should be paid to communicating results, and especially to connecting risk analyses to management (Bossenbroek et al. 2005). The main problems are the lack of reporting probabilities, and failure to report the type of model used to estimate the probabilities (Nelson et al. 2007). One common approach to communicating risk is to rank order locations from greatest risk to least risk for invasion (e.g., Schneider et al. 1998; location A is more likely to be invaded than location B under the assumption that the sites are independent). Alternatively, it may be possible to estimate the relative risk of invasion, where the relative risk is a ratio of invasion probabilities for two locations (e.g., Herborg et al. 2007; location A is 10 times more likely to be invaded than location B). While both rank orders and relative risk measures are formulated from probabilities and represent a risk analysis, they do not directly communicate the probability of invasion needed to apply risk management as proposed by Leung et al. (2002). Therefore, it is unclear if ranked and relative risk measurements provide sufficient information to make risk management decisions (Nelson et al. 2007; see also chapter 7). For some systems, it may be necessary, due to data limitations, and reasonable, given simplifying assumptions, to apply relative risk estimators to perform a risk analysis (Jerde and Lewis 2007), and the use of ranks and relative measures may be useful for sampling design and monitoring. As discussed, however, this needs further development and careful interpretation.

Risk management includes evaluating the potential influence of management actions on reducing the invasion probability (Suter 1993; Leung et al. 2002). Experiments that estimate this influence are undoubtedly useful but may be costly. Alternatively, models of dispersal and spread can be analyzed using sensitivity analysis. Sensitivity analysis evaluates how perturbations of model parameters influence the probability of invasion (Suter 1993). Small changes in model parameters may substantially change the calculated probability of invasion. Identifying which parameters contribute most to reductions in the probability of invasion may provide insights into the effectiveness of management actions that aim to reduce invasion risk.

CONCLUSIONS

The ability to quantify dispersal and population spread is a key component in NIS management because reductions in species dispersal can reduce the likelihood of species establishment. A strength of dispersal models such as gravity and RUM models is their flexible modeling framework. Both of these model types are well suited for anthropogenic dispersal and can provide a mechanistic basis for multiple dispersal vectors, in general. RUM models are further refined than are gravity models, because RUM models take into consideration an individual's behavior in terms of minimizing economic costs. Current challenges for gravity and RUM models are to quantify the actual propagule pressure of NIS for various dispersal mechanisms and to tie in NIS dispersal with models of habitat suitability and biological resistance to invasions.

For models of population spread, strengths of ID models include being able to incorporate non-Gaussian distribution kernels and being able to be tailored for species in which particular life stages are dispersed. One challenge is to apply models of NIS spread to heterogeneous environments such as networks of lakes and streams, islands, or corridors in a terrestrial landscape.

Models of dispersal or population spread are able to guide management depending on whether the goal is to predict and prevent new invasions or to control established NIS populations. For predicting and preventing new invasions, model predictions, when developed using historical abundance or presence/absence data, can then be cast into a probabilistic framework for use in risk analysis. For curtailing the spread of established populations, these models can be used to identify an optimal management strategy when used as a basis for what-if scenarios.

Acknowledgments This chapter was substantially improved by the editors of this book and three reviewers. This material is based on work supported by the Integrated Systems for Invasive Species project (D. M. Lodge, principal investigator) funded by the National Science Foundation (DEB 02-13698). Support was provided by Ontario Graduate Fellowships (J.R.M.), the University of Notre Dame Center for Aquatic Conservation (A.M.B.), the U.S. Department of Agriculture Program of Research on the Economics of Invasive Species Management (J.M.B.), and a National Sciences and Engineering Research Council of Canada collaborative research grant to H. J. MacIsaac and M. A. Lewis (C.L.J.). This is publication 2009-02 from the University of Toledo Lake Erie Center.

References

Andow, D. A., P. M. Kareiva, S. A. Levin, and A. Okubo. 1990. Spread of invading organisms. Landscape Ecology 4:177–188.

Avise, J. C. 1994. Molecular markers, natural history and evolution. Chapman and Hall, New York.

Aylor, D. E. 2003. Spread of plant disease on a continental scale: role of aerial dispersal of pathogens. Ecology 84:1989–1997.

Boag, D. A. 1986. Dispersal in pond snails: potential role of waterfowl. Canadian Journal of Zoology 64:904–909.

Bobeldyk, A. M., J. M. Bossenbroek, M. A. Evans-White, D. M. Lodge, and G. A. Lamberti. 2005. Secondary spread of zebra mussels (*Dreissena polymorpha*) in coupled lake-stream systems. Ecoscience 12:339–346.

Bockstael, N. E., W. M. Hanemann, and I. E. Strand. 1986. Measuring the benefits of water quality improvements using recreation demand models. Report presented to the Environmental Protection Agency under cooperative agreement CR-811043-01-0. Washington, DC.

Bossenbroek, J. M., L. E. Johnson, B. Peters, and D. M. Lodge. 2007. Forecasting the expansion of zebra mussels in the United States. Conservation Biology 21:800–810.

Bossenbroek, J. M., C. E. Kraft, and J. C. Nekola. 2001. Prediction of long-distance dispersal using gravity models: zebra mussel invasion of inland lakes. Ecological Applications 11:1778–1788.

Bossenbroek, J., J. McNulty, and R. Keller. 2005. Can ecologists heat up the discussion on invasive species risk? Risk Analysis 25:1595–1597.

Bubb, D. H., T. J. Thom, and M. C. Lucas. 2004. Movement and dispersal of the invasive signal crayfish *Pacifastacus leniusculus* in upland rivers. Freshwater Biology 49:357–368.

Buchan, L. A. J., and D. K. Padilla. 1999. Estimating the probability of long-distance overland dispersal of invading aquatic species. Ecological Applications 9:254–265.

Cáceres, C. E., and D. A. Soluk. 2002. Blowing in the wind: a field test of overland dispersal and colonization by aquatic invertebrates. Oecologia 131:1432–1939.

Corrigan, J. R., K. J. Egan, and J. A. Downing. In Press. Aesthetic values of lakes and rivers. *In* G. Likens, editor. The encyclopedia of inland waters. Elsevier Science, London.

Crooks, J. A. 2002. Characterizing ecosystem-level consequences of biological invasions: the role of ecosystem engineers. Oikos 97:153–166.

Davis, H. G., C. M. Taylor, J. C. Civille, and D. R. Strong. 2004. An Allee effect at the front of a plant invasion: *Spartina* in a Pacific estuary. Journal of Ecology 92:321–327.

DeVlaming, V., and V. W. Proctor. 1968. Dispersal of aquatic organisms: viability of seeds recovered from the droppings of captive killdeer and mallard ducks. American Journal of Botany 55: 20–26.

Ferrari, M. J., O. N. Bjornstad, J. L. Partain, and J. Antonovics. 2006. A gravity model for the spread of a pollinator-borne plant pathogen. American Naturalist 168:294–303.

Fisher, R. A. 1937. The wave of advance of advantageous genes. Annals of Eugenics 7:355–369.

Flather, C. H., and M. Bevers. 2002. Patchy reaction-diffusion and population abundance: the relative importance of habitat amount and arrangement. American Naturalist 159:40–56.

Garbelotto, M., P. Svihra, and D. M. Rizzo. 2001. Sudden oak death syndrome fells 3 oak species. California Agriculture 55:9–19.

Garrett, K. A., and R. L. Bowden. 2002. An Allee effect reduces the invasive potential of *Tilletia indica*. Phytopathology 92:1152–1159.

Gilbert, M., S. Guichard, J. Freise, J. C. Gregoire, W. Heitland, N. Straw, C. Tilbury, and S. Augustin. 2005. Forecasting *Cameraria ohridella* invasion dynamics in recently invaded countries: from validation to prediction. Journal of Applied Ecology 42:805–813.

Gilpin, M. 1990. Ecological prediction. Science 248:88–89.

Green, A. J., and Figuerola, J. 2005. Recent advances in the study of long-distance dispersal of aquatic invertebrates via birds. Diversity and Distributions 11:149–156.

Greene, D. F., and E. A. Johnson. 1989. A model of wind dispersal of winged or plumed seeds. Ecology 70:339–347.

Hanski, I. 1999. Metapopulation ecology. Oxford University Press, New York.

Hastings, A., K. Cuddington, K. F. Davies, C. J. Dugaw, S. Elmendorf, A. Freestone, S. Harrison, M. Holland, J. Lambrinos, U. Malvadkar, B. A. Melbourne, K. Moore, C. Taylor, and D. Thomson. 2005. The spatial spread of invasions: new developments in theory and evidence. Ecology Letters 8:91–101.

Hebert, P. D. N., and M. E. A. Cristescu. 2002. Genetic perspectives on invasions: the case of the Cladocera. Canadian Journal of Fisheries and Aquatic Sciences 59:1229–1234.

Hengeveld, R. 1989. Dynamics of biological invasions. Chapman and Hall, New York.

Herborg, L.-M., C. L. Jerde, D. M. Lodge, G. M. Ruiz, and H. J. MacIsaac. 2007. Predicting invasion risk using measures of introduction effort and environmental niche models. Ecological Applications 17:663–674.

Higgins, S. I., and D. M. Richardson. 1996. A review of models of alien plant spread. Ecological Modelling 87:249–265.

Holmes, E. E., M. A. Lewis, J. E. Banks, and R. R. Veit. 1994. Partial differential equations in ecology: spatial interactions and population dynamics. Ecology 74:17–29.

Horvath, T. G., G. A. Lamberti, D. M. Lodge, and W. L. Perry. 1996. Zebra mussel dispersal in lake-stream systems: source-sink dynamics. Journal of the North American Benthological Society 15:564–575.

Iverson, L. R., A. Prasad, J. Bossenbroek, D. Sydnor, and M. W. Schwartz. In press. Modeling potential movements of an ash threat: the emerald ash borer. *In* J. Pye and M. Raucher, editors. Advances in threat assessment and their application to forest and rangeland management.

Jarnagin, S. T., B. K. Swan, and W. C. Kerfoot. 2000. Fish as vectors in the dispersal of *Bythotrephes cederstroemi*: diapausing eggs survive passage through the gut. Freshwater Biology 43:579–589.

Jenkins, D. G., and M. O. Underwood. 1998. Zooplankton may not disperse readily in wind, rain, or waterfowl. Hydrobiologia 338:15–21.

Jerde, C. L., and M. A. Lewis. 2007. Waiting for invasions: a framework for the arrival of non-indigenous species. American Naturalist 170:1–9.

Jeschke, J. M., and D. L. Strayer. 2005. Invasion success of vertebrates in Europe and North America. Proceedings of the National Academy of Sciences of the United States of America 102:7198–7202.

Jung, C. L., and B. A. Croft. 2001. Aerial dispersal of phytoseiid mites (Acari: Phytoseiidae): estimating falling speed and dispersal distance of adult females. Oikos 94:182–190.

Kendall, D. G. 1957. In discussion on Bartlett, M. S.: measles periodicity and community size. Journal of the Royal Statistical Society Series A (General) 120:48–70.

Kolar, C. S., and D. M. Lodge. 2001. Progress in invasion biology: predicting invaders. Trends in Ecology and Evolution 16:199–204.

Kot, M., M. A. Lewis, and P. van den Driessche. 1996. Dispersal data and the spread of invading organisms. Ecology 77:2027–2042.

Krkošek, M., J. S. Lauzon-Guay, and M. A. Lewis. 2007. Relating dispersal and range expansion of California sea otters. Theoretical Population Biology 71:401–407.

Leibold, M. A., M. Holyoak, N. Mouquet, P. Amarasekare, J. M. Chase, M. F. Hoopes, R. D. Holt, J. B. Shurin, R. Law, D. Tilman, M. Loreau, and A. Gonzalez. 2004. The metacommunity concept: a framework for multi-scale community ecology. Ecology Letters 7:601–613.

Leung, B., J. M. Bossenbroek, and D. M. Lodge. 2006. Boats, pathways, and aquatic biological invasions, estimating dispersal potential with gravity models. Biological Invasions 8: 241–254.

Leung, B., and D. G. Delaney. 2006. Managing sparse data in biological invasions: a simulation study. Ecological Modelling 198:229–239.

Leung, B., J. M. Drake, and D. M. Lodge. 2004. Predicting invasions: propagule pressure and the gravity of Allee effects. Ecology 85:1651–1660.

Leung, B., D. M. Lodge, D. Finnoff, J. F. Shogren, M. A. Lewis, and G. A. Lamberti. 2002. An ounce of prevention or a pound of cure: bioeconomic risk analysis of invasive species. Proceedings of the Royal Society of London Series B Biological Sciences 269:2407–2413.

Lewis, M. A. 1997. Variability, patchiness, and jump dispersal in the spread of an invading population. Pages 46–69 *in* D. Tilman and P. M. Kareiva, editors. Spatial ecology: the role of space in population dynamics and interspecific interactions. Princeton University Press, Princeton, NJ.

Lewis, M. A., M. G. Neubert, H. Caswell, J. S. Clark, and K. Shea. 2006. A guide to calculating discrete-time invasion rates from data. Pages 169–192 *in* M. W. Cadotte, S. M. McMahon, and T. Fukami, editors. Conceptual ecology and invasion biology: reciprocal approaches to nature. Springer, Dordrecht, The Netherlands.

Liebhold, A. M., and J. Bascompte. 2003. The Allee effect, stochastic dynamics and the eradication of alien species. Ecology Letters 6:133–140.

Linneman, H. V. 1966. An econometric study of international trade flows. North-Holland, Amsterdam.

Lockwood, J. L., M. F. Hoopes, and M. P. Marchetti. 2007. Invasion Ecology. Blackwell, Oxford.

Lonsdale, W. M. 1993. Rates of spread of an invading species—*Mimosa pigra* in Northern Australia. Journal of Ecology 81:513–521.

MacArthur, R. H., and E. O. Wilson. 1963. The Theory of Island Biogeography. Princeton University Press, Princeton, NJ.

MacIsaac, H. J., J. Borbely, J. R. Muirhead, and P. Graniero. 2004. Backcasting and forecasting biological invasions of inland lakes. Ecological Applications 14:773–783.

Mack, R. N., and W. M. Lonsdale. 2001. Humans as global plant dispersers: getting more than we bargained for. BioScience 51:95–102.

Macpherson, A. J., R. Moore, and B. Provencher. 2006. A dynamic principal-agent model of human-mediated aquatic species invasions. Agricultural and Resource Economics Review 35: 144–154.

Madsen, J. D., L. W. Eichler, and C. W. Boylen. 1988. Vegetative spread of Eurasian watermilfoil in Lake George, New York. Journal of Aquatic Plant Management 26:47–50.

Muirhead, J. R., B. Leung, C. van Overdijk, D. W. Kelly, K. Nandakumar, K. R. Marchant, and H. J. MacIsaac. 2006. Modelling local and long-distance dispersal of invasive emerald ash borer *Agrilus planipennis* (Coleoptera) in North America. Diversity and Distributions 12:71–79.

Murray, J. D., E. A. Stanley, and D. L. Brown. 1986. On the spatial spread of rabies among foxes. Proceedings of the Royal Society of London Series B Biological Sciences 229:111–150.

Myers, J. A., M. Vellend, S. Gardescu, and P. L. Marks. 2004. Seed dispersal by white-tailed deer: implications for long-distance dispersal, invasion, and migration of plants in eastern North America. Oecologia 139:35–44.

Nathan, R., G. Perry, J. T. Cronin, A. E. Strand, and M. L. Cain. 2003. Methods for estimating long-distance dispersal. Oikos 103:261–273.

Nelson, W., A. Potapov, M. A. Lewis, A. Hundsdorfer, and F. He. 2007. Balancing ecological complexity in predictive models: a reassessment of risk models in the mountain pine beetle system. Journal of Applied Ecology doi:10.1111/j.1365-2664.2007.01374.x.
Neubert, M. G., and H. Caswell. 2000. Demography and dispersal: calculation and sensitivity analysis of invasion speed for structured populations. Ecology 81:1613–1628.
Noble, J. V. 1974. Geographic and temporal development of plagues. Nature 250:726–729.
Okubo, A. 1980. Diffusion and ecological problems: mathematical models. Springer-Verlag, Berlin.
Olden, J. D., N. L. Poff, M. R. Douglas, M. E. Douglas, and K. D. Fausch. 2004. Ecological and evolutionary consequences of biotic homogenization. Trends in Ecology and Evolution 19:18–24.
Puth, L. M., and T. F. H. Allen. 2005. Potential corridors for the rusty crayfish, *Orconectes rusticus*, in northern Wisconsin (USA) lakes: lessons for exotic invasions. Landscape Ecology 20:567–577.
Rahel, F. J. 2002. Homogenization of freshwater faunas. Annual Review of Ecology and Systematics 33:291–315.
Reilly, W. J. 1931. The law of retail gravitation. Knickerbocker Press, New York.
Ricciardi, A. 2006. Patterns of invasion in the Laurentian Great Lakes in relation to changes in vector activity. Diversity and Distributions 12:425–433.
Roy, J. R., and J. C. Thill. 2004. Spatial interaction modelling. Papers in Regional Science 83:339–361.
Schneider, D. W., C. D. Ellis, and K. S. Cummings. 1998. A transportation model assessment of the risk to native mussel communities from zebra mussel spread. Conservation Biology 12:788–800.
Sharov, A. A., and A. M. Liebhold. 1998. Bioeconomics of managing the spread of exotic species with barrier zones. Ecological Applications 8:833–845.
Shigesada, N., and K. Kawasaki. 1997. Biological invasions: theory and practice. Oxford University Press, New York.
Shigesada, N., and K. Kawasaki. 2002. Invasion and the range expansion of species: effects of long-distance dispersal. Pages 350–373 *in* J. M. Bullock, R. E. Kenward, and R. S. Hails, editors. Dispersal ecology. Blackwell, Oxford.
Shigesada, N., K. Kawasaki, and Y. Takeda. 1995. Modeling stratified diffusion in biological invasions. American Naturalist 146:229–251.
Shurin, J. B., and J. E. Havel. 2002. Hydrologic connections and overland dispersal in an exotic freshwater crustacean. Biological Invasions 4:431–439.
Skarpaas, O., K. Shea, and J. M. Bullock. 2005. Optimizing dispersal study design by Monte Carlo simulation. Journal of Applied Ecology 42:731–739.
Skellam, J. G. 1951. Random dispersal in theoretical populations. Biometrika 38:196–218.
Soons, M. B. 2006. Wind dispersal in freshwater wetlands: knowledge for conservation and restoration. Applied Vegetative Science 9:271–278.
Stoeckel, J. A., D. W. Schneider, L. A. Soeken, K. D. Blodgett, and R. E. Sparks. 1997. Larval dynamics of a riverine metapopulation: implications for zebra mussel recruitment, dispersal, and control in a large-river system. Journal of the North American Benthological Society 16:586–601.
Stohlgren, T., and J. Schnase. 2006. Risk Analysis for biological hazards: what we need to know about invasive species. Risk Analysis 26:163–173.
Suarez, A. V., D. A. Holway, and T. J. Case. 2001. Patterns of spread in biological invasions dominated by long-distance jump dispersal: insights from Argentine ants. Proceedings of the National Academy of Sciences of the United States of America 98:1095–1100.
Suter, G. W. 1993. Ecological risk assessment. Lewis Publishers, Chelsea, MI.
Taylor, R. A. J., L. S. Bauer, D. L. Miller, and R. A. Haack. 2004. Emerald ash borer flight potential. Research and Technology Development Meeting, Romulus, Michigan.
Thomas, R. W., and R. J. Huggett. 1980. Modeling in geography: a mathematical approach. Harper and Row, London.
Train, K. 2003. Discrete choice methods with simulation. Cambridge University Press, Cambridge, UK.
Trakhtenbrot, A., R. Nathan, G. Perry, and D. M. Richardson. 2005. The importance of long-distance dispersal in biodiversity conservation. Diversity and Distributions 11:173–181.
Turchin, P. 1998. Quantitative analysis of movement. Sinauer Associates, Sunderland, MA.

Turchin, P., and W. T. Thoeny. 1993. Quantifying dispersal of southern pine beetles with mark-recapture experiments and a diffusion model. Ecological Applications 3:187–198.

Veit, R. R., and M. A. Lewis. 1996. Dispersal, population growth, and the Allee effect: dynamics of the house finch invasion of eastern North America. American Naturalist 148:255–274.

Viboud, C., O. N. Bjornstad, D. L. Smith, L. Simonsen, M. A. Miller, and B. T. Grenfell. 2006. Synchrony, waves, and spatial hierarchies in the spread of influenza. Science 312:447–451.

Wiens, J. A. 2001. The landscape concept of dispersal. Pages 96–109 *in* Clobert, J., E. Danchin, A. A. Shondt, and J. D. Nichols, editors. Dispersal. Oxford University Press, New York.

Wonham, M. J., J. T. Carlton, G. M. Ruiz, and L. D. Smith. 2000. Fish and ships: relating dispersal frequency to success in biological invasions. Marine Biology 136:1111–1121.

Zipf, G. K. 1946. The P1P2/D hypothesis: on the intercity movement of persons. American Sociological Review 11:677–686.

7

Uncertain Invasions: A Biological Perspective

Christopher L. *Jerde and Jonathan* M. *Bossenbroek*

In a Clamshell

When there is variability in predictions or when mechanisms of how systems or processes work are unknown, there is uncertainty. From medicine to engineering, and biology to economics, uncertainty is an important consideration when testing hypotheses and predicting outcomes. In this chapter, we explore the role of uncertainty on explaining and predicting invasions from a biological perspective. We begin by reviewing a classification of uncertainty that conforms to how many biologists and statisticians perceive the role of uncertainty in their scientific explorations. We then narrow our focus by evaluating the uncertainty in explanations and predictions from a gravity model of zebra mussel invasion into the western United States. This includes providing bootstrapped confidence intervals on parameter estimates from survey data and evaluating the predictive performance of the gravity model on a subset of economically and ecologically valuable lakes using probability theory and receiver operator characteristic curves. Many of these evaluations of uncertainty are uncommon in current gravity model applications to invasive species and are generalizable to other modeling approaches and the larger concern of predicting successful invasions.

In the absence of perfect knowledge about how a system or process works, there is unexplained variability in observations and predictions, or uncertainty. Uncertainty is the reason biologists experience difficulty identifying the properties of successful invaders and predicting invasions (Kolar and Lodge 2001). Prior to and following Elton's (1958) treatise, invasion biology received mainly disparate scientific interest, and it has only recently gained general ecological popularity (Davis et al. 2001; Puth and Post 2005) because of threats to biodiversity and financial losses (Pimentel et al. 2000, 2005). Now scientists are faced with the challenge of predicting and ideally preventing invasions without fully understanding the mechanisms that lead to successful invasion. Some have argued that predicting invasions in light of such uncertainty is futile (Gilpin 1990). Nevertheless, there has been considerable progress in predictive methods to identify species likely to invade and the locations likely to become invaded, which have emerged from the study of biological invasions (Reichard and Hamilton 1997; Schneider et al. 1998; Kolar and Lodge 2001; Rouget et al. 2004) and the uncertainty of invasions has been reduced.

Uncertainty is an interdisciplinary subject and consequently has varied usage. In economics, uncertainty is regularly mentioned in the same context of risk, where risk is a product of the probability of an event occurring and the impact of that event (Knight 1921). As long as the probability of the event is not zero or one, there is uncertainty. Biologists generally associate uncertainty with unexplained processes and mechanisms by which variability in observations arise (Taper and Lele 2004). This uncertainty is often presented as confidence intervals, standard errors, or posterior distributions on model parameters (Lewin-Koh et al. 2004; Cumming et al. 2007). However, there is considerable overlap between these two generalized perspectives of uncertainty (e.g., Dovers and Norton 1996). With an interdisciplinary approach to biological invasions, it is therefore unsurprising to find multiple perspectives regarding the importance and role of uncertainty. Because of the breadth of perspectives regarding uncertainty, in this chapter we provide a biological perspective of uncertainty and invasions.

Our biological perspective of uncertainty begins by reviewing a general taxonomy of uncertainty (Regan et al. 2002) and applying this taxonomy to biological invasions. Our list of sources of uncertainty is not comprehensive, but we believe it highlights some areas of invasion biology that, if emphasized in future studies, will strengthen our explanatory and predictive capabilities. We then focus our study of uncertainty and biological invasions to deconstructing a national gravity model of zebra mussel invasion. Specifically, we demonstrate a bootstrapping method of survey data used to parameterize a gravity model of the United States and assess the influence of uncertainty on the risk of invasion for two environmentally and economically valuable lakes, Lake Mead (Arizona and Nevada, USA) and Lake Roosevelt (Washington State, USA). We then evaluate the risk of invasion for 13 uninvaded lakes and evaluate the predictive performance of the gravity model on 15 lakes recently invaded. In closing, we discuss the challenges of prediction and validation of predictions in the face of uncertain invasions.

SOURCES OF UNCERTAINTY

Even in the discipline of biology there are multiple perspectives on sources of uncertainty (e.g., Shaffer 1981; Burnham and Anderson 2002). Many of these treatments focus on a particular type of uncertainty (e.g., measurement error: Thomas et al. 1993) or a particular methodology of assessing uncertainty (e.g., natural variability: Harwood and Stokes 2003; Cumming et al. 2007). However, a broad perspective of classifying uncertainty was taken by Regan et al. (2002) and is generally consistent with biological and statistical perspectives of sources of uncertainty. We adopt this classification for its completeness as a taxonomy with which to discuss the uncertainty related to biological invasions.

Linguistic Uncertainty

The general classification system starts by separating uncertainty into linguistic and epistemic uncertainty (Regan et al. 2002). Linguistic uncertainty is associated with communicating ideas and definitions. Chapter 5 contains a discussion about the linguistic uncertainty associated with defining when an invader is said to have established. Establishment is defined both by an abundance of the invader and by a persistence of the invader in a new landscape. In theoretical treatments of the invasion process, establishment is rarely defined with precise, numerical thresholds, but for empirical applications, these thresholds are required to define whether a species has established in a system and are necessary to perform statistical analyses. The variability in thresholds used to define establishment represents one form of linguistic uncertainty.

Within invasion biology, linguistic uncertainty has received some attention. Richardson et al. (2000) discuss the vagueness of the terms "naturalized" and "invasive." They also attempt to provide a clear vocabulary for discussing the invasion process. The vagueness, context dependence, and ambiguity (Regan et al. 2002) of invasion biology terms is pervasive throughout the entire invasion process from transport and arrival of invaders to the establishment, spread, and impact of those invasions (Sakai et al. 2001; Colautti and MacIsaac 2004).

Another example of linguistic uncertainty is in the assessment of whether an invasive species has had an impact. Invasion biologists have variably defined impact as the presence of any nonindigenous species (NIS), an NIS that has spread, and an NIS that has produced harmful environmental changes, particularly to native species (Ricciardi and Cohen 2007). Although attempts have been made to provide clear working definitions, such as the Parker et al. (1999) formulation stating that impact is the product of species range (area), abundance, and per-unit (or biomass) effect. These formulations may be entirely appropriate for considering impact in an ecological context, but this ignores the economic impact.

The simplest solution to linguistic uncertainty is to provide precise definitions that can be agreed upon by the scientific community (Richardson et al. 2000; Regan et al. 2002). However, finding consistent terminology has been, and continues to be, a problem for exploring the patterns and processes of biological invasion (Shrader-Frechette 2001) and also for communicating invasion biology and risk to the public and policy makers (Bossenbroek et al. 2005; Hodges 2008). The evolution of invasion biology into an objective discipline will likely be tied to the preciseness of the definitions employed in its description and application (Colautti and MacIsaac 2004). Another solution to the linguistic uncertainty of invasion biology is to provide syntheses of work on a particular subject, such as propagule pressure (Lockwood et al. 2005), that makes connections across multiple definitions and inferences.

Epistemic Uncertainty

Epistemic uncertainty is more closely related to data, models, and the methods of scientific inquiry, which can be secondarily separated into uncertainty due to measurement error, systematic error, natural variation, inherent randomness, and subjective judgment (Regan et al. 2002). Epistemic uncertainty is often accounted for by reporting quantitative measures such as confidence intervals, prediction intervals, probability distributions, or *p*-values.

Measurement error and systematic error are associated with errors in the recording of data by either human observation or errors from measurement devices, where measurement error is an unbiased measurement and systematic error is a biased measurement (Thomas et al. 1993). Natural variation is uncertainty due to spatial or temporal differences in the values of model parameters (e.g., Lele et al. 1998). Even when model parameters are accurately and precisely estimated, some ecological processes are inherently (effectively) random because it is infeasible to account for all the processes that influence model outcomes, nor is it possible to know the initial conditions with such precision as to make deterministic predictions (Gilpin 1990). Model uncertainty occurs when a biological process is described using a mathematical representation (Jonzen et al. 2002), and model uncertainty applies to not only variability in approaches (e.g., deterministic and stochastic modeling) but also to interpolation and extrapolation of model results beyond the coverage of the data that can be used to support such modeling-based inferences. An example of the variability in approaches for modeling the dispersal of invasive species is discussed in chapter 6. Lastly, subjective judgment emerges as a form of uncertainty throughout scientific inquiry, from evaluating the quality of data and choosing a modeling approach, to interpreting results and making decisions (Harwood and Stokes 2003). Table 7.1 provides the taxonomy of epistemic uncertainty with examples from biological invasions.

Many of the epistemic uncertainty issues in invasion biology are common problems in general ecology. For example, some species are able to invade at low population densities and remain undetected at low densities, only later experiencing population growth and spread (Christian and Wilson 1999). Detecting species at low abundance is the same problem community ecologists face when attempting to detect the presence of rare species (Longino and Colwell 1997; Costello and Solow 2003). Similarly, in conservation biology, demographic stochasticity and minimum viable population size (Lande 1993) are directly related to propagule pressure in invasion biology (Lockwood et al. 2005; Drake and Lodge 2006).

Invasion biology, however, faces some unique sources of uncertainty. In community ecology, the rare species are usually known to occur within the area being searched, and the observer has some search recognition pattern. This is often not true in invasion biology, where many invaders are surprise discoveries (Solow and Costello 2004). Similarly, population parameters such as growth rate, survivorship, fecundity, and reproductive value may be known for a species in its native range, but unless the destination has very similar habitat characteristics, these parameters will be different and will remain unknown until the invasion has occurred and population

TABLE 7.1. Uncertainty in biological invasions.

Type of uncertainty	Mechanism
Measurement error (ME)	Estimates of propagule pressure (species abundance) Estimates of boat registrations Estimates of lake attractiveness
Systematic error (SE)	Measurement of failed invasions Detection of invasive species Ignorance of important vectors and pathways of introduction
Natural variability (NV)	Fluctuations in donor region populations Changes in pathways of introduction Changes in recipient region environment Seasonality of species life cycle
Model uncertainty (MU)	Presence of Allee effects Population growth models Choosing parameters in an ecological niche model
Inherent randomness (IR)	Demographic stochasticity of introduction and survival Genetic bottlenecks due to small founder populations Predator avoidance in new locations
Subjective judgment (SJ)	Choice of species to study Including expert and public opinion into risk management and policy Use of survey data designed for other purposes to validate model

Biological invasions have a mixture of common ecological and unique sources of uncertainty. Each source of uncertainty could potentially hinder our ability to accurately predict successful invasions. Some examples are specific to dispersal of NIS by recreation boaters.

data collected. As a consequence, and in part to sidestep added uncertainty due to population dynamics, there has been a call to develop robust methods and models of invasive species risk assessment that do not require specific details regarding population dynamics (Simberloff 2003).

One solution to both epistemic and linguistic uncertainty is through mathematical modeling of biological (Taper and Lele 2004) and economic processes (Leung et al. 2002). For epistemic uncertainty, the modeling solution is straightforward. Models represent hypotheses about how a system or process works. Confronting models with data allows us to perform hypothesis testing and model selection among competing hypotheses (Hilborn and Mangel 1997; Burnham and Anderson 2002; Lewin-Koh et al. 2004). Alternatively, models of these processes can be evaluated on their predictive performance. Both strategies require accounting for uncertainty, that is, quantifying the explanatory (e.g., goodness of fit) and predictive (e.g., receiver operating characteristic [ROC] curves) capabilities (Hosmer and Lemeshow 2000). Methods to assess the explanatory and predictive performance of models are demonstrated in the following section.

Less obvious is the role mathematical modeling plays in reducing linguistic uncertainty. When theory and experimentation meet, there is necessarily a measurable quantity to evaluate from a model that is evaluated with data (hypothesis testing or model selection: Lewin-Koh et al. 2004). Returning to the example of defining establishment (chapter 5), empiricists must specify thresholds of abundance and/or persistence above which establishment is said to have occurred. It is possible, although not likely, that invasion biologist, mathematical modelers, and economists could reach an agreement regarding the threshold of establishment. It is more likely that we will continue to have a variety of thresholds. Arguably, the most common thresholds will be determined by the data available (e.g., presence or absence vs. count data) and the models used (chapter 6) to assess the process of establishment. Although some linguistic uncertainty will remain, depending on the data collected and modeling approach, mathematical models will force clearly delineated (at least mathematically) definitions of establishment that theory likely would not evolve to if left in a conceptual form. Byers and Goldwasser (2001) provide one such example where thresholds are defined in order to assess a variety of invasion biology issues.

We now transition to an example of estimating the risk of invasion and accounting for uncertainty by modeling the transport of zebra mussels through the use of a gravity model produced by Bossenbroek et al. (2007). The purpose of this specific example is to demonstrate how uncertainty in explanation and prediction can be accounted for. While gravity models are regularly applied to aquatic invasions, particularly zebra mussels, previous studies have only haphazardly quantified uncertainty (but see Bossenbroek et al. 2001; Leung et al. 2006). Here we show a bootstrapping routine that accounts for uncertainty in parameter estimates used in the gravity model, and we evaluate the predictive power of the gravity model on the order of lake invasions.

ZEBRA MUSSELS AND A GRAVITY MODEL OF ARRIVAL

The dreissenid mussel invasion of the Great Lakes began about 1986, with the successful establishment of the zebra mussel, *Dreissena polymorpha*, in Lake St. Clair via ships' ballast water (Hebert et al. 1989), causing extensive ecological and economic impacts. The present North American range of *D. polymorpha* includes much of northeastern and north central North America, including more than 400 inland lakes. In 2008, zebra mussels were also discovered in one reservoir in Colorado and one in California. A second species of dreissenid, the quagga mussel (*D. bugensis* (= *D. rostriformis bugensis* [Andrusov (1897)])), was found in the Erie Canal and Lake Ontario in 1991 (May and Marsden 1992) and is now common in lakes Erie and Ontario (Diggins et al. 2004). This species was also discovered in Lake Mead on the Colorado River in January 2007 and now extends to several other reservoirs up and downstream of Lake Mead. The range expansion of dreissenid mussels in North America to date has resulted from a combination of processes, involving the dispersal within and between water bodies. The primary pathways of dreissenid dispersal include shipping routes in the United States, natural downstream dispersal, and overland dispersal by human vectors, such as recreational boaters (see chapter 12 for a more thorough discussion of these processes). Gravity models of recreational boater movement patterns have been used to forecast the overland dispersal of *D. polymorpha* (Schneider et al. 1998; Bossenbroek et al. 2001, 2007; Leung et al. 2006).

Gravity Model Formulation

Gravity models use formulations, analogous to Newton's laws of attraction, to estimate the arrival of invaders into discrete patches. The variables and parameters susceptible to epistemic uncertainty are listed in table 7.2, and an overview of the mathematics of gravity models can be found in Bossenbroek et al. (2001) or Leung et al. (2006). The formulation used here follows the national gravity model for zebra mussel dispersal (Bossenbroek et al. 2007). The critical result useful for predicting invasions from the gravity model is the number of arriving boaters that previously visited waters infected with zebra mussel (Q). Bossenbroek et al. (2007) report Q as the proportion of all boaters traveling to a destination from a zebra mussel source. Here Q is the number of infested boaters arriving into a destination.

Estimating the number of arriving boaters starts by first modeling the distribution of all boaters from watershed, i, to destination, j, between N watersheds,

$$T_{i,j} = \frac{O_i W_j c_{ij}^{-\alpha}}{\sum_{j=1}^{n} W_j c_{ij}^{-\alpha}}, \qquad \text{for all } i,$$

where O_i is the number of licensed boaters in their watershed of origin, W_j is the area of lakes in the destination watershed, c_{ij} is the Euclidean distance between the source and destination watersheds, and

$$c_{ii} = \delta \min_{j \neq i} (c_{ij}) c_{ii}.$$

The sum in the denominator is a balancing factor that ensures all boaters that leave a source arrive at a destination. The national gravity model for zebra mussels

TABLE 7.2. Uncertainty in gravity models.

Symbol	Description	Examples of uncertainty[a]
Gravity model input		
O_i	Number of boats at the source	Temporal changes in the number of boats (NV) Only licensed boats (SE)
W_j	Attractiveness	Alternative measure to area (SJ, MU)
c_{ij}	Distance	Euclidian versus travel distance (SJ, MU)
$\delta\alpha$	Distance multiplier Distance coefficient	Change in gas prices (NV) Selective sampling (SE) Sampling error (ME)
Gravity model output		
Q	Number of boats entering a destination with NIS	Functional relationship between Q and the probability of establishment (ME)

The parameters α and δ are subject to uncertainty common to estimation methods and sampling. The variables O, W, c, and Q are also subject to uncertainty. One concern is whether attractiveness is rightly associated with the area of the lake.

[a] Uncertainty abbreviations are as defined in table 7.1.

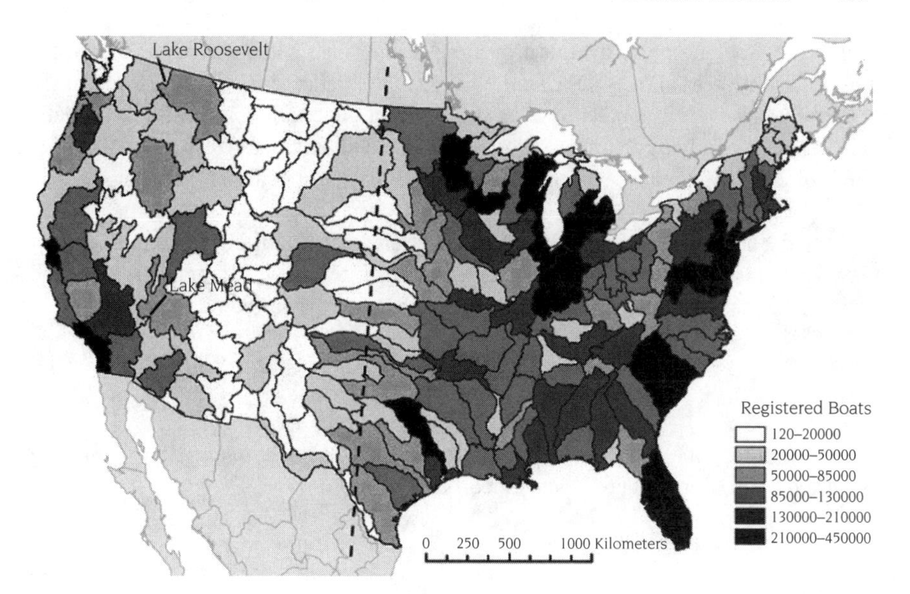

FIGURE 7.1.
Reservoirs and watersheds. Lakes Roosevelt and Mead receive boater traffic from watersheds east of the 100th meridian (dashed line) that have zebra mussels present. These lakes have been the focus of managers as likely locations for zebra mussels to be introduced in the western United States. Each watershed is shaded based the number of licensed boats found within the watershed. The Great Lakes region, which has the highest density of waters invaded by zebra mussels, also has a high density of licensed recreational boaters. Adapted from Bossenbroek et al. (2007), with permission of the Society for Conservation Biology.

(Bossenbroek et al. 2007) has two notable differences from previous zebra mussel studies (Bossenbroek et al. 2001; Leung et al. 2006). First, the locations are delineated by watersheds (figure 7.1) rather than counties, and second, it is possible for boaters to redistribute themselves within the same watershed ($i = j$). This leads to a need to define how far boaters travel within their own watershed, because a distance value of 0 would render a gravity model useless. A typical convention is to have the distance traveled within a watershed be some proportion (δ) of the distance to the next nearest possible destination (Thomas and Hugget 1980). The parameter α is a distance coefficient that describes the deterrent effect of distance upon a boater. The variables O, W, and c are properties of each watershed, while the parameters α and δ are estimated from data.

Survey data for the 100th Meridian Initiative were used to parameterize the national gravity model. The 100th Meridian Initiative is a cooperative effort by state, federal, and provincial agencies to prevent the spread of zebra mussels and other aquatic nuisance species into western North America. Surveys were conducted at 20 reservoirs throughout the Great Plains region (i.e., roughly along the 100th meridian) that recorded the distance traveled by recreational boaters between sources and destinations. The 13 reservoirs that had more than 50 completed surveys were used for the parameterization, and this procedure is consistent with the analysis performed in Bossenbroek et al. (2007). Using boater survey data from only reservoirs in the Great Plains states is subject to several sources of uncertainty, including systematic error, because only a portion of the country is used to parameterize a national model, and subjective judgment, because the surveyors may be specifically targeting boaters from long distances.

In the national gravity model, the number of boaters carrying zebra mussels (Z_i) is assumed to be proportional to the area of lakes infested with zebra mussels found within watershed i. This results in the redistribution of infested boaters,

$$R_{i,j} = \frac{Z_i}{O_i} T_{ij},$$

where Z_i/O_i is the proportion of boaters carrying zebra mussels. The number of infested boaters arriving to a watershed j is

$$Q_j = \sum_{i=1}^{n} R_{ij}.$$

The number of boaters visiting a watershed that previously visited watersheds with invaded lakes (Q) is purportedly an indicator of invasion risk and can be interpreted as the dose in a dose–response application of biological invasions (chapter 5; Bossenbroek et al. 2001; Leung et al. 2004; Lockwood et al. 2005).

The 100th Meridian Initiative was initially established to stop or slow the spread of zebra mussels into the western United States, but the funding was grossly inadequate relative to what would be required to meet this goal. No guidance currently exists about how best to allocate funding to alternative methods of prevention and control. The national gravity model was developed in part to inform this management dilemma (see chapter 12). We focus our analysis hereafter on accounting for the uncertainty in the gravity model, because this is a necessary component for implementing management actions (Leung et al. 2002). How the measured uncertainty, dealt with here, feeds into economic considerations and resulting management decisions is more thoroughly treated in chapter 8. We begin our treatment of uncertainty in gravity models by first considering the invasion of two popular destinations for boaters across the United States, lakes Mead and Roosevelt.

Order of Invasion

So which of these two lakes is most likely to be invaded first? Jerde and Lewis (2007), using invasion waiting times (chapter 5), formulated the order of invasion for two locations. This ordering, applied to outputs from gravity models, requires three assumptions: the survivorship of individuals is small, the survivorship at both locations is approximately the same, and the gravity scores (Q_m, Lake Mead; Q_r, Lake Roosevelt) are proportional to the actual propagule pressure. Under these conditions, the probability of Lake Mead being invaded before Lake Roosevelt is $(1 + Q_r/Q_m)^{-1}$, and the probability of Lake Roosevelt being invaded before Lake Mead is the complement, $(1 + Q_m/Q_r)^{-1}$.

Based on Bossenbroek et al. (2007), the probability of Lake Mead being invaded before Lake Roosevelt is 0.797, and the probability of Lake Roosevelt being invaded before Lake Mead is 0.203. These probabilities are based on the gravity model structure, the measured variables, the estimated parameters, and the model assumptions—all of which are subject to the influences of uncertainty. We now turn to quantifying the uncertainty in the parameters α and δ that are estimated from the survey data regarding the distance boaters traveled to get to reservoirs in the Midwest of the United States.

Bootstrapping Confidence Intervals

To investigate the change in the probabilities of ordered invasion, due to uncertainty in the estimates of α and δ, we performed a Monte Carlo simulation by the following:

1. For a single reservoir, we estimated the probability an arriving boater would be from a specific watershed using the survey data used to parameterize the gravity model.
2. We then drew from a multinomial distribution with these probabilities as parameters and recorded the distance this boater traveled to arrive at the destination (c_{ij}).
3. This was repeated at a single reservoir for the number of surveys recorded at that reservoir.
4. Steps 1–3 were then repeated for each reservoir.
5. From steps 1–4, the parameters δ and α were estimated following Bossenbroek et al. (2007) and recorded.
6. Q_m, Q_r, and the estimated probabilities of ordered invasion were recorded.
7. Lastly, steps 2–6 were repeated 1,000 times.

This procedure is a bootstrapping routine that accounts for the uncertainty in and from the survey samples. The uncertainty arises not from the number of boaters arriving to the reservoir but from the variability in the contribution of observed sources to the estimates. The list of replicates resulting from the bootstrapping routine can be used to build 95% confidence intervals on α, δ, and the resulting uncertainty in the ordered probabilities of invasion (Efron and Tibshirani 1993).

The distance coefficient (α) estimate has fairly tight confidence intervals, while the distance multiplier (δ) has broader confidence intervals (table 7.3). This is expected after inspecting the sums-of-squares surface provided in figure 1 of Bossenbroek et al. (2007). The relatively small range of the confidence intervals on the parameters and minimal change in the probabilities of ordered invasion indicate that the uncertainty in the boater surveys due to variability in the source of boater movements is negligible. A subsequent sensitivity analysis (Bossenbroek et al. 2007) indicated that a 25% reduction in α could lead to approximately an 8% decrease in

TABLE 7.3. Confidence intervals (CIs) on parameters and probability estimates.

Parameter	Lower 95% CI	Upper 95% CI
α	2.49	2.63
δ	0.70	1.35
Pr(Mead before Roosevelt)	0.795	0.799
Pr(Roosevelt before Mead)	0.201	0.205

There appears to be very little variation in the probability estimates due to variability in the parameters estimated from the survey data. However, the parameter estimates may have some temporal variability from changes in driving habits, such as increased gas prices, that are not reflected in the confidence intervals.

the proportion of boats arriving to a location. However, as demonstrated here, this does little to change the predicted ordered probabilities of invasion. Taken together, the reduction in α would likely increase the expected invasion waiting time for both lakes but does not change the order in which the invasion would likely occur (Jerde and Lewis 2007).

Bootstrapped confidence intervals account for uncertainty in the parameters that can be quantified from the survey data, and this encompasses many of the common sources of uncertainty surrounding surveys, such as sample size, randomness, and completeness (Barnett 2002). Bootstrapping does not account for any bias, such as interviewing boaters with only out-of-state plates, nor does this bootstrapping account for any temporal variability in the values of α or δ.

Gravity Models, Evaluating Variables, and Model Selection

Uncertainty may also influence the variables in a model (e.g., the gravity variables O, W, and c) and the model structure. Assessing model structure and the inclusion or exclusion of different variables is generally referred to as model selection. In statistical practice, there are multiple ways to perform model selection, such as likelihood ratio tests and Akaike's information criterion (Burnham and Anderson 2002). But these methods have yet to be applied to invasion gravity models because the data are insufficient and, due to the expenses involved in continuously monitoring boater traffic at multiple uninvaded locations, unlikely to ever be collected.

As an alternative to directly applying a model selection approach on Q, the estimated number of boaters arriving is used as an explanatory variable and then the probability of establishment is modeled from presence or absence of an invasion into a lake. This may be accomplished using logistic regression or a functional form of the response curve (Bossenbroek et al. 2001; Leung et al. 2004, 2006). Statistically, this approach is problematic because the explanatory variable, in this case Q, is usually assumed fixed and known (Hosmer and Lemeshow 2000), but here the number of arriving boaters is an estimate from the gravity model and therefore contains some variability, or uncertainty. Specifically, there is measurement error and/or systematic error in the explanatory variable. Measurement error can change the observed mean structure, the variance structure, and obscure significant covariates (Thomas et al.

1993). This in turn will lead to poor predictions and model fit from a dose-response curve. Proper model selection using this approach would require accounting for the measurement error. Assessing the gravity model structure and accounting for measurement error in gravity scores are important future directions for invasion biology research. In the following section, we offer a few prescriptions for where gravity model research may proceed in order to account for uncertainty as it relates to predictive performance.

UNCERTAINTY AND PREDICTION

Predicting establishment with accuracy is an aspiration of invasion biology (Kolar and Lodge 2001), and of gravity models in particular (Bossenbroek et al. 2001; Leung et al. 2006). Predictive accuracy is critical for proactive management to prevent invasions (chapter 1). However, it is still unclear whether models of species arrival and spread (chapter 6) developed thus far are sufficient to offer reliable guidance about when and where to intervene against invasive species (Gilpin 1990; but see chapter 9). Here, we demonstrate three related predictive insights related to gravity models. First, we look at the pairwise probabilities of 13 uninvaded lakes. Second, we estimate the pairwise probabilities of 15 recently invaded lakes and show graphically the relationship between Q and the pairwise probabilities of invasion given the distribution of Q from the invaded lakes. Third, we apply logistic regression, knowing full well there is uncertainty in the explanatory variable Q, to demonstrate how uncertainty confounds predictive performance.

Making Predictions and Gravity Models

Gravity scores, Q, are known to be positively correlated with successful invasions (MacIsaac et al. 2004). Moreover, invasion order probability, as demonstrated for lakes Mead and Roosevelt, can be calculated. For the 13 lakes of interest presented in Bossenbroek et al. (2007), the pairwise probabilities are provided in table 7.4. All lakes in the table were uninvaded by zebra mussels at the onset of this project. Since then, however, the Lake of the Ozarks, just downstream from H. S. Truman Reservoir, was reported to contain zebra mussels in June 2006, Lake Mead was reported to contain quagga mussels in January 2007, and Lake Perry in Kansas was reported to contain zebra mussels in October 2007.

As apparent from gravity score, Q, H. S. Truman Reservoir is the most likely to become invaded, and each paired probability is greater than 0.75. This later result indicates that we should not expect many, if any, lakes in this group to become invaded before H. S. Truman Reservoir. Analytically, the probability of r lakes becoming invaded before some time, t, while H. S. Truman Reservoir remains uninvaded is

$$\Pr(R = r) = \binom{N}{r} \left[1 - (1 - p_u)^t\right]^r \left[(1 - p_u)^t\right]^{N-r} (1 - p_i)^t$$

TABLE 7.4. Pairwise order of invasion for currently uninvaded lakes.

	Q	H. S. Truman Reservoir	Lake Oahe	Barren River Lake	Martin Lake	Austin Lake	Lake Mead	Upper Saranac Lake	Perry Lake	Chesuncook Lake	Roosevelt Lake	Amistad Reservoir	Goose Lake	Conchas Lake
H. S. Truman Reservoir	3145.4	—	0.778	0.782	0.820	0.839	0.902	0.910	0.958	0.963	0.972	0.986	0.987	0.995
Lake Oahe	899.1	0.222	—	0.506	0.565	0.599	0.724	0.744	0.867	0.881	0.910	0.953	0.955	0.982
Barren River Lake	876.5	0.218	0.494	—	0.559	0.593	0.719	0.739	0.865	0.879	0.907	0.952	0.952	0.981
Martin Lake	691.1	0.180	0.435	0.441	—	0.534	0.669	0.691	0.834	0.851	0.886	0.940	0.943	0.976
Austin Lake	602.2	0.161	0.401	0.407	0.466	—	0.637	0.660	0.814	0.833	0.871	0.932	0.935	0.973
Lake Mead	342.6	0.098	0.276	0.281	0.331	0.363	—	0.525	0.714	0.739	0.797	0.886	0.891	0.953
Upper Saranac Lake	309.6	0.090	0.256	0.261	0.309	0.340	0.475	—	0.693	0.719	0.776	0.876	0.881	0.949
Perry Lake	137.4	0.042	0.133	0.135	0.166	0.186	0.286	0.307	—	0.532	0.606	0.758	0.766	0.891
Chesuncook Lake	120.9	0.037	0.119	0.121	0.149	0.167	0.261	0.281	0.468	—	0.575	0.733	0.742	0.878
Roosevelt Lake	89.3	0.028	0.090	0.093	0.114	0.129	0.203	0.224	0.394	0.425	—	0.670	0.680	0.842
Amistad Reservoir	44.0	0.014	0.047	0.048	0.060	0.068	0.114	0.124	0.242	0.267	0.330	—	0.511	0.724
Goose Lake	42.0	0.013	0.045	0.046	0.057	0.065	0.109	0.119	0.234	0.258	0.320	0.489	—	0.715
Conchas Lake	16.8	0.005	0.018	0.019	0.024	0.027	0.047	0.051	0.109	0.122	0.158	0.276	0.285	—

Each lake has a probability, estimated from the gravity score (Q), of being invaded before a different lake. For example, the probability of H. S. Truman Reservoir being invaded before Lake Oahe is 0.778. The complement, the probability of Lake Oahe being invaded before H. S. Truman Reservoir is 0.222. Probabilities near 0.5 indicate lake pairs with similar invasion risk based on the gravity score. For example, Lake Mead and Upper Saranac Lake have similar gravity scores (Q = 342.6 and 309.6 respectively). These lakes are the largest lakes within their watershed and are considered the most likely to become invaded.

where there are N lakes in the group of uninvaded lakes (table 7.4; $N = 12$ when excluding H. S. Truman Reservoir). The parameter p_u is the probability at each discrete time step that an uninvaded lake transitions to become invaded, and p_i is the probability that H. S. Truman Reservoir becomes invaded; $[1 - (1 - p_u)^t]^r$ is the probability of r invaded lakes, $[1 - (1 - p_u)^t]^{N-r}$ is the probability of $N - r$ invaded lakes, and $(1 - p_i)^t$ is the probability of H. S. Truman Reservoir remaining uninvaded. This formulation rests on the assumptions that each lake in the group has the same p_u and that both p_i and p_u are known. Unfortunately, neither of these assumptions is easily justified or met for the gravity scores presented in table 7.3. We know the gravity scores, Q, are very different between the group of uninvaded lakes, and this can reflect different probabilities of invasion (Von Holle and Simberloff 2005; see table 7.4). The pairwise probabilities (table 7.4) of Lake Oahe, Barren River Lake, Martin Lake, and Austin Lake all have order pairings close to 0.5, implying that either lake could be invaded before the other, and these lakes form a group with similar likelihoods of invasion. Given a larger sample than just four lakes with similar pairing, scores may be more useful for producing the group of reference lakes.

For an example, we could assume that the group of uninvaded lakes were similar to Lake Oahe, the next most likely invaded lake based on the gravity scores. Then, using the relative probability formulation (presented in Jerde and Lewis 2007), the relative probability of H. S. Truman Reservoir transitioning to invaded is $p_i = (Q_{\text{H.S.Truman}})(Q_{\text{Oahe}})^{-1}(p_u) = 3.5p_u$. This can be inserted into the equation above, and we are left with one parameter to estimate, p_u. Alternatively, we can insert $(899.1/3145.4)p_i$ for p_u, and then we are similarly left with p_i to estimate. However, neither of these parameters is known.

Figure 7.2A is a plot of the probability of observing one or more of the 12 reference lakes becoming invaded (given $p_u = (899.1/3145.4)p_i$) as a function of probability that H. S. Truman Reservoir becomes invaded, and figure 7.2B is a plot of the probability of observing five or more of the 12 reference lakes becoming invaded under the same conditions. The probability of observing one or more lakes invaded ranges from zero to approximately 0.5 for the possible values of p_i and for $t = 1$, 5, and 10 time step periods of observation. With almost no better than a fair coin's chance of one or more of the reference lakes becoming invaded and the uncertainty due to p_i being unknown, observing a few of the reference lakes becoming invaded before H. S. Truman would not invalidate the predictions of the gravity model. However, observing five or more lakes invaded before H. S. Truman (figure 7.2B) has only a 0.1 or less probability of being observed by chance for all time periods and all values of p_i. This observation provides a robust rule of thumb for evaluating the performance of predictions gleaned from the gravity scores. If we observe five or more lakes in the reference group become invaded before H. S. Truman Reservoir becomes invaded, we should be skeptical of the gravity model's predictive capabilities.

Evaluating Predictions from Gravity Models

Validating gravity models through testing of predictions is desperately needed. Gravity models for invasive species have a relatively young history compared to other

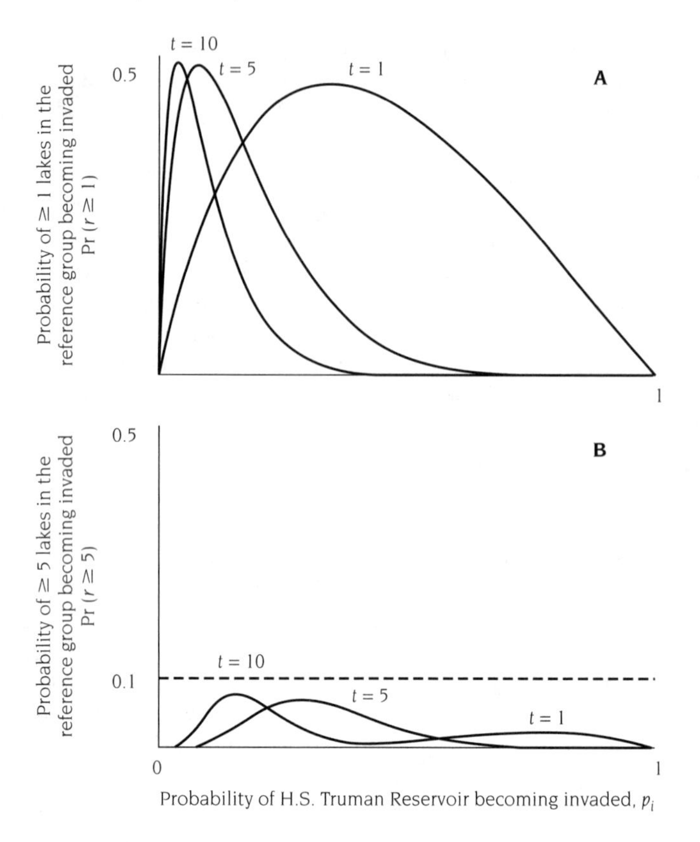

FIGURE 7.2.
The probability that one or more (A) or five or more (B) of the reference lakes ($n = 12$) becoming invaded by time t as a function of the probability that H. S. Truman Reservoir becomes invaded, p_i. Uncertainty is captured in the range of probability values (height of the curves). Because p_i is unknown, there is considerable uncertainty about the probability of one or more lakes becoming invaded while H. S. Truman remains uninvaded, even over 10 time steps. However, for all values of p_i and $t = 1, 5$, and 10, the probability of observing five or more lakes invaded is low [$\Pr(r \geq 5) < 0.1$].

models of species spread (Schneider et al. 1998; Bossenbroek et al. 2001, 2007; Leung et al. 2004, 2006; MacIsaac et al. 2004). One evaluation of predictive performance is to compare the gravity scores of recently invaded locations to uninvaded locations (Leung et al. 2004, 2006; MacIsaac et al. 2004). As mentioned above, logistic regression is used with presence/absence data to test the significance of the gravity score. In the studies thus far, the gravity score is shown to be positively correlated with successful invasions and is a significant explanatory variable. This is the usual extent to which diagnostics of model fit and predictive power are performed (but see MacIsaac et al. 2004). However, more diagnostics exist, and some of these diagnostics are more useful for evaluating the predictive capability of logistic regression models.

The receiver operating characteristic (ROC) curve and the resulting area under the curve (AUC) are one such diagnostic. The purpose of such curves is to evaluate how well the logistic model is able to discriminate invaded lakes and uninvaded lakes based on the gravity score. The curve is a reflection of the sensitivity and specificity. Sensitivity accounts for the correct discrimination of successful invasions—meaning it is a proportion of lakes that are predicted and observed to be invaded. Similarly, specificity accounts for correctly discriminating lakes predicted and observed to be uninvaded. Both sensitivity and specificity are proportions and range from 0 to 1. The ROC uses 1-specificity. This reflects the proportion of misclassified lakes that are predicted to be invaded but are actually uninvaded. A more thorough discussion of ROC curves, sensitivity, and specificity can be found in Hosmer and Lemeshow (2000).

The heuristic measure of the ROC is the AUC. When the AUC is between 0.9 and 1, the model does an excellent job of discriminating between invaded and uninvaded. In contrast, when the AUC is close to 0.5, discriminating between invaded and uninvaded lakes is really no better than flipping a fair coin to predict whether it is invaded.

Table 7.5 provides the gravity scores for 15 lakes invaded just prior to construction of the national gravity model (Bossenbroek et al. 2007). Applying logistic regression to these lakes and the gravity scores of the uninvaded lakes found in table 7.4 produces the ROC and AUC shown in figure 7.3. With an AUC of 0.63, we can conclude the logistic regression model with the gravity score as an explanatory variable does a relatively poor job of discriminating, and hence predicting, successful invasions. However, a few caveats are in order. First, this is a small subsample of the lakes in the national gravity model, and it would be unfair to say that the failure of this subset to provide a larger AUC is evidence for a failure of the entire system of lakes and the gravity model. Second, the logistic regression showed that Q was not a significant explanatory variable. Using a likelihood ratio test between a constant only and constant with parameter for the Q variable, the p-value was 0.57. This leads us to select the constant-only model even though we used the model with the added parameter for estimating the AUC. Poor model fit is often, although not necessarily, associated with poor discrimination (Hosmer and Lemeshow 2000). As mentioned above, one of the consequences of measurement error is the failure to detect significant covariates (Thomas et al. 1993). Future studies may consider the approach and subsequent improvements on this concern suggested by Wacholder et al. (1993) or Lele and Allen (2006).

The purpose of discussing ROC and AUC as diagnostics is not to call into question the national gravity model for zebra mussels. Rather, ROC and AUC diagnostics represent a tool to evaluate the predictive performance of the model. ROC and AUC are not exclusive to logistic regression and may be useful with other applications, and therefore should be the subject of future studies that seek to demonstrate the predictive capabilities of invasion models.

Pairwise probabilities can also be used as a visual diagnostic to assess uncertainty in model predictions. Table 7.5 contains the pairwise probabilities of invaded (rows) and uninvaded (columns) lakes. The table itself is difficult to decipher, as

TABLE 7.5. Probabilities of invaded lakes being invaded before currently uninvaded lakes.

		Uninvaded lakes												
Invaded lakes	Q	H. S. Truman Reservoir	Lake Oahe	Barren River Lake	Martin Lake	Austin Lake	Lake Mead	Upper Saranac Lake	Perry Lake	Chesuncook Lake	Roosevelt Lake	Amistad Reservoir	Goose Lake	Conchas Lake
Lake of the Ozarks	2717.6	0.464	0.751	0.756	0.797	0.819	0.888	0.898	0.952	0.957	0.968	0.984	0.985	0.994
Grand Lake O' the Cherokees	1024.5	0.246	0.533	0.539	0.597	0.630	0.749	0.768	0.882	0.894	0.920	0.959	0.961	0.984
Hamilton Lake	629.09	0.167	0.412	0.418	0.476	0.511	0.647	0.670	0.821	0.839	0.876	0.935	0.937	0.974
Lake Margarethe	608.92	0.162	0.404	0.410	0.468	0.503	0.640	0.663	0.816	0.834	0.872	0.933	0.935	0.973
Clear Lake	356.32	0.102	0.284	0.289	0.340	0.372	0.510	0.535	0.722	0.747	0.800	0.890	0.895	0.955
Rock Lake	214.22	0.064	0.192	0.196	0.237	0.262	0.385	0.409	0.609	0.639	0.706	0.830	0.836	0.927
Prairie River Lake	85.83	0.027	0.087	0.089	0.110	0.125	0.200	0.217	0.385	0.415	0.490	0.661	0.671	0.837
Horicon National Wildlife Refuge	82.27	0.025	0.084	0.086	0.106	0.120	0.194	0.210	0.375	0.405	0.479	0.652	0.662	0.831
Lake George	76.86	0.024	0.079	0.081	0.100	0.113	0.183	0.199	0.359	0.389	0.462	0.636	0.647	0.821
Winfield City Lake	68.63	0.021	0.071	0.073	0.090	0.102	0.167	0.181	0.333	0.362	0.434	0.610	0.620	0.804
Cass Lake	56.19	0.018	0.059	0.060	0.075	0.085	0.141	0.154	0.290	0.317	0.386	0.561	0.572	0.770
Big Bradford Lake	54.46	0.017	0.057	0.059	0.073	0.083	0.137	0.150	0.284	0.311	0.379	0.553	0.565	0.765
Hartwick Lake	52.92	0.017	0.056	0.057	0.071	0.081	0.134	0.146	0.278	0.304	0.372	0.546	0.558	0.759
Ess Lake	23.62	0.007	0.026	0.026	0.033	0.038	0.065	0.071	0.147	0.164	0.209	0.350	0.360	0.585
Base Lake[a]	2.90	0.001	0.003	0.003	0.004	0.005	0.008	0.009	0.021	0.023	0.031	0.062	0.065	0.149

The regular invasion of lakes with small paired probabilities of invasion relative to more susceptible lakes would suggest that the model is not adequate to predict invasions.

[a] Base Lake is inside the confines of a military base and was likely invaded through a different mechanism or pathway than is modeled by a gravity model based on recreational boaters.

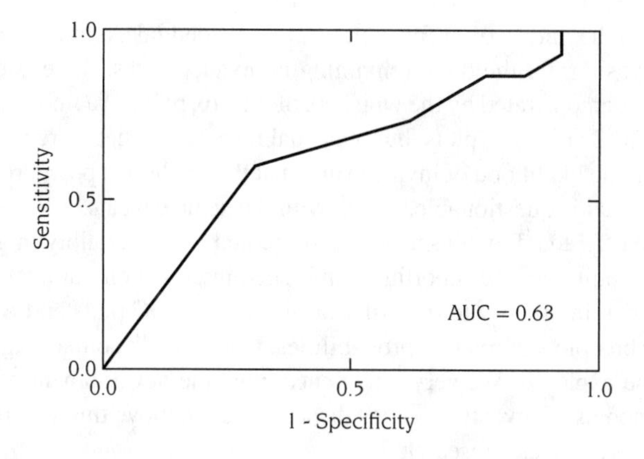

FIGURE 7.3.
Receiver operating characteristic (ROC) curve for the logistic regression of invaded ($n = 13$) and uninvaded lakes ($n = 15$) based on the gravity scores. The AUC is 0.63 and indicates there is poor discrimination between invaded and uninvaded lakes. As a diagnostic, the ROC and resulting AUC indicate that the gravity score for this subsample of lakes has poor predictive capabilities.

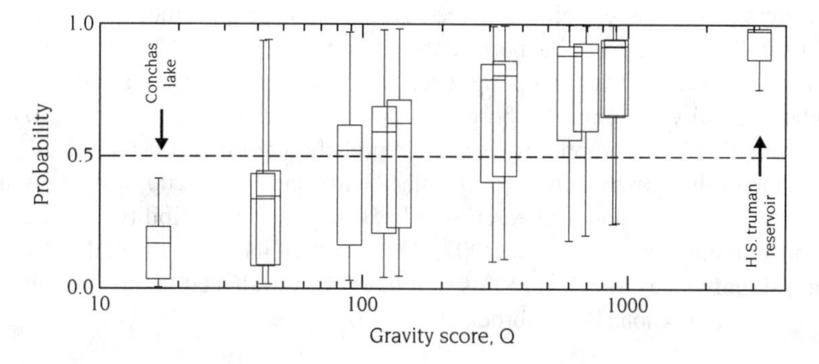

FIGURE 7.4.
The pairwise probability distributions of invaded lakes for each uninvaded lake. Each box plot is one of the uninvaded lakes from table 7.4, positioned at its respective gravity score, Q. The distribution of the box plot is all of the pairwise probabilities with invaded lakes. Most uninvaded lakes have expectations greater than 0.5, which implies they are more susceptible, based on the gravity score, to invasion than are lakes already invaded.

would be the ROC and AUC diagnostics without some familiarity. However, figure 7.4 contains 13 box plots, one for each uninvaded lake, that show the distribution of the pairwise probabilities to the group of invaded lakes. The box plots are ordered by the gravity score, Q, such that H. S. Truman Reservoir, with the largest gravity score, is the rightmost box plot and Conchas Lake, with the smallest gravity score, is the leftmost plot. The spread of each box represents the uncertainty associated with each uninvaded lake given the observed group of invaded lakes. Interestingly, 10 of the 13 uninvaded lakes have means greater than 0.5, indicating that relative to the values of Q for the recently invaded lakes, there is a reasonable chance of invasion. Of the three remaining uninvaded lakes, there is considerable variability, as demonstrated by the whiskers of the box plots. We may conclude, similarly to the ROC and AUC plots, but by visual inspection, that there is considerable uncertainty in the likelihood of invasion and that the predictive power resulting from the gravity model is questionable, again, with the same caveats.

We have offered a few prescriptions to account for uncertainty in gravity models, including building and reporting confidence intervals on parameter estimates, applying probability theory to the order of invasions, ROC plots and AUC metrics, and building box plots of relative probabilities. Undoubtedly, other diagnostics exist and should be explored. We very much encourage the development of diagnostics for gravity models of invasive species dispersal, and believe this will be an avenue of future invasive species research.

From the small collection of invaded and uninvaded lakes and the analyses performed here, one may conclude that gravity models are left wanting. We believe this is not a fair conclusion. Indeed, there appears to be predictive performance issues with gravity models. However, this is less an indictment of gravity models and more of a guidepost of where our research needs to go. For example, there are likely groups of lakes with similar susceptibility of being invaded where there are substantial differences in susceptibility between groups. This phenomenon was observe in the suitability of U.S. shipping ports to the potential invasion of Chinese mitten crab (*Eriocheir sinensis*) when coupling a relative measure of propagule pressure and habitat suitability mapping (Herborg et al. 2007). The result of comparing two locations with different susceptibility is that the probability of invasion will be different even though the gravity scores are similar (Jerde and Lewis 2007). A likely next objective in gravity model research will be to include susceptibility in evaluating the model predictions (Muirhead 2007). In the case of the uninvaded lakes listed in table 7.4, only lakes Mead and Roosevelt have been evaluated for susceptibility of zebra mussel invasion (Bossenbroek et al. 2007).

DISCUSSION

The study of biological invasions is plagued by uncertainty. From identifying the characteristics of successful invaders (Goodwin et al. 1999) or detecting new invaders at a location (Costello et al. 2007), to predicting when and where the next lake in the western United States is going to be invaded by zebra mussels (Bossenbroek et al. 2007), there are few topics that are deterministic (Gilpin 1990). This includes not only the epistemic uncertainty emphasized in this chapter, but also the linguistic uncertainty of the terminology used in the biological and bioeconomic research of invasive species (Colautti and MacIsaac 2004; Shrader-Frechette 2001). Here, we have provided a biological and statistical perspective of uncertainty with emphasis on the role of making and evaluating predictions, in particular, to a subset of invaded and uninvaded lakes with scores from a gravity model of zebra mussel dispersal (Bossenbroek et al. 2007). We have offered a few specific prescriptions for evaluating the predictive capability of gravity models, but much more needs to be done to account for the uncertainty in invasion biology. Listing all the ways we may account for uncertainty would be an arduous task indeed. Instead, we offer examples from the literature, following the order of the sections of this chapter, that we hope will guide researchers in future studies and investigations of biological invasions.

Identifying sources of uncertainty for any biological process is a critical step that should be done in concert with formulating hypotheses and models (Hilborn and Mangel 1997; Lewin-Koh et al. 2004). Carlton (1996) provides an exemplary overview of the sources of uncertainty for predicting the arrival of invasive species into new locations. He identified six important sources of uncertainty: changes in donor regions, new donor regions, changes in recipient region, invasion windows, stochastic inoculation events, and dispersal vector changes. These processes have become the subject of empirical investigations (e.g., Wonham et al. 2000) and theoretical frameworks (e.g., Jerde and Lewis 2007). More generally, overviews of the study of biological invasions (e.g., Vermeij 1996; Puth and Post 2005) highlight areas in the field that need more attention, due in large part to the uncertainty of specific processes, such as predicting invasions (Kolar and Lodge 2001).

Reducing linguistic uncertainty has been accomplished, so far, by evaluating the terminology used in the literature. To this end, there are many notable examples of how to proceed (e.g., Colautti and MacIsaac 2004; Hodges 2008). When recognizable differences between definitions for the same term are detected, it may be time for a critical review that attempts to bridge and clarify the discrepancy. One such term, from a bioeconomic perspective, is the term “risk.” For biologists, risk is often associated with a probability of some unwanted event occurring, such as invasion (Suter 1993; Jerde and Lewis 2007). However, economists generally associate risk with the probability of an event occurring times the loss accrued because of that event occurring. Undoubtedly interdisciplinary approaches to biological invasions will uncover similar disparities and will require some attention.

Reporting parameters with some measure of variability is a common method for dealing with uncertainty (Cumming et al. 2007). One approach to mathematical modeling is to produce a model and then search the literature for the parameter values of a particular species and/or system from which to make predictions. This is actually quite difficult because often only point estimates (i.e., means), and not measures of variability, are reported, especially for parameters such as growth and predation rates. A measure of variability in parameter estimates can be used to analyze qualitative differences between model predictions (e.g., Wonham et al. 2006) from perturbation analysis (Caswell 2001). This problem can easily be overcome with diligence in the reporting of descriptive statistics for point estimates (e.g., the variance or standard deviation) in future empirical studies of biological invasions. A good review of the appropriate error bars to produce for point estimates is presented by Cumming et al. (2007).

Assessing gravity model structure and the variables to include in a gravity model has received limited attention. The differences in models are usually determined by the available data, such as production constrained or doubly constrained gravity models (Leung et al. 2006; see also chapter 6). That said, Leung et al. (2004) provided an example of how to detect Allee effects from gravity scores. Yet many questions remain for gravity model research. For example, the attraction coefficient, Wj (table 7.2), the area of the destination lake or the area of lakes within a water shed (Bossenbroek et al. 2007), has been shown to be a reasonable measure of how attractive a lake is to boaters (e.g., Reed-Andersen et al. 2000). But should measures of recreational fishing opportunities, or distance to population centers, or availability of facilities (see Reed-Andersen et al. 2000), or water quality, or water skiing also be used? Similarly, watersheds that have big reservoirs, such as lakes Mead, Roosevelt, and Oahe, can probably be assumed to be the main attractor of these watersheds, and it seems reasonable that the proportion of boaters coming to these big reservoirs is related to the overall proportion of water in the watershed. For watersheds in Michigan, for example, with a lot of small lakes, the spatial interactions within the watershed, including the distribution of people living within the watershed, may be a critical consideration not currently captured in the gravity model dynamics. With the predictive performance observed within this chapter, it is likely that assessing the model structure of gravity models will be a fruitful area for reducing the uncertainty of invasions.

Likely the most pressing issue for invasion biologists with respect to epistemic uncertainty is that of predicting successful invasions—successfully. This is why we focused much of our attention here on gravity model diagnostics of predictive performance. To date, little has been presented regarding the predictive performance of gravity models other than to show significant correlation between the gravity scores and observed invasions (Leung et al. 2004; MacIsaac et al. 2004). But it bears repeating that correlation is not necessarily an indicator of good predictive performance (Hosmer and Lemeshow 2000).

Applying the probability theory of waiting times (Drake et al. 2005; Jerde and Lewis 2007) appears to be one avenue for evaluating predictions from gravity models with a lot of potential. With the specific question, "what is the probability of observing r lakes invaded before H. S. Truman Reservoir," we were able to find a robust threshold; that is to say, there is a probability of less than 0.1 of observing five lakes invaded within 1-, 5-, and 10-year time periods. Similarly, there appears to be some usefulness of formulating relative waiting times (Jerde and Lewis 2007), but this likely needs to be expanded into formulations that include multiple lakes or groups of lakes with similar invasion susceptibility (Herborg et al. 2007).

One of the drawbacks of predictive formulations based on waiting times is that in order to validate or invalidate the predictions, we must wait for invasions to occur. This is likely a problem for the impatient and, more important, for managing the spread of invasive species. Alternatively, there are diagnostics such as ROC and AUC that assess predictive performance based on the model's ability to discriminate between invaded and uninvaded lakes based on the gravity score and can be conducted on existing data without having to wait for future invasions. However, the ability to discriminate between invaded and uninvaded can be sensitive to measurement error (Thomas et al. 1993). Therefore, it appears that gravity scores will need to be calibrated with census data about the number of boaters arriving to lakes that are infested with zebra mussels, but some of the necessary information may already be available (e.g., Johnson et al. 2001).

We have focused on the biological processes of invasive species and the uncertainty in our understanding of these processes. The next step is to translate these measures into decision-making frameworks based on the economic realities of managing invasive species. In transition to chapter 8, which deals with some of the bioeconomic issues of uncertainty, one last uncertainty issue should be raised, and that is the uncertainty of uncertainty. As mentioned above, risk, in economics, is the product of the probability of an event occurring and the loss accrued due to that event occurring. Certainty is defined by the probability of an event occurring and is equal to 0 or 1, and uncertainty in the outcome occurs everywhere in between. Much of the uncertainty we have been emphasizing is in the estimate of the probability of that event occurring. As a result, the biological uncertainty we have in a processes will in large part also influence our ability to make any economic decision.

Acknowledgments This chapter was substantially improved by the editors of this book and reviews by A. Potapov, J. Rothlisberger, and three anonymous reviewers. This material is based on work supported by the Integrated Systems for Invasive Species project (D. M. Lodge, principle investigator) funded by the National Science Foundation (DEB 02-13698). C.L.J. was supported by a National Sciences and Engineering Research Council of Canada Collaborative Research Opportunities Grant to M. A. Lewis and the Canadian Aquatic Invasive Species Network. J.M.B. was supported by a National Sea Grant and U.S. Fish and Wildlife Service award to D. M. Lodge. This is publication 2009-03 from the University of Toledo Lake Erie Center.

References

Barnett, V. 2002. Sample survey: principles and methods, 3rd edition. Hodder Arnold Press, London.

Bossenbroek, J., C. Kraft, and J. Nekola. 2001. Prediction of long-distance dispersal using gravity models: zebra mussel invasion of inland lakes. Ecological Applications 11:1778–1788.

Bossenbroek, J., J. Ladd, B. Peter, and D. Lodge. 2007. Forecasting the expansion of zebra mussels in the United States. Conservation Biology 21:800–810.

Bossenbroek, J., J. McNulty, and R. Keller. 2005. Can ecologists heat up the discussion on invasive species risk? Risk Analysis 25:1595–1597.

Burnham, K., and D. Anderson. 2002. Model selection and multimodel inference: a practical information-theoretic approach, 2nd edition. Springer, New York.

Byers, J., and L. Goldwasser. 2001. Exposing the mechanism and timing of impact of nonindigenous species on native species. Ecology 82:1330–1343.

Carlton, J. 1996. Pattern, process, and prediction in marine invasion ecology. Biological Conservation 78:97–106.

Caswell, H. 2001. Matrix population models, 2nd edition. Sinauer, Sunderland, MA.

Christian, J., and S. Wilson. 1999. Long-term ecosystem impacts of an introduced grass in the northern Great Plains. Ecology 80:2397–2407.

Colautti, R., and H. MacIsaac. 2004. A neutral terminology to define invasive species. Diversity and Distributions 10:135–141.

Costello, C., J. Drake, and D. Lodge. 2007. Evaluating an invasive species policy: ballast water exchange in the Great Lakes. Ecological Applications 17:655–662.

Costello, C., and A. Solow. 2003. On the pattern of discovery of introduced species. Proceedings of the National Academy of Sciences of the United States of America 100:3321–3323.

Cumming, G., F. Fidler, and D. Vaux. 2007. Error bars in experimental biology. Cell Biology 177:7–11.

Davis, M., K. Thompson, and J. Grime. 2001. Charles S. Elton and the dissociation of invasion ecology from the rest of ecology. Diversity and Distributions 7:97–102.

Diggins, T., M. Weimer, K. Stewart, R. Baier, A. Meyer, R. Forsberg, and M. Goehle. 2004. Epiphytic refugium: are two species of invading freshwater bivalves partitioning spatial resources? Biological Invasions 6:83–88.

Dovers, S., and T. Norton. 1996. Uncertainty, ecology, sustainability and policy. Biodiversity and Conservation 5:1143–1167.

Drake, J., P. Baggenstos, and D. Lodge. 2005. Propagule pressure and persistence in experimental populations. Biology Letters 1:480–483.

Drake, J., and D. Lodge. 2006. Allee effects, propagule pressure and the probability of establishment: risk analysis for biological invasions. Biological Invasions 8:365–375.

Efron, B., and R. Tibshirani. 1993. An introduction to the bootstrap. Chapman and Hall/CRC, Boca Raton, FL.

Elton, C. 1958. The ecology of invasion by animals and plants. Methuen, London.

Gilpin, M. 1990. Review: ecological prediction. Science 248:88–89.

Goodwin, B., A. McAllister, and L. Fahrig. 1999. Predicting invasiveness of plant species based on biological information. Conservation Biology 13:422–426.

Harwood, J., and K. Stokes. 2003. Coping with uncertainty in ecological advice: lessons from fisheries. Trends in Ecology and Evolution 18:617–622.

Hebert, P., B. Muncaster, and G. Mackie. 1989. Ecological and genetic studies on *Dreissena polymorpha* (Pallas): a new mollusc in the Great Lakes. Canadian Journal of Fisheries and Aquatic Sciences 46:1587–1591.

Herborg, L., C. Jerde, D. Lodge, G. Ruiz, and H. MacIsaac. 2007. Predicting the North American distribution of Chinese mitten crabs (*Eriocheir sinensis*) using measures of propagule pressure and environmental niche models. Ecological Applications 17:663–674.

Hilborn, R., and M. Mangel. 1997. The ecological detective: confronting models with data. Princeton University Press, Princeton, NJ.

Hodges, K. 2008. Defining the problem: terminology and progress in ecology. Frontiers in Ecology and the Environment 6:35–42.

Hosmer, D., and S. Lemeshow. 2000. Applied logistic regression, 2nd edition. Wiley Interscience, New York.

Jerde, C., and M. Lewis. 2007. Waiting for invasions: a framework for the arrival of non-indigenous species. American Naturalist 170:1–9.

Johnson, L., A. Ricciardi, and J. Carlton. 2001. Overland dispersal of aquatic invasive species: a risk assessment of transient recreational boating. Ecological Applications 11:1789–1799.

Jonzen, N., P. Lundberg, E. Ranta, and V. Kaitala. 2002. The irreducible uncertainty of the demography-environment interaction in ecology. Proceedings of the Royal Society of London Series B Biological Sciences 269:221–225.

Knight, F. 1921. Risk, uncertainty, and profit. Houghton Mifflin, Boston.

Kolar, C., and D. Lodge. 2001. Progress in invasion biology: predicting invaders. Trends in Ecology and Evolution 16:199–204.

Lande, R. 1993. Risks of population extinction from demographic and environmental stochasticity and random catastrophes. American Naturalist 142:911–927.

Lele, S., and K. Allen. 2006. On using expert opinion in ecological analyses: a frequentist approach. Environmetrics 17:683–704.

Lele, S., M. Taper, and S. Gage. 1998. Statistical analysis of population dynamics in space and time using estimating functions. Ecology 79:1489–1502.

Leung, B., J. Bossenbroek, and D. Lodge. 2006. Boats, pathways, and aquatic biological invasions: estimating dispersal potential with gravity models. Biological Invasions 8:241–254.

Leung, B., J. Drake, and D. Lodge. 2004. Predicting invasions: propagule pressure and the gravity of Allee effects. Ecology 85:1651–1660.

Leung, B., D. Lodge, D. Finnoff, J. Shogren, M. Lewis, and G. Lamberti. 2002. An ounce of prevention or a pound of cure: bioeconomic risk analysis of invasive species. Proceedings of the Royal Society of London Series B Biological Sciences 269:2407–2413.

Lewin-Koh, N., M. Taper, and S. Lele. 2004. A brief tour of statistical concepts. Pages 3–16 *in* M. L. Taper and S. R. Lele, editors. The nature of scientific evidence: statistical, philosophical, and empirical considerations. University of Chicago Press, Chicago.

Lockwood, J., P. Cassey, and T. Blackburn. 2005. The role of propagule pressure in explaining species invasions. Trends in Ecology and Evolution 20:223–228.

Longino, J., and R. Colwell. 1997. Biodiversity assessment using structured inventory: capturing the ant fauna of a tropical rain forest. Ecological Applications 7:1263–1277.

MacIsaac, H., J. Borbely, J. Muirhead, and P. Graniero. 2004. Backcasting and forecasting biological invasions of inland lakes. Ecological Applications 14:773–783.

May, B., and J. Marsden. 1992. Genetic identification and implications of another invasive species of dreissenid mussel in the Great Lakes. Canadian Journal of Fisheries and Aquatic Science 49: 1501–1506.

Muirhead, J. 2007. Forecasting and validating the establishment of aquatic nonindigenous species. Ph.D. thesis, University of Windsor.

Parker, I., D. Simberloff, W. Lonsdale, K. Goodell, M. Wonham, P. Kareiva, M. Williamson, B. Von Holle, P. Moyle, J. Byers, and L. Goldwasser. 1999. Impact: toward a framework for understanding the ecological effects of invaders. Biological Invasions 1:3–19.

Pimentel, D., L. Lach, R. Zunign, and D. Morrison. 2000. Environmental and economic costs of non-indigenous species in the United States. BioScience 50:53–67.

Pimentel, D., R. Zuniga, and D. Morrison. 2005. Update on the environmental and economic costs associated with alien-invasive species in the United States. Ecological Economics 52:273–288.

Puth, L., and D. Post. 2005. Studying invasion: have we missed the boat? Ecology Letters 8:715–721.

Reed-Andersen, T., E. Bennett, B. Jorgensen, G. Lauster, D. Lewis, D. Nowacek, J. Riera, B. Sanderson, and R. Stedman. 2000. Distribution of recreational boating across lakes: do landscape variables affect recreational use? Freshwater Biology 43:439–448.

Regan, H., M. Colyvan, and M. Burgman. 2002. A taxonomy and treatment of uncertainty for ecology and conservation biology. Ecological Applications 12:618–628.

Reichard, S., and C. Hamilton. 1997. Predicting invasions of woody plants introduced into North America. Conservation Biology 11:193–203.

Ricciardi, A., and J. Cohen. 2007. The invasiveness of an introduced species does not predict its impact. Biological Invasions 9:309–315.

Richardson, D., M. Pysek, M. Rejmanek, M. Barbour, F. Panetta, and C. West. 2000. Naturalization and invasion of alien plants: concepts and definitions. Diversity and Distributions 6:93–107.

Rouget, M., D. Richardson, S. Milton, and D. Polakow. 2004. Predicting invasion dynamics of four alien *Pinus* species in a highly fragmented semi-arid shrubland in South Africa. Plant Ecology 152:79–92.

Sakai, A., F. Allendorf, J. Holt, D. Lodge, J. Molofsky, K. With, S. Baughman, R. Cabin, J. Cohen, N. Ellstrand, D. McCauley, P. O'Neil, I. Parker, J. Thompson, and S. Weller. 2001. The population biology of invasive species. Annual Reviews of Ecology and Systematics 32:305–332.

Schneider, D., C. Ellis, and K. Cummings. 1998. A transportation model assessment of the risk to native mussel communities from Zebra Mussel spread. Conservation Biology 12:788–800.

Shaffer, M. 1981. Minimum population sizes for species conservation. BioScience 31:131–134.

Shrader-Frechette, K. 2001. Non-indigenous species and ecological explanation. Biology and Philosophy 16:507–519.

Simberloff, D. 2003. How much population biology is needed to manage introduced species? Conservation Biology 17:83–92.

Solow, A., and C. Costello. 2004. Estimating the rate of species introductions from the discovery record. Ecology 85:1822–1825.

Suter, G. 1993. Ecological risk assessment. Lewis Publishers, Chelsea, Michigan.

Taper, M., and S. Lele. 2004. Dynamical models as paths to evidence in ecology. Pages 275–297 *in* M. L. Taper and S. R. Lele, editors. The nature of scientific evidence: statistical, philosophical, and empirical considerations. University of Chicago Press, Chicago.

Thomas, D., D. Stram, and J. Dwyer. 1993. Exposure measurement error: influence on exposure disease relationships and methods for correction. Annual Reviews of Public Health 14:69–93.

Thomas, R. W., and R. J. Hugget. 1980. Modeling in geography. Barnes and Noble Books, Totowa, New Jersey.

Vermeij, G. 1996. An agenda for invasion biology. Biological Conservation 78:3–9.

Von Holle, B., and D. Simberloff. 2005. Ecological resistance to biological invasion overwhelmed by propagule pressure. Ecology 86:3212–3218.

Wacholder, S., B. Armstrong, and P. Hartge. 1993. Validation studies using an alloyed gold standard. American Journal of Epidemiology 137:1251–1258.

Wonham, M., J. Carlton, G. Ruiz, and L. Smith. 2000. Fish and ships: relating dispersal frequency to success in biological invasions. Marine Biology 136:1111–1121.

Wonham, M., M. Lewis, J. Renclawowicz, and P. van den Driessche. 2006. Transmission assumptions generate conflicting predictions in hostvector disease models: a case study in West Nile virus. Ecology Letters 9:706–725.

8

Economic Valuation and Invasive Species

Christopher R. *McIntosh, David* C. *Finnoff, Chad Settle, and Jason* F. *Shogren*

In a Clamshell

Invasive species policy can be better informed if we understand how people value reductions in the risks posed by invasive species. Eliciting these values, however, presents a challenge to research because invasive species have both market and nonmarket impacts on people. While decades of research has been devoted to valuing market and nonmarket impacts on the environment, minimal effort has been expended on valuing damages due to invasive species. This is changing—more research is focusing on valuing the reduced human and environmental damages posed by invasive species so we can better understand the net benefits of prevention, control, and eradication efforts. This chapter presents two applications of nonmarket valuation methods to estimate the value of invasive species prevention and control. Application 1 looks at the value of delaying the inevitable risks posed by aquatic invaders in freshwater in the United States; application 2 examines the value of control lake trout in Yellowstone Lake, Wyoming.

Invasive species impose costs on society by disrupting scarce ecosystem services, changing unique natural landscapes, and requiring people to reallocate scarce wealth to remove or eradicate unwanted species. *Scarcity* is the key word here. Economic value emerges from the idea that economics is a discipline of scarcity. The fact that resources are scarce means that using up resources in one way (e.g., ignore or control invasive species) incurs an *opportunity cost*—the cost of forgoing the next best alternative use. Economists use opportunity costs to capture how people value any scarce good or service, including protection from or eradication of invasive species (see, e.g., Nunes and van den Bergh 2004; Born et al. 2005).

Opportunity costs are relevant to invasive species management—allowing invaders to take over an ecosystem implies an opportunity cost of, say, lost or reduced native wildlife benefits. Alternatively, complete eradication of an invasive species may mean increasing public expenditures to such a degree we forgo other valuable public policy opportunities for, say, health care or education. Deciding how to manage invasive species in either way entails a sacrifice, which is the benefits we could have gotten by using these scarce resources in some other way. If society is considering whether to implement stricter invasive species policies (e.g., increased border inspections), people must decide whether the benefits are worth the costs of the policy—as measured by the forgone best alternative opportunity.

This chapter begins by briefly illustrating what we mean by economic value and why this differs in function and form from recent estimates of "economic value" provided in some well-publicized papers. We then consider two examples of our recent work in which we estimate the economic value of invasive species protection, one at the national level and one for Yellowstone Lake. The reader interested in general overviews of nonmarket valuation and stated preference methods could consult, for instance, Bateman et al. (2002), Champ et al. (2003), Adamowicz and Deshazo (2006), and Hanley et al. (2007).

Before continuing, we believe it is useful to discuss one overarching criticism that can be leveled against stated preference survey work in general, and our work on invasive species is no exception. *Hypothetical bias* exists when survey respondents give valuation answers to survey questions that do not match up with a real economic commitment—how they would actually value the good or service if they were spending their own money. Hypothetical bias arises whenever elicited preferences are different depending on whether the elicitation method has monetary consequences. The accumulated evidence, mainly from lab experiments, leads many to conclude that hypothetical bias exists, which undercuts the basic foundations of popular state preference valuation methods used in cost–benefit analyses. The gap between hypothetical statements of value and real economic commitments is a common problem that has long troubled work on stated preference (see, e.g., Murphy et al. 2005). Other work finds evidence to suggest that cultural differences might explain why hypothetical bias is not observed in every nation. For example, Ehmke et al. (2008) implemented the same referendum lab valuation experiment in China, France, Niger, and the United States. They found that U.S. subjects (Indiana and Kansas) exhibit a significant hypothetical "positive" bias, subjects in China and Niger are likely to exhibit a "negative" bias, and French subjects from Grenoble are the least prone to the bias. Other work has focused on ex ante or ex post methods to correct for or reduce hypothetical bias with varying degrees of success (see, e.g., Cummings and Taylor 1999). This all said, many economists, including ourselves, have long argued that stated preference valuation exercises should be best viewed from a relative perspective, not an absolute perspective—the relative value of option A versus option B (see, e.g., Hanley et al. 2007)

WHAT IS ECONOMIC VALUE?

Opportunity costs are measured by what economists call *total surplus*—the difference between the maximum each buyer is willing to pay for a good and the market price (*consumer surplus*) added to the difference between the market price and the

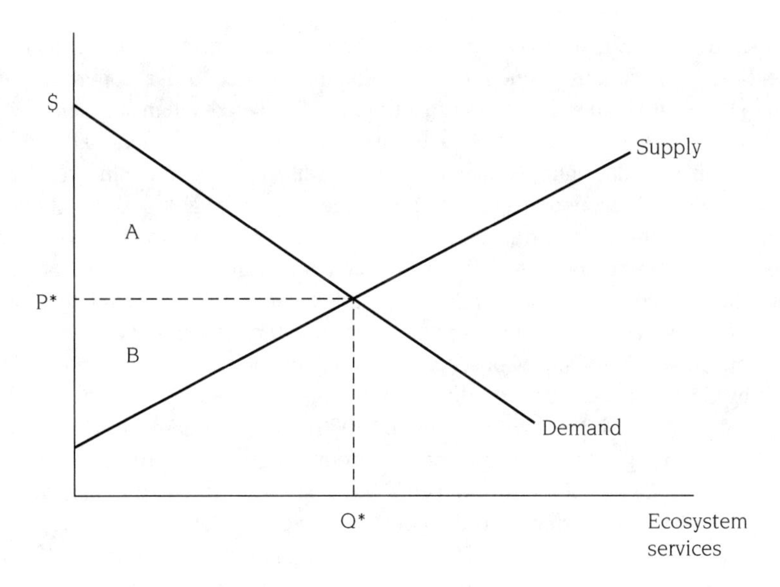

FIGURE 8.1.
Supply and demand for ecosystem services. Area A is consumer surplus; area B is producer surplus; P* and Q* are market equilibrium price (P) and quantity (Q).

minimum each seller is willing to accept for the good (*producer surplus*). You may recall learning in your microeconomics course about supply and demand and the market equilibrium. When supply equals demand at a market-clearing price, both consumers and producers benefit from the exchange. Removing this ability to trade for whatever reason, including invasive species, removes these benefits as measured by total surplus—which is the opportunity cost to society (Hanley et al. 2007).

Let us make this more concrete. Consider the supply and demand of some local ecosystem service, such as pollination of apple trees by neighboring bees, as illustrated in figure 8.1. On the vertical axis we have dollars; on the horizontal we have ecosystem services. We have the demand (D) for ecosystem services fall as price increases; we have the supply (S) of services increasing as price increases. Loosely speaking, the demand curve represents each buyer's maximum willingness to pay (WTP), arranged from highest to lowest. The supply curve reflects each seller's minimum willingness to accept (WTA), arranged from lowest to highest. The intersection of supply and demand reveals the market-clearing price, P^*, and quantity sold, Q^*.

In figure 8.1, area A represents consumer surplus, that is, the benefits to consumers of this pollination service. Consumer surplus captures the aggregate difference between the maximum WTP of each buyer and the market price, P^*. This is also the opportunity costs to consumers if some invasive species enters the ecosystem and somehow disrupts this service. Area B represents producer surplus—the benefit to the producers from making the trade as captured by the aggregate difference between the market price and the minimum WTA for each seller, again, the opportunity cost suffered by producers if an invader disrupts this service. Together, areas A and B represent the total economic benefit of the pollination service, or the opportunity cost to society if an invader affects the functioning of this ecosystem. Remember, the term *economic benefit* can include both market and nonmarket goods and services; economic benefit does not just mean financial gains to people in the marketplace.

Applying this economic approach to capture economic value can be a challenge because it requires measuring these surplus measures for more invasive species protection. While tempting to fall back on as a guide, market prices do not reflect this value in a consistent manner. Rather, market prices capture the relative rate at which the "market" is willing to exchange one good for another, not the social value of the good itself. This is why most economists reject popular press estimates such as those by Costanza et al. (1997) on the costs of replacing the earth's services, or those provided by the biologists Pimentel et al. (2000) on the costs of invasive species. Although at first glance this may seem an intuitively correct approach, economists recognize that these estimates of benefits and costs are derived from a summation of replacement costs, which, on the contrary, have no relationship to the idea of total surplus (see Bockstael et al. 2000). Even though market prices are used in both calculations, the replacement costs and surplus measures are two entirely different concepts. Also frequently found in engineering studies of financial costs and benefits, these replacement cost numbers are constructed by the simple rubric of multiplying market price times quantity needed, rather than making use of the standard surplus welfare measures that economists have developed over two centuries of research.

Figure 8.2 illustrates the simple analytics behind why the recent attempts to value the earth and its services via replacement costs can be misleading. Figure 8.2a reproduces figure 8.1—the supply and demand for pollination. Figure 8.2b illustrates a private market for a bee keeper who might be hired to replace the ecosystem service if some invasive species has disrupted services, that is, the replacement market. We know that total economic value is reflected by the two areas, A and B, in figure 8.2a. So protecting these ecosystem services from an invasive species would yield an economic value of A + B, which can be estimated in dollars. Alternatively, allowing this ecosystem to be lost to an invasive species would cost A + B.

In contrast, one could use the replacement market as a guide (figure 8.2b). Here one would multiply the market price, P^{**}, as the basis for value times equilibrium quantity, Q^{**}, to get the total expenditures for this ecosystem service. Total value would be represented by the area C1 + C2. Comparing the welfare triangle area A + B with the $P \times Q$ rectangle area C1 + C2, we see there is no fundamental relationship between the two. The only way A + B would equal C1 + C2 is by coincidence. In addition, one can not say a priori that A + B will be greater than or less than C1 + C2 without knowing more about the slopes of the demand and supply curves for the ecosystem services.

For instance, if there is no substitute for this ecosystem service (e.g., if you tried to replace a lake in the Sahara), the demand curve will be perfectly vertical (inelastic demand), and then area A will be substantial. This no-substitution possibility is what triggered Michael Toman's (1998) comment that the Costanza et al. (1997) valuation estimate of the earth was a "serious underestimate of infinity" (area A $\rightarrow \infty$). If

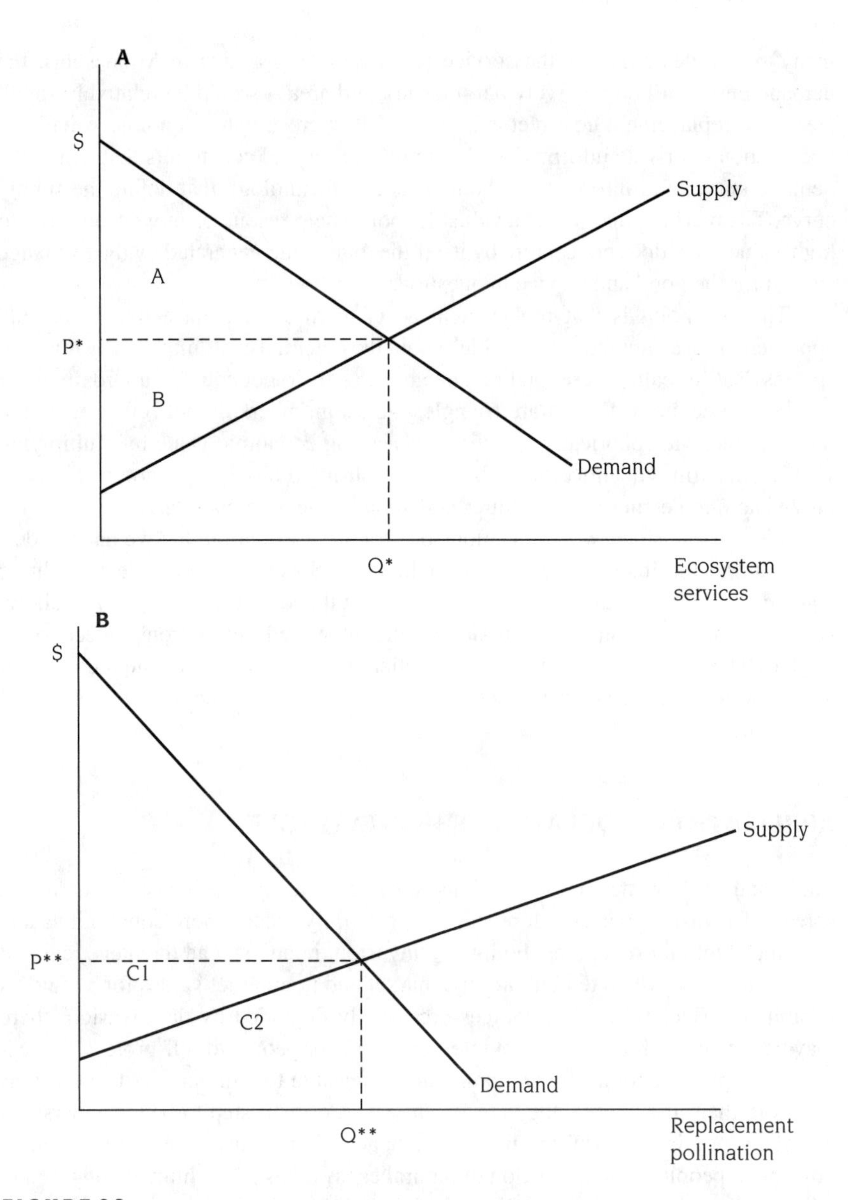

FIGURE 8.2.
Supply and demand for ecosystem service and replacement market. (A) The supply and demand for pollination. (B) A private market for a bee keeper who might be hired to replace pollination ecosystem services if they are disrupted by an invasive species. P** and Q** are market equilibrium price (P) and quantity (Q); C1 + C2 represents total revenue (or P × Q).

many substitutes exist for the service (e.g., replacing a lake in Minnesota), the demand curve will be relatively flat (elastic), and area A would be relatively small. Here, the replacement cost method could well overestimate economic value. We need to obtain private information about preferences and constraints that define the demand curve, and information about costs and technology that define the supply curve. The market price, while a valuable tool to help resources move from low- to high-value uses, does not capture by itself the total value generated by the exchange system for the good and service in question.

The key point is that replacement costs seem at first glace like a "useful" approach to get a "number on the table" about the potential costs imposed by invasive species. But in reality these total revenue rectangles do not equal or approximate or parallel the economist's welfare triangles these numbers do not reflect real economic value. Metaphorically speaking, measuring economic value by multiplying visible price times quantity would be like measuring biological phenotype by using only visible molecules and structures coded by the genetic material.

We now consider two applications of economic valuation that we have undertaken during the Integrated Systems for Invasive Species (ISIS) project—valuing aquatic species protection at a national level and the lake trout invaders in Yellowstone Lake (see chapter 2's discussion on the integrated Yellowstone bioeconomic model). The goal of the economist's analytical framework is to create a consistent and rigorous baseline against which we can judge the net value of any given set of policy options.

APPLICATION 1: DELAYING THE INEVITABLE

Trade and trade routes among regions are known to be a primary vector in the spread of invasive species. More shipping generally means more aquatic invaders introduced into these regions. Following invasion, inland spread threatens regional lakes and rivers, which tend to provide market and nonmarket values for a region's inhabitants. Because these resources are usually degraded by the invasion, there may be incentives for government intervention or for people to self-protect. Since it is in large part economically and politically infeasible to eliminate trade, invasions are more than just likely—they are inevitable. A critical step to better understand and gauge public support for policies to combat biological invasions is to determine how much people value protection of natural ecosystems given limited budgets and other public policy challenges. Given the inevitability of invasion, perhaps the most pertinent question becomes: What are people willing to pay to maintain (temporarily) the current (high) level of environmental quality if environmental degradation from the invasive species is guaranteed at some future point?

The WTP measure we consider is for the marginal delay in invasion damages, given that these damages will occur with certainty at some point in time. Determining this WTP requires a dynamic model, in which a person maximizes lifetime utility over good and bad states and faces a budget constraint. Box 8.1 presents the formal details of the model. During periods in the good state, the individual receives constant utility from consumption and the corresponding environmental quality. When the bad state is realized, utility then depends on consumption minus some market damage and the lower level of environmental quality. Our model closely follows Rosen (1988). Full versions can be found in Shogren et al. (2006) and McIntosh et al. (2007).

BOX 8.1. The Economic Value of Delaying the Inevitable

For a detailed description of methods of environmental valuation, the reader is directed to Hanley et al. (2007). Most of the methods focus on estimation of static (with time period) values and proceed from an assumption that an individual derives utility (or satisfaction) from environmental quality Q and market goods and services $x = (x_1, x_2, \ldots, x_n)$, where utility is given by the twice differentiable, concave utility function $U(x, Q)$. Individuals are presumed to have their expenditures constrained by their income M and take prices of goods and services as given $p = (p_1, p_2, \ldots, p_n)$. Within the framework, the choice problem is for people to choose their consumptions of each good and service to maximize their utility, taking their income, prices, and the level of environmental quality as given. The values of changes in environmental quality can be determined by looking at the money metrics of changes in optimized utility due to the changes in environmental quality.

One commonly employed method approaches the problem from the dual formulation, one of an individual cost minimization. In this setting, individuals choose levels of market goods and services to minimize their expenditures, given a fixed level of utility $\bar{U}$ and the given level of environmental quality Q^0. The result of the optimization problem is an expenditure function $e(p, Q^0, \bar{U})$ that is a function of prices, environmental quality, and the fixed level of utility each of which is exogenous to the individual. If environmental quality is changed from Q^0 to Q^1, the individual's maximum WTP (formally known as the compensating surplus) for the change is the difference in expenditure functions $WTP = e(p, Q^0, \bar{U}) - e(p, Q^1, \bar{U})$.

The changes in environmental quality related to invasive species, however, will occur over time rather than instantaneously. People also have an opportunity to delay the timing of the environmental change. We extend the valuation framework by letting a representative person's lifetime utility be written as

$$\bar{U} = \int_0^{\tau} \bar{U}^0 e^{-\rho t} dt + \int_{\tau}^{T} \bar{U}^1 \left[c(t) - \alpha D[x(t) + \tilde{x}(t)], Q^1\right] e^{-\rho t} \, dt, \qquad (1)$$

where $\bar{U}^0$ is a constant utility in the good state, ρ is the rate of time preference, $\bar{U}^1$ is the utility in the bad state, c is consumption in period t, α represents the proportion of damages faced by the person, D is the damage function, x are the monetary contributions to invasion control, $\tilde{x}$ represents all contributions to invasion control by other parties, Q^1 is environmental quality in the bad state, τ reflects the invasion time, and T is the time of death.

Assume either ρ or T is sufficiently large such that $e^{-\rho T} \to 0$. To determine the relevant budget constraint, assume the person is endowed with wealth W. He or she confronts a pure-consumption-loans market at interest rate r and cannot die in debt. All capital is consumed in the lifetime, so the choice of consumption path, $c(t)$, and contributions toward lowering market damages, $x(t)$, is constrained by

$$W = \int_0^T [c(t) + x(t)]\, e^{-rt}\, dt.$$

Assume the person receives no utility in the good state from contributions and that contributions cannot be saved into a rainy-day fund to reduce market damages in the bad state. This provides no incentive for the person to contribute to damage protection in the good state and creates a new budget constraint:

$$W = \int_0^\tau c(t) e^{-rt}\, dt + \int_\tau^T [c(t) + x(t)]\, e^{-rt}\, dt$$

Since good state utility is constant and determined by consumption and the unchanging environmental state, utility from consumption in each good state period should be the same (as discounting is multiplicative). Because consumption in the good state is equal over periods in the good state and discounted at the rate of interest, the simplified budget constraint is derived as

$$W = \left[\frac{c(0)}{r}\right] * \left(1 - e^{-r\tau}\right) + \int_\tau^T [c(t) + x(t)] e^{-rt}. \tag{2}$$

The Lagrangian for this problem is

$$\begin{aligned} \underset{c,x}{L} = & \int_\tau^T \left[\bar{U}^1(c(t) - \alpha D(x(t) + \tilde{x}(t)), Q^1) - \bar{U}^0\right] e^{-\rho t}\, dt \\ & + \lambda \left[W - \left[\frac{c(0)}{r}\right] * \left(1 - e^{-r\tau}\right) - \left[\int_\tau^T (c(t) + x(t))\, e^{-rt} dt\right]\right] \end{aligned} \tag{3}$$

Differentiating equation 3 with respect to consumption and control expenditures and letting $M^1 = c(t) - \alpha D\,(x(t) + \tilde{x}(t))$ and $\hat{X} = x + \tilde{x}$ leads to the following first-order conditions:

$$c(t){:}\ \bar{U}^1_{M^1}\left[c\,(t) - \alpha D\,(x\,(t) + \tilde{x}\,(t))\,, Q^1\right] e^{-\rho t} + \lambda\left(-e^{-rt}\right) = 0$$

$$\text{for } \tau \le t \le T \tag{4}$$

$$x(t){:}\ -\alpha \bar{U}^1_{M^1}\left[c\,(t) - \alpha D\,(x\,(t) + \tilde{x}\,(t))\,, Q^1\right] D_{\hat{X}} e^{-\rho t} - \lambda e^{-rt} = 0,\ D_{\hat{X}} \le 0$$

$$\text{for } \tau \le t \le T \tag{5}$$

Simultaneously solving the optimality conditions 4 and 5 to solve for λ gives

$$\lambda = \frac{U^1_{M^1}\left(1 - \alpha D_{\hat{X}}\right) e^{-\rho t}}{2e^{-rt}}. \tag{6}$$

The WTP is found as the value of a change in the time of transition from good to bad states, τ. Indirect utility is a function of W and τ (and the other parameters held at their original levels) and defines (W, τ) indifference curves. Holding utility constant at the original level and viewing wealth, W, as a function of the exogenous probability of the good state, τ, allows the slope of the indifference curve (MRS) to be found as

$$-\frac{dW}{d\tau} = \frac{\partial L/\partial \tau}{\partial L/\partial W}. \tag{7}$$

Applying the envelope theorem to equation 3 allows for calculating the partial derivatives necessary to determine the WTP value, V:

$$V = -\frac{dW}{d\tau} = \frac{\partial L/\partial \tau}{\partial L/\partial W}$$

$$= \frac{-\left[U^1\left(c\left(\tau\right) - \alpha D\left(x\left(\tau\right) + \tilde{x}\left(\tau\right)\right), Q^1\right) - \bar{U}^0\right] e^{-\rho\tau} + \lambda\left[\left(-c(0) + c(\tau) + x(\tau)\right) e^{-r\tau}\right]}{\lambda}$$

Allowing $c(0) = c(\tau)$, holding c and x at their optimum levels (given by equations 4 and 5), and using the dual optimality condition 6 to simplify allows the value at $t = \tau$ to be determined:

$$V = \frac{\left[\bar{U}^0 - \bar{U}^1\left(c\left(\tau\right) - \alpha D\left(x\left(\tau\right) + \tilde{x}\left(\tau\right)\right), Q^1\right)\right] e^{-rt}}{\left[\bar{U}^1_{M^1}(c(\tau) - \alpha D(x(\tau) + \tilde{x}(\tau)), Q^1)^*(1 - \alpha D_{\hat{X}})\right]/2} + x(\tau)e^{-r\tau} \tag{8}$$

Assuming $\bar{U}^0 > \bar{U}^1$ and $D_{\hat{X}} < 0$, V is always positive and equal to the difference in utility from the bad state to the good state over the average of marginal utility and the derivative of the utility function with respect to individual contributions, plus monetary contributions to invasion control at time τ with all terms discounted to time τ. The first term is similar to a value of a statistical life result. The optimal monetary contributions at time $x(\tau)$ are added since the amount can be used to extend the time in the good state instead of the alternative of spending it on controlling damages in the bad state.

We tested the analytical model using a survey designed to elicit subject's WTP to delay the inevitable. The study was conducted at the University of Wyoming and administered in economics courses. The average respondent completed the survey in less than 15 minutes. The instrument initially defined lakes and rivers, the respondent's region, invasive species, impact of invasive species, and impact severity levels (described as either low or high). This introductory section was created to inform the respondents of the terminology used in the valuation questions. The survey used the contingent valuation method of eliciting demand for delaying the inevitable, that is, WTP to delay the impacts of aquatic invasive species in regional lakes and rivers.

Table 8.1 illustrates the invasive species impact chart, which describes the type of impacts that can occur for a specific invasion. Impacts are categorized by lake aesthetics, risks to biodiversity health, risks to human health, economic production, navigation, and recreation. The aquatic species are clustered into four categories: fishes, crustaceans, mollusks, and aquatic plants. The impact chart is provided to subjects to help them understand the potential pros and cons associated with aquatic species. To test for scoping effects, whether respondents consider protection against more impacts more valuable or provide the same WTP regardless of the impacts faced, some respondents received information for only one of the four species categories. This information set up the valuation questions, which were given as three scenarios:

Scenario 1: What is the most you would be willing to pay to keep all lakes and rivers in your region *not invaded* (no impacts) from *all groups* for *one year*? These groups will cause *low* impacts after one year for the foreseeable future.

Scenario 2: Imagine your lakes and rivers have been invaded. What is the most you would be willing to pay to keep all lakes and rivers in your region at *low* impacts from *all groups* for *one year*? These groups will cause HIGH impacts after one year for the foreseeable future.

Scenario 3: Similar to scenario 2 only prevention lasts for *ten years*?

We collected 120 surveys; 106 were included in the statistical analysis. There were 26 completed surveys for fish species invasions and 80 for all species. Since each respondent answered three WTP questions, there were 318 total observations in the data set. Ten surveys were excluded because of missing WTP estimates for one of the three valuation questions. Three additional surveys were excluded because reported WTP exceeded the reported annual household income (see Freeman 1993). One more survey was excluded due to very high reported WTP of \$20,000/year, \$7,500/year, and \$7,500/year for 1 year of protection from low impacts, 1 year of protection from high impacts, and 10 years of protection from high impacts from all species, respectively. While these WTP values were less than reported household income (\$80,000–99,000), they are large outliers that likely greatly overstate actual WTP (hypothetical bias).

All statistics discussed here are based on respondents WTP values per year for the given scenarios.[1] Table 8.2 presents summary statistics for each of the six WTP question responses. Mean WTP/year from fish species was \$57 for 1 year of protection from low impacts, \$70 for 1 year of protection from high impacts,

TABLE 8.1. Impact chart: Types of impacts that will occur if the given group(s) (e.g., fishes) invades.

Invasive species group	Lake aesthetics	Risks to biodiversity health	Risks to human health	Economic production	Navigation	Recreation
Fishes (e.g., round goby, carp)	Reduce lake clarity	Reduce native fish and aquatic plants		Reduce commercial fish		Reduce sport fish
Mollusks (e.g., snails, mussels)	Improve lake clarity	Reduce native mollusks; can kill wildlife that eats them	Cut feet; can make people sick	Clog pipes; reduce filtering; stick to boat hulls; reduce commercial fish	Clog locks, dams, and canals	
Crustaceans (e.g., spiny water flea, rusty crayfish)		Reduce native aquatic animals and plants; cross-breed with native species			Improve by reducing native plants	Stick to fishing lines and nets; reduce sport fish
Aquatic plants (e.g., Eurasian water milfoil)	Lake can look full of weeds	Reduce native aquatic plants	Increase mosquitoes and swimmer's itch	Clog irrigation and water treatment intake pipes	Stick to boat propellers	Stick to fishing lines and nets; reduce sport fish; reduce swimming areas

TABLE 8.2. Summary statistics for WTP question responses.

	All[a]	FZL1Y[b]	FLH1Y[c]	FLH10Y[d]	AZL1Y[e]	ALH1Y[f]	ALHZL10Y[g]
n[h]	318	26	26	26	80	80	80
Mean, $/year	131	57	70	35	108	146	213
Standard deviation, $/year	286	107	111	35	219	289	420
Minimum, $/year	0	0	0	0	0	0	0
Median, $/year	28	18	33	28	28	28	43
Maximum, $/year	2,000	505	505	100	1,000	1,250	2,000

[a] Summary results when all WTP question responses are grouped.
[b] WTP for 1 year (1Y) of protection from low damages (ZL) from fish species (F).
[c] WTP for 1 year (1Y) of protection from high damages (LH) from fish species (F).
[d] WTP per year for 10 years (10Y) of protection from high damages (LH) from fish species (F).
[e] WTP for 1 year (1Y) of protection from low damages (ZL) from all species (A).
[f] WTP for 1 year (1Y) of protection from high damages (LH) from all species (A).
[g] WTP per year for 10 years (10Y) of protection from high damages (LH) from all species (A).
[h] Number of WTP question responses used in the statistical analysis.

and $35/year for 10 years of protection from high impacts (since this is a per-year measure, the mean WTP for all 10 years of high impacts is $350). Mean WTP per year from all species was $108 for 1 year of protection from low impacts, $146 for 1 year of protection from high impacts, and $213/year for 10 years of protection from high impacts (indicating a mean WTP of $2,130 for 10 years). We see a large heterogeneity across people and their WTP to delay the inevitable.

The WTP numbers seem reasonable when compared to estimates in Nunes and van den Bergh (2004). Their estimates of recreational costs (travel costs) for beach closure due to harmful algal blooms result in average values of approximately $45/year. In addition, they implement a contingent valuation survey to test for non-market benefits of a ballast water monitoring and treatment program. They found average values of these nonmarket benefits (associated with beach recreation, human health, and marine ecosystem impacts) of approximately $62/year to prevent these invasions. Their total of about $107/annually is matched to the total here of $57/year for fish species protection from low impacts for 1 year (similar to preventing this invasion by continuing to have no impacts to regional lakes and rivers) when one considers their surveys were administered geographically close to the beach resort where the invasion will occur (making it more likely that respondents will have recreational values).

We used these survey data to explore the determinates of WTP to delay the inevitable. The model we chose was a multiplicative heteroskedastic ordinary least squares (OLS) estimator:

$$
\begin{aligned}
&WTP/Yr \\
&\quad = \beta_1 + \beta_2 FishZL1Y + \beta_3 FishLH1Y + \beta_4 FishLH10 + \beta_5 AllLH1Y \\
&\qquad + \beta_6 AllLH10 + \beta_7 FishZL1YIncH + \beta_8 FishLH1YIncH + \beta_9 FishLH10IncH \\
&\qquad + \beta_{10} AllZL1YIncH + \beta_{11} AllLH1YIncH + \beta_{12} AllLH10IncH + \varepsilon(i,t)
\end{aligned}
$$

The dependent variable, *WTP/Yr*, is WTP per year. Respondents were asked their WTP per year for either invasive protection from fish species or all species groups. The *Fish* prefix classifies a dummy variable for observations from fish species protection, while *All* indicates observations with all species groups invading. The two severity levels are represented as binary variables: *ZL*, defined as impacts starting at zero and then increasing to low intensity, and *LH*, defined as impacts starting at low then increasing to high intensity. There are two time frames for the length of invasion delay, 1 year ($1Y$) and 10 years (10). Finally, respondents were divided into two income categories, and *IncH* classifies observations corresponding to people with high household incomes (defined as $45,000 or more for the previous year). For example, $FishZL1Y$ is an observation in which a person was asked his or here WTP for protection from fish species creating low impacts (from zero) for 1 year. Similarly, $FishZL1YIncH$ is an observation with the same WTP question but from a person with a reported high household income.

Each parameter of the model describes a key relationship. The constant captures WTP in the low-income category for all species groups creating low impacts for 1 year. It is expected to be positive since delaying impacts to environmental quality presumably has some value (such that a known invasion time suggests some consequence). This is the baseline case and provides the comparative benchmark for all other coefficients in the dummy variable expressions. In what follows, we discuss predictions over signs of coefficients and make conclusions based on the relationship between the variable and the benchmark.

$FishZL1Y$ is for a similar WTP question as the constant but regarding protection from fish species only. Since invasive fish have fewer types of impacts (than the baseline of all species), this coefficient is predicted to be negative. The sign of the coefficient on $FishLH1Y$ is difficult to predict. Because it reflects just fish, the value is expected to be lower than for all species (as captured by the constant). But the coefficient also incorporates a change in quality, from low to high impacts. The trade-off is whether people place a higher value on keeping lakes and rivers unspoiled (zero to low impacts) or prefer protecting them from significant damages (low to high). It is expected that the value of keeping areas pristine is higher than the value of protecting against increased impacts of damaged goods, reflected in a negative coefficient when compared to the benchmark case.

The severity of impacts assists in predicting the signs of the rest of the low-income variables: $FishLH10$, $AllLH1Y$, and $AllLH10$. One difference is that two of these variables capture WTP per year for a 10-year delay versus 1 year. If participants discount the future, it is likely, on a per-year basis, that WTP will be higher in a 1-year

TABLE 8.3. Multiplicative heteroskedastic OLS estimates.

Variable	Prediction	Coefficient	*p*-Value
Constant	+	57.01	0.000
*Fish*ZL1Y	−	−10.63	0.563
*Fish*LH1Y	−	1.74	0.935
*Fish*LH10	−	−21.38	0.182
*All*LH1Y*r*	−	10.17	0.580
*All*LH10Y	−	67.75	0.027
FZL*Inc*H	?	46.96	0.588
FLHI*nc*H	?	48.75	0.557
FLHTI*nc*H	?	−3.88	0.820
*All*ZL1YI*nc*H	+	145.85	0.008
*All*LH1YI*nc*H	+	226.57	0.027
*All*LH10I*nc*H	+	252.03	0.042

Note: $n = 318$; adjusted $R^2 = 0.11$.

scenario. All three of these variables are predicted to be negative in comparison to the benchmark.

The final six variables (*FishZL*1*YIncH*, *FishLH*1*YIncH*, *FishLH*10*IncH*, *AllZL*1*YIncH*, *AllLH*1*YIncH*, *AllLH*10*IncH*) are observations from high-income subjects. It is expected that delaying impacts is a normal service, so their coefficients should be greater than their corresponding low-income coefficients. Since income was divided into only two groups, it seems reasonable to think that these people may have significantly larger WTP. These variables are expected to be positive compared to the benchmark.

Table 8.3 shows the results of the multiplicative heteroscedastic estimation model.[2] The regression model is significant at the 1% level. The constant is positive and significant, as predicted. It is clear that the baseline case influenced the statistical results through the constant, which is the mean WTP for all species/low income/1-year delay. Notice it is statistically different than zero. The rest of the coefficients must be considered in relation to this result. For example, the coefficient for *FishZL*1*Y* (−10.63) suggests that the WTP is less than that for the baseline. The *p*-value tells us it is not statistically different than the baseline, and further tests confirmed that the WTP for *FishZL1Y* is statistically positive. Tests indicated 10 of 12 individual coefficients to be significantly different than zero.

Two of the three low-income fish species terms are negative (as expected), but all three are insignificantly different than the base case. The low-income all-species terms are positive (opposite of expected) but, again, are insignificantly different from the benchmark. The high-income terms were predicted to be positive; the fish species terms are insignificant, while the all-species terms are positive and significantly different than the baseline case of all species/low income/1-year delay.

Two key results emerge based on the regression analysis. First, we find that delaying impacts is valuable to our sampled population. We reject the null hypothesis at the 1% significance level that WTP is zero, and WTP is positive. To study the results in greater detail, we tested each of the three scenarios. Joint tests incorporating both income groups (low, high) for each of the three scenarios (WTP per year for delaying low damages for 1 year, delaying high damages for 1 year, and delaying high damages for 10 years) are significant at the 1% level. Null hypotheses suggesting delaying impacts are not valuable are all rejected. This result supports the notion that delaying environmental degradation is valuable, as anticipated from the terms of WTP in the analytical model.

Second, since validity is a common concern with stated preference surveys and hypothetical scenarios, we conducted more tests on the validity of the WTP estimates. Overall, we found that the relative characteristics of the WTP estimates are consistent with economic theory. We reject the null hypothesis at the 5% significance level that total WTP for 10 years of protection from high impacts is no more than the WTP for 1 year of protection from high impacts. This result indicates that respondents are not saturated in invasive species protection after 1 year; they value and are willing to pay for addition protection past 1 year.

We also reject the null hypothesis that protection from impacts from all species groups is no more valuable than protection from impacts from invasive fishes only (rejected at the 1% significance level). Since all-species invasions are defined as having a larger number of impacts than fish species invasions, the rejection of this hypothesis suggests that respondents considered these additional impacts in their valuation and were not just giving WTP based on general preferences for environmental concerns (i.e., scoping effects/surrogate bidding; see, e.g., Diamond et al. 1993). Finally, we reject the hypothesis that subjects classified as having high incomes have no greater WTP for invasive species protection (rejected at the 5% significance level).

In summary, we developed a dynamic framework to address how people value the inevitable risks to environmental quality posed by invasive species. We tested the theory by developing a survey instrument that measured whether people who experienced utility loss from environmental degradation have a positive WTP for the temporary elimination of risk and damages. Consistent with our model, we determined from the pilot survey results that delaying impacts to regional lakes and rivers is valuable.

Policy makers and resource managers may be hesitant to make investments when the project is destined to fail. Yet our survey results indicate that expenditures to postpone market and nonmarket impacts may be justifiable. While longer delays were more valuable, even 1-year gains lead to positive WTP (on average). In terms of public policy toward invasive species, these results suggest that it may be reasonable to fight what is ultimately a losing battle.

A practical extension to our pilot study is to test our results with a larger and more diverse survey population. We enlisted the help of the Wyoming Survey and Analysis Center to simplify the language and format of the survey, conduct focus groups, and perform a nationwide mailing. This will help us determine if the results of our pilot study are applicable nationally and allow us to extend our analysis to consider other factors (e.g., spatial differences) that may affect the value of delaying the inevitable.

We now consider a second valuation application, one that does address a national audience but provides a more complicated valuation task.

APPLICATION 2: YELLOWSTONE RATIONALITY SPILLOVER

Many bioeconomic models have economic parameters that can be defined based on market data (e.g., the cost of fishing vessels, relative prices of fish). The strength of using market data to parameterize a model is that it is based on choices and decisions of people in active exchange institutions. These prices really exist—they are not just conjured up in people's minds—and they serve to force people to make real choices given binding budget constraints.

But bioeconomic models also have parameters that cannot be valued using market data, and require surveys as we discussed in application 1. These surveys have been criticized, however, because they are based on hypothetical choices outside the marketplace. This complaint rests on the idea that economic value is based on the idea that people can make "rational" choices. Here, rationality usually means "consistent" choices and values—people are assumed to make consistent choices in economic representations of choice and value. A rational consumer has experience in markets, has experience with the available bundles of goods, and has clearly defined preferences over those bundles. Such rationality was presumed for our first application—people valuing a delay in the inevitable will value it consistently.

But this rationality has been called into question in the literature for a variety of reasons, one being preference reversals—the inconsistent choices consumers make when choosing preference between two bundles of goods and then being asked for their WTP for the two bundles (Grether and Plott 1979). A preference reversal occurs when a person says he prefers one apple to an orange but then puts a greater dollar value on the orange—if he prefers the apple, he should be willing to pay more for the apple. The phenomenon of preference reversals shows a potential failure in economic theory. Constraints on individuals' cognition and humankind's physiological limits on cognition play a role in the ultimate ability to process information (see Simon 1955; Heiner 1983). But the limitations that reduced cognition puts on the types of rationality assumed in economics may or may not be a binding constraint. It is possible that economic agents could learn from market experience and that inexperience, rather than cognition, is the limiting factor. Do these reversals lend credence to the theory that consumers are not rational? Can we achieve rationality in this context, or are preference reversals persistent?

The observed differences between assumed optimizing behavior in economic theory and the actual behavior of people can be affected through repeated market transactions with large enough sums of money at stake for the individual (see Smith 1989; Shogren 2006). Recent research has shown that rationality can be increased in situations in which preference reversals are prevalent (Cherry et al. 2003). If preference reversals are a sign of irrationality, if rationality can be learned through market experience, and if rationality *can* spill over from one market to another, it is the experience in the market that makes consumers rational. Market experience can then be used as a tool to help consumers make more rational decisions. While the work of Cherry et al. (2003) has shown that rationality spillovers exist in lab experiments, these lab experiments may need to be applied to specific problems. The lab experiments have shown that rational behavior can spill over from one market to another, from a market context to a hypothetical context with a low-probability, high-severity event, such as an environmental good. Applying this method to a specific problem requires that we take the theory from the lab to the field and target people who are interested in these nonmarketed goods. If the interactive survey can be used in the field to elicit preferences for and values of bundles of environmental goods, not only can we elicit values, but also those values will come from a more rational consumer—the values may well be closer to the consumer's true WTP.

This research is an attempt to take the rationality spillover design from the lab into the field. We wish to determine preferences for and values of seeing species in and around Yellowstone Lake that might be affected from the introduction of an exotic species into Yellowstone Lake: lake trout. Lake trout are an exotic species to Yellowstone Lake and are a predator of the native and popular species, cutthroat trout. Cutthroat trout not only are important to anglers who come to Yellowstone Lake to fish for cutthroat, but also are an important food source for grizzly bears, osprey, white pelicans, river otter, and many other species in Yellowstone National Park. Not only are cutthroat trout expected to decline in number, but these other species relying on cutthroat trout for food may decline, as well.

We combined integrated models and valuation to measure preferences for reduced risks to the native species cutthroat trout in Yellowstone National Park (for complete details, see Settle and Shogren 2006). The survey was designed to explore the degree to which people are concerned about protecting a native species such as cutthroat trout within the park against threats from invasive species such as lake trout. The goal was to apply into the field the insight from the past decade of laboratory valuation work and to use our empirical results to parameterize the composite visitor's demand within an integrated bioeconomic model. Using a seven-step experimental strategy, we created a Yellowstone interactive survey. (See appendix 8.1 for an example of the survey instrument.)

The first step is to develop wildlife lotteries to which people can assign an economic value.[3] These wildlife lotteries represent the probability that a visitor experiences a species and the core attractions. Due to the large number of possible permutations for the set of wildlife lotteries, we limited our probability distributions of species to those most important to parameterize the integrated bioeconomic model. The 90 lottery pairs capture the reality of most environmental policy by defining many of our wildlife lotteries as low-probability/high-outcome lotteries. The second step is to use marketlike arbitrage as a disciplining device in valuation. Asking people to value low-probability/high-outcome lotteries introduces the possibility that people might not act as rationally as expected utility theory presumes. The fear was that people could have inconsistent preferences over the wildlife lotteries and therefore state inconsistent values. Rational valuation is defined by consistency between valuation and preference. One solution is to introduce arbitrage into the experimental design (see Cherry et al. 2003). Over 10 rounds, each participant was presented with two lottery markets side by side. One market is a real market; people played money lotteries for actual cash. For these real lotteries, we also used a low probability of a high payout and a high probability of a low payout. The second market is the hypothetical wildlife lotteries; people play these lotteries but do not get paid in cash.

The third step is to construct the actual interactive valuation survey questions such that they match up with the integrated model. The instructions explained each stage in the survey: Each person is presented two separate situations, each with two options (options A and B). They say which option they prefer in each situation (A or B), and then they state a dollar value for the two options in both situations. They were aware that a computer market may buy, sell, or trade with them in situation 1 according to indicated preferences and values. At the end of a round, the next round appears with a new initial money balance in situation 1 and new options for both situations. Once they completed the last round, we determined their earnings for the survey.

The fourth and fifth steps were the actual implementation of the survey inside the park and over the Internet. In the park sessions, we set up tables outside the Visitors Center at the south entrance to Yellowstone and Teton National Parks. Visitors were asked if they would be willing to participate in an experiment taking about 30 minutes on laptop computers set up on the tables. We ran the survey over the Internet and attracted participants through a paid advertisement run over 5 weeks on the *New York Times* website (www.nytimes.com). Our sample focuses on people already interested in the park, which further suggests that our valuation estimates reflect an upper bound on the composite person's value for cutthroat.

The sixth step is to evaluate the valuation statements. For tractability, assume the average respondent's WTP function is separable in each argument—e.g., the value of catching a cutthroat trout does not depend on whether you have seen the core attractions of the park. We use this assumption to back out the value of each probability from a large number of different types of lottery pairs. Using this method, we estimated values for each of the species in the park.

The final step is to add the estimated values as parameters in the integrated model. We next determined the value for each of the probabilities of seeing/catching each species and used these estimates to parameterize the value to see/catch each species in our composite visitor's welfare function. If a visitor valued a cutthroat trout at $4 and had a 25% chance of catching a cutthroat trout, the visitor valued that 25% chance in the lottery at $1. The values for each species were included in our welfare function used in the simulations to measure the welfare to visitors of visiting Yellowstone National Park.

We implemented the experiment over the Internet by recruiting participants from two sources—newsgroups from the Internet and the *New York Times*—with the goal to broaden our sample of people. The first source was done by posting announcements to environmental newsgroups on the Internet. The number of participants was too low from the newsgroups, and more data needed to be collected, which led to the second source, an advertisement. We ran a paid ad on the *New York Times* website to gather participants. Approximately 250 people were gathered from the *New York*

Times website, as compared with fewer than 20 from the newsgroups. The hit rate from the banner advertisement was well above the industry standard of 0.5%, nearly doubling that to 0.9%. Possible explanations for the high hit rate were the $20 that was average earnings for participation and the cleverness of the banner ad. The campaign ran from June 7 to July 6, 2000, at various locations on the *New York Times* website.

A total of 269 people completed the Internet experiment. Of these 269, 82 actually visited Yellowstone National Park at least once before. The first important distinction to make is to differentiate anglers from all other visitors. While about 1% (about 25,000 out of approximately 3 million total annual visitors to Yellowstone National Park) of the people who visit Yellowstone National Park go to the park to fish, 5.6% of all participants in the experiment classified themselves as anglers. If we consider only participants who actually visited Yellowstone National Park, the percentage of these participants who classified themselves as anglers jumps to 17.1%.

For the 269 participants, the data contained information on valuation, composite probabilities of seeing or catching the various species (including the park attractions), individual characteristics, and a unique participant identification number. Eighty-two of 269 had actually visited Yellowstone National Park. Valuations were collected for both lotteries placed before the participant (both lotteries A and B) over 10 rounds for each participant. We used 10 rounds per participant, 269 participants, and 2 lotteries per round, to obtain the total of 5,380 observations.

We used a two-way fixed-effects model to account for time effects and participant effects. Time-specific fixed effects account for the impact of arbitrage across rounds. Participants who exhibited irrational behavior lost money as a direct result of their irrational behavior. In both the Cherry et al. (2003) lab experiment and our survey, participants learned to stop being inconsistent—but rather than changing preferences, they instead began reducing their stated valuations of low-probability events. Preferences for the lotteries were fixed, but the stated valuations fell with arbitrage. We use player-specific fixed effects to determine if certain players have a higher value for all environmental goods examined in this experiment.

The regression equation is $Value_i = \alpha + \beta_1 X_{1i} + \beta_2 X_{2i} + \beta_{3i} X_{3i} + u_i$, where X_{1i} are the dummy variables for individual effects, X_{2i} are the different dummy variables for time effects (a dummy for each of the rounds of the experiment) with the dummy variable for the first round being excluded, and X_{3i} are the variables for valuation—percent chance of seeing or catching each species (cutthroat trout, lake trout, grizzly bear, bird of prey, core attractions) and the square root of the percent chance of seeing each species.

Table 8.4 shows the results of the two-way fixed effects model. The coefficients are jointly statistically significant even if the individual parameters are not. The percent chance of seeing each species is included to capture the notion of diminishing marginal utility. Diminishing marginal utility is usually captured by squaring the variable. But here, since the probability of seeing each species is bounded between 0 and 1, the probability needs to be a square root to allow for diminishing marginal utility. Also, the information from the regression shows that the most significant

TABLE 8.4. Regression results from the Yellowstone experiment.

Variable	Parameter estimate	*t*-Value	Probability
Time 2	−0.062	−0.108	0.914
Time 3	−0.203	−0.354	0.724
Time 4	0.068	0.118	0.906
Time 5	−0.019	−0.333	0.739
Time 6	−0.167	−0.290	0.772
Time 7	−0.062	−0.109	0.913
Time 8	0.406	0.707	0.479
Time 9	0.640	1.116	0.264
Time 10	−0.564	−0.983	0.326
ct	−2.951	−1.138	0.255
*ct*2	1.948	1.211	0.226
lt	−0.506	−0.200	0.841
*lt*2	−0.057	−0.036	0.971
b	0.879	0.343	0.732
*b*2	0.060	0.038	0.970
g	4.767	1.914	0.056
*g*2	−2.090	−1.328	0.184
ca	5.614	1.443	0.149
*ca*2	−3.459	−1.752	0.080
Constant	4.169	7.900	0.000

Abbreviations: *time n* = round of play $n = 2, \ldots, 10$; *ct*, cutthroat trout; *lt*, lake trout; *b*, birds of prey; *g*, grizzly bears; *ca*, core attractions. $N = 5{,}380$, $R^2 = 0.485$, $F = 0.90$. Note the F-value has already netted out the player-specific fixed effects. The F-value for the player-specific fixed effects is 17.815.

component of determining a person's WTP for a visit to Yellowstone is not a slight change in the probability of seeing species or catching trout in the lake, but rather who is the individual person, since some people value a basket of goods highly when compared to other people regardless of the percentage chance of seeing species or catching species in the park. That is, the parameter capturing individual fixed effects is the most significant component (some individuals have a higher value for any individual basket than do other people).

The valuation estimates show that the composite visitor has a positive value for seeing a cutthroat trout, seeing a bird of prey, and seeing the core attractions of the park. Viewing grizzly bears is borderline, and catching a lake trout is an economic bad. The composite visitor needs to be compensated for catching a lake trout, which matches our initial expectations.

The final simulation results from the combination of bioeconomic integration and valuation were interesting. While integrating economics and biology was worth the effort for predicting physical changes, we found it did not matter for welfare estimates. We estimate a trivial difference between the present value of net benefits between the best- and worst-case scenarios. The average person surveyed cared more about improving road quality than protecting cutthroat trout. Their bumper sticker would read "fix the roads, forget the fish." This suggests that if park managers want to justify cutthroat trout protection based on visitor preferences, they would benefit from better educational tools to close the gap between today's average visitor and the farsighted ecotourist. Regardless, the point is that the valuation exercise was guided by the demands of an integrated model, and the structure of the integrated model was affected by what needed to be valued. Future work exploring this second level of integration seems most worthwhile.

CONCLUDING REMARKS

Protecting human and ecosystem services from invasive species generates both costs and benefits to society. Understanding how to estimate these costs and benefits in a consistent manner is essential so that decision makers can compare policy options to achieve more protection at less cost. Nonmarket valuation is the basic tool economists use to elicit the value associated with controlling exotic invasives. There are many examples of nonmarket valuation in the economics literature for many forms of environmental protection and ecosystem services. Less work, however, has been devoted to understanding how people value reduced risks posed by invasive species. This is a significant gap in the literature given the potential importance that invasive species could have on the world's ecological and economic systems.

This chapter reviews two of our applications on how economists define economic value from invasive species protection. First, we consider the economic value of delaying inevitable environmental damage due to aquatic invasive species, a problem especially relevant to tropical and subtropical regions. We developed an analytical framework and tested it using a stated preference survey. Results suggest that delaying impacts is valuable to people. Other tests reveal characteristics of the WTP estimates are consistent with economic theory, suggesting that people are making consistent choices when thinking about how an invasive species might affect them.

Second, we consider how to use valuation methods to measure preferences for risks to cutthroat trout in Yellowstone Lake. Valuing natural resources within Yellowstone National Park sets a visible target given that Yellowstone is the world's first national park and is considered the crown jewel of the National Park System. If people were unconcerned about the risks that invasive species posed on unambiguous resources such as cutthroat trout within the park, they are probably less likely to value them outside the park. Our results here suggest that the valuation estimates for park visitors were positive and provides an upper bound on the average person's value for cutthroat trout survival.

APPENDIX 8.1. INTERNET YELLOWSTONE EXPERIMENT

The design of the experiment follows seven steps. The following descriptions are taken directly from the experiment to show how each of the seven steps were explained.

Step 1: Options and Situations

Options

Options are uncertain outcomes. Each option has an outcome and the likelihood of realizing that outcome. You will face two types of options: *money* and *wildlife*.

- *Money option*: An example of a money option is having a 50% chance of winning \$10 and a 50% chance of losing \$5.
- *Wildlife option*: An example of a wildlife option is having a 50% chance of seeing a grizzly bear.

Situations

There are two situations: 1 and 2. The screen is split down the middle with Situation 1 on the left side of the screen and Situation 2 on the right side of the screen. Though the two situations are presented together, they are separate. In both situations, you are presented two options. Thus, the layout of the options and situations will be as shown in table A8.1.

Situation 1. The two options in situation 1 are money options. For the example in table A8.2, option A1 has an 86% chance of winning \$5.25 and a 14% chance of losing \$1.25. Option B1 has a 28% chance of winning \$9.00 and a 72% of losing \$1.75.

Situation 2. The two options in situation 2 are wildlife options. The wildlife options will contain one or more of the animals in table A8.3.

Wildlife Categories and Symbols

Reference Sheet. A reference sheet to help you remember what the wildlife abbreviations represent is available on your browser. When you are running the survey, a link to this reference sheet is located just below the survey.

Number of Sightings and/or Catches. We show the number of each wildlife sighting or catch in an option by the number listed prior to the wildlife symbol. For

TABLE A8.1. Internet Yellowstone experiment situations.

Situation 1	Situation 2
Option A1	Option A2
Option B1	Option B2

TABLE A8.2. Options for situation 1 in the Yellowstone experiment.

Option	Outcome
A1	86% 5.25, 14% – 1.25
B1	28% 9.00, 72% – 1.75

TABLE A8.3. Wildlife categories and symbols for the Yellowstone experiment.

Wildlife categories and symbols	Description
Birds of prey (B)	Including birds of prey means that you may see an osprey or a pelican, which prey on trout
Cutthroat trout (CT)	Including cutthroat trout means that you may catch a cutthroat trout, which are the native trout species in Yellowstone Lake
Lake trout (LT)	Including lake trout means that you may catch a lake trout, which are nonnative to Yellowstone Lake
Grizzly bears (G)	Including grizzly bear means that you may see a grizzly bear without harm
Core attractions (CA)	Including core attractions means that you may see "roadside" sites, such as Old Faithful or Yellowstone Falls

TABLE A8.4. Examples of lotteries for the Yellowstone experiment.

Option	Example 1	Example 2
A2	80% 1CT	86% 1CA
B2	39% 1B, 1CA	28% 1G, 2B

example, "2G" means that you may see two grizzly bears, and "3CT, 1LT" means that you may catch three cutthroat trout and one lake trout.

Options in Situation 2

Table A8.4 gives an example of how the options are displayed. In example 1, option A2 has an 80% chance of catching one cutthroat trout (1CT) and option B2 has a 39% chance of seeing one bird of prey and the core attractions (1B, 1CA). In example 2, option A2 has an 86% chance of seeing the core attractions (1CA), and option B2 has a 28% chance of seeing one grizzly bear and two birds of prey (1G, 2B).

The screen will be split down the middle. Situation 1 is on the left; situation 2 is on the right. The two situations are presented simultaneously but are separate. The presentation of both situations looks like that shown in table A8.5.

Beginning Money Balance

For Situation 1. You are given an amount of money. You may use part or all of your money to buy options. If you purchase an option, your money balance will decrease by the amount of the purchase price, and you will own the option. Once you own an option, it may be bought from you—increasing your money balance by the selling price. You keep any unspent portion of your beginning balance.

For Situation 2. You are not given money, and no buying or selling will take place in situation 2. Recall, situation 1 and 2 are separate.

TABLE A8.5. Situations 1 and 2 for the Yellowstone experiment.

Situation 1	Situation 2
A1	A2
88%	84%
5.00	1CT, 2LT
12%	
−1.25	
B1	B2
25%	35%
9.00	1B, 1CA, 1G
75%	
−1.75	

Step 2: Binding Contracts for Choices

Indicating Your Preference

You are first asked to state which option you prefer in each situation.

Beginning with situation 1, you are asked "*which option would you prefer, A1 or B1?*" In other words, if you had to choose one of the two options to face, which would it be? Indicate the option you prefer by typing either "A1" or "B1."

By typing in your preferred option, you agree to a binding contract. For example, suppose you indicate that you prefer option A1. If you are holding option B1, you may have to trade the option you hold (B1) for the option you prefer (A1)—accept A1 and give up B1.

This will be repeated for situation 2, where you will indicate which option you prefer—A2 or B2. Indicate the option you prefer by typing either "A2" or "B2." In situation 2, you are not agreeing to a binding contract by typing in your preferences.

Step 3: Binding Contracts for Values

Indicating Your Values

Next, you are asked how much you value each option. Your valuation of each option should be the dollar and cent amount that you would be willing to buy or sell the option, in other words, how much you are willing to pay to face the option OR how much you are willing to receive to give up facing the option.

Beginning in situation 1, you are asked "*What is your value for option A1?*" Indicate your value by typing the amount using dollars and cents (e.g., 6.29). Then you are asked "*What is your value for option B1?*" Again, indicate your value by typing the amount using dollars and cents.

By entering a dollar amount, you agree to a binding contract. For example, suppose you indicate that your value for option A1 is $X.XX. The computer market may buy or sell option A1 for $X.XX.

This will be repeated for situation 2, in which you will be asked your value for options A2 and B2. Indicate your value of facing the wildlife option by typing the amount using dollars and cents. In situation 2, you are not agreeing to a binding contract by typing in a dollar amount.

Step 4: Buying, Selling, and Trading Options

After you indicate (1) which option you prefer in each situation and (2) your value for each option, the computer market may buy, sell, and/or trade options with you in situation 1 according to your binding contracts.

For example, suppose you agreed to the following binding contracts for situation 1:

- You preferred option A1 to B1
- Your value for option A1 is $X.XX
- Your value for option B1 is $Y.YY

The computer market may do one or more of the following:

- Buy or sell option B1 for your value of $Y.YY
- Buy or sell option A1 for your value of $X.XX
- Trade option A1 for option B1 (give up A1 for your B1)

The computer market will act only according to your binding contracts. Thus, the computer market only buys and sells at your indicated values, and only trades according to your preferences.

The computer market will buy, sell, and/or trade with you only if it is beneficial. If your selling price is too high, the market will not buy from you. If your buying price is too low, the market will not sell to you.

The computer market will only buy, sell, and/or trade in situation 1. There is no buying, selling, and trading in situation 2.

Your Holdings of Options

If the computer decides to sell you an option, the transaction will be indicated on the screen. For example, suppose the market sells you option A1 for $4.00. The following message will appear: "*The market sells you option A1 for $4.00.*" Also, the screen will indicate your money balance decreased by the purchase price to $6.00 and that you now hold option A. This is shown in table A8.6.

There are no transactions, and thus no balance or holding, in situation 2.

Your Best Strategy

It is in your best interest to be accurate in your *preferences* and *values*; that is, the best thing you can do is to be honest.

If your preferences are incorrect, you are passing up opportunities that you prefer and face situations that you do not prefer. Suppose you prefer option A1 to B1, but you indicate that you prefer option B1 over option A1; the computer market may trade A1 for B1 even though you would rather keep A1.

If your values are incorrect, you are passing up opportunities that you prefer. For example, if you overstate your value for an option, the computer market may

TABLE A8.6. Situations 1 and 2 for the Yellowstone experiment, after one transaction.

Situation 1	Situation 2
A1	A2
85%	75%
5.25	1CT, 2LT
15%	
−1.25	
B1	B2
35%	22%
9.00	1B, 1CA, 1G
65%	
−1.75	
Balance: −4.00, Holding A	

sell you the option for the overstated price. Thus, you may pay a price more than you wish. If you understate your value for an option, the computer market may buy the option for the understated price. Thus, you may sell for a price lower than you wish. Again, the best strategy is to be honest and accurate in your preferences and valuations.

Step 5: Playing the Options

After any transactions in situation 1, you may or may not be holding an option in situation 1. If you hold an option, the outcome will be determined at this time. The outcome will be determined by a random draw.

For example, suppose you hold the option "78%, 4.75; 22%, –1.00."

The random draw will determine the outcome where you have a 78% chance of winning $4.75 and a 22% chance of losing $1.00.

Step 6: End of Round

Ending Money Balance

After the outcomes of the options held in situation 1 are determined from the random draws, your money balance will be adjusted by the amount of your earnings or losses.

Thus, your ending balance will be

- The beginning balance,
- *Less* the amount spent on buying options,
- *Plus* the amount received from selling options,
- *Plus* the earnings received from any winning option,
- *Less* the losses incurred from any losing option.

The ending balance for situation 1 is presented on the screen. This concludes the first round of the experiment. Additional rounds will follow. The subsequent rounds will be identical, but will have a new beginning balance in situation 1 and different options in both situations. There will be 10 rounds.

Step 7: End of Survey

Total Earning from Survey
Your total earnings will be the sum of the ending balances for all rounds.

Summary

1. You are presented two separate situations simultaneously (situation 1 and 2). Each situation will present two options (options A and B).
2. You first indicate which option you prefer in each situation (A or B).
3. You then indicate your value for the two options in both situations.
4. The computer market may buy, sell, or trade with you in situation 1 according to your indicated preferences and values.
5. The outcome of any option you may own in situation 1 is determined with your money balance being adjusted according to your winnings or losses.
6. End of round, next round will repeat with a new initial money balance in situation 1 and new options for both situations.

After completion of the final round, your balance from each round will be totaled and converted to real cash—this is your earnings for the survey.

Acknowledgments This chapter draws on material from McIntosh et al. (2007) and Settle et al. (2008).

Notes

1. The conversion of total WTP in each scenario to WTP per year affects only scenario 3, in which households gave upfront WTP values for 10 years. Total WTP for 10 years was reported as a per-year measure to simplify the entire results discussion.
2. Given that individuals answered three WTP questions each, it would seem that fixed or random effects would be more appropriate to account for individual effects. Using fixed effects, however, does not allow for independent variables that do not vary across individual observations (income), and models without income were not significant. After including the income terms, there was no longer a significant random error component, which implies that OLS is not rejected. After checking plots of OLS residuals against the independent variables, it appeared that several of the interaction dummy terms were candidates for group heteroskedasticity. This problem can be addressed using a general heteroskedastic form or specifying the responsible variables. Since the plots indicated using a group specification would be reasonable, we used Harvey's multiplicative heteroskedastic model to produce the OLS results (maximum likelihood estimation results were very similar). White's (1980) adjusted estimator allows for a more general form and yielded very similar results. Other variables were included and determined insignificant, including lake and

river visits, ex ante familiarity of invasive species, age, sex, race, marital status, earned college or technical school degree, membership to an environmental organization, survey clarity variables, and survey treatment dummies (three versions of survey were given). Given that the subjects were students in economics classes, there was little variation in many of these variables. We intentionally excluded these variables to increase the model's significance and keep the number of independent variables to a reasonable total given the number of observations.

3. The lotteries were designed to be probabilistic in nature just as a visit to a national park is a lottery. No one visitor is guaranteed to see or catch any one species; rather, the visitor has a chance to see or chance to catch various species in the park each visit.

References

Adamowicz, W., and J. R. Deshazo. 2006. Frontiers in stated preferences methods: an introduction. Environmental and Resource Economics 34:1–6.

Bateman, I. J., R. T. Carson, B. Day, M. Hanemann, N. Hanley, T. Hett, M. Jones-Lee, G. Loomes, S. Mourato, E. Özdemiroglu, D. W. Pearce, R. Sugden, and J. Swanson. 2002. Economic valuation with stated preference techniques: a manual. Edward Elgar, Cheltenham, UK.

Bockstael, N., A. M. Freeman, R. Kopp, P. Portney, and V. K. Smith. 2000. On measuring economic values for nature. Environmental Science and Technology 34:1384–1389.

Born, W., F. Rauschmayer, and I. Brauer. 2005. Economic evaluation of biological invasions—a survey. Ecological Economics 55:321–336.

Champ, P., K. J. Boyle, and T. C. Brown, editors, 2003. A primer on nonmarket valuation. Kluwer/Springer, Berlin.

Cherry, T., T. Crocker, and J. Shogren. 2003. Rationality spillovers. Journal of Environmental Economics and Management 45:63–84.

Costanza, R., R. d'Arge, R. de Groot, S. Fraber, M. Grasso, B. Hannon, K. Limburg, S. Naeem, R. O'Neill, and J. Paruelo. 1997. The value of the world's ecosystem services and natural capital. Nature 387:253–260.

Cummings, R. G., and L. O. Taylor. 1999. Unbiased value estimates for environmental goods: a cheap talk design for the contingent valuation method. American Economic Review 89:649–665.

Diamond, P. A., J. A. Hausman, G. K. Leonard, and M. A. Denning. 1993. Does contingent valuation measure preferences? Experimental evidence. Pages 41–85 *in* J. A. Hausman, editor. Contingent valuation: a critical assessment. North-Holland, Amsterdam.

Ehmke, M. D., J. L. Lusk, J.A. List. 2008. Is hypothetical bias a universal phenomenon? A multinational investigation. Land Economics 84: 489–500.

Freeman, A. M., III. 1993. The measurement of environmental and resource values: theory and methods. Resources for the Future Press, Washington, DC.

Grether, D. M., and C. R. Plott. 1979. Economic theory of choice and the preference reversal phenomenon. American Economic Review 69:623–638.

Hanley, N., J. Shogren, and B. White. 2007. Environmental economics in theory and practice, 2nd edition. Palgrave Macmillan, New York.

Heiner, R. 1983. The origins of predictable behavior. American Economic Review 73:560–595.

McIntosh, C., J. Shogren, and D. Finnoff 2007. Invasive species and delaying the inevitable: results from a pilot valuation experiment. Journal of Agricultural and Applied Economics 36:81–93.

Murphy, J. J., P. G. Allen, T. H. Stevens, and D. Weatherhead. 2005. A meta-analysis of hypothetical bias in stated preference valuation. Environmental and Resource Economics 30:313–325.

Nunes, P., and J. van den Bergh. 2004. Can people value protection against invasive marine species? Evidence from a joint TE-CV survey in the Netherlands. Environmental and Resource Economics 28:517–532.

Pimentel, D., L. Lach, R. Zuniga, and D. Morrison. 2000. Environmental and economic costs of non-indigenous species in the U.S. BioScience 50:53–67.

Rosen, S. 1988. The value of changes in life expectancy. Journal of Risk and Uncertainty 1:285–304.

Settle, C., T. Cherry, and J. Shogren. 2008. Rationality spillovers in Yellowstone. Pages 383–394 *in* T. Cherry, S. Kroll, and J. Shogren, editors. Environmental economics, experimental methods. Routledge, London.

Settle, C., and J. F. Shogren. 2006. Does integrating economic and biological systems matter for public policy? The case of Yellowstone Lake. Topics in Economic Analysis and Policy 6(1) (www.bepress.com/bejeap/topics/vol6/iss1/art9/).

Shogren, J. F. 2006. A rule of one. American Journal of Agricultural Economics 88:1147–1159.

Shogren, J. F., D. C. Finnoff, C. McIntosh, and C. Settle. 2006. Integration-valuation nexus in invasive species policy. Agricultural and Resource Economics Review 35:11–20.

Simon, H. 1955. A behavioral model of rational choices. Quarterly Journal of Economics 69:99–118.

Smith, V. L. 1989. Theory, experiment and economics. Journal of Economic Perspectives 3:151–169.

Toman, M. 1998. Why not to calculate the value of the world's ecosystem services and natural capital. Ecological Economics 25:57–60.

White, H. 1980. A heteroscedasticity-consistent covariance matrix estimator and a direct test for heteroscedasticity. Econometrica 48:817–838.

9

Modeling Integrated Decision-Making Responses to Invasive Species

Mark A. Lewis, Alexei B. Potapov, and David C. Finnoff

In a Clamshell

This chapter integrates ecological and economic models and theory into mathematical models of optimization and optimal control for invasive species. The primary goal from this is to design management and policy that accounts for invader population dynamics, control and eradication measures, cost-benefit analyses, and methods for optimal decision making. Because the outcomes from invasive species control depend strongly on how control programs are implemented, we begin by discussing some major economic models of behavioral choice: static, dynamic, deterministic, and stochastic. We then consider model implementations for the problem of lake invasions. It appears that, besides control efficiency and invasion cost, the optimal policy is strongly influenced by the choice of control horizon and discounting factor. Additionally, the efficiency of prevention measures is strongly related to population growth rates when the species is at low density. When growth rates of small populations are low or negative, the optimal control policy may be to prevent invader spread. We expand our models to consider the optimal eradication policy for an invader that has high growth rates at low population densities. Applications to controlling zebra mussels in lake systems are made, and how our modeling approach can be applied to other habitats and species is explained.

In this chapter we develop and analyze bioeconomic models for decision making in the control of invasive species. The costs associated with impacts of invasive species (chapter 1) are such that control of the invader is an option that must be considered. While designed to confer a benefit to the ecosystem and/or economy, controls also incur a cost. It is through finding a balance between the costs and benefits of control that the optimal control policy is determined. Understanding the balance necessitates bioeconomic cost-benefit analysis.

Mathematical models for bioeconomics come to the fore in areas such as control of invasive species, where there are complex nonlinear interactions between the economics and the ecology (chapter 2). Here, intuition can be of limited use—history tells us that unexpected outcomes can arise, even with simple bioeconomic interactions (Clark 1990). In the area of invasive species research, the disciplines of economics and ecology have developed in parallel, with few interactions. Mathematical models require that we carefully build a bridge between the economics and the ecology. They also allow us to quantitatively formalize the joint contributions from economics and ecology in new, synthetic ways. While the area of mathematical bioeconomics is, itself, an established field (Clark 1990), the application of bioeconomics to invasive species is a recent development.

Control measures for single-species invasions can be broadly grouped into *prevention*, which attempts to stem invasive spread, and *eradication*, which attempts to remove invasive species from regions where they are present (see examples in chapter 1). Thus, regulations for the movement of ballast water by cargo ships, the quarantining of beetle-infested firewood, and the cleaning of recreational boats as they move from one lake to the next are all prevention measures. By way of contrast, physical removal of invasive crayfish from lakes, the chemical spraying of weed species, and the biocontrol of invasive insects are all eradication measures, whether or not they are entirely successful in the eradication. (History tells us that they seldom are [Rejmanek and Pitcairn 2002].) Elsewhere in the invasion literature, eradication by itself is sometimes referred to as *control*. However, in this chapter we use the term "control" to refer to both eradication and prevention measures. This allows us to consider an appropriate combination of eradication and prevention measures using the theory of optimal control.

Overall, models for integrated decision-making responses to invasive species can have a variety of goals. First, they can define general principles and strategies that characterize optimal responses to invasive species. We may ask:

- How should a producer respond to an invasive pest by adapting input labor or capital?
- How does the time horizon for control (e.g., political vs. ecological) or economic characteristics of control (e.g., discount rate or preferences toward risk) affect optimal control strategies?

Models to answer such questions frequently use the power of the abstract to simplify highly complex situations to a level where they can be understood and analyzed. Specific details of the model are simplified to the point where it is possible to tease out the underlying principles and strategies.

By way of contrast, tactical models may incorporate specific details of invasion economics and ecology, with a view to determining optimal responses to specific invasive species. With these models we may ask:

- How much should we spend on preventing species invasions?
- How, where, and when should we specifically apply control effort?

Most of the models we consider in this chapter are of the strategic form. However, we are now at the stage where we are starting to formulate and analyze tactical models for specific invasion processes. In this chapter, our tactical focus will be on the control of zebra mussels (*Dreissena polymorpha*) in networks of freshwater lakes.

Developing the framework for decision-making responses to invasive species requires incorporation of both ecological and economic components. We start this chapter by outlining models of behavioral choice, based on economic theory (see "Models of Behavioral Choice," below). Methods for determining optimal behavioral choice are then illustrated for the case of a power plant operator dealing with zebra mussel invasion in a single lake ecosystem. Both static optimization and stochastic dynamic programming approaches are employed (see also chapter 12).

We then introduce basic dynamic ecological models for invasive species in networks of patches or lakes, with effects of control actions included. In the last three sections of this chapter, we show these can be coupled to the economic models and how the bioeconomic approach can be used to deduce optimal control strategies.

To illustrate the bioeconomic approach in networks, we consider a number of models for the process of an aquatic invader spread in a lake network system, and for the control of this spread. It is known that invasions occur due to transportation of invader propagules from lake to lake along with recreational boats (Johnson and Carlton 1996; Johnson et al. 2001; MacIsaac et al. 2004). Invaders include zebra mussels and spiny water fleas (*Bythotrephes longimanus*). Processing of such boats after they leave an invaded lake, or before they arrive at an uninvaded lake, can reduce the number of transported invaders and hence slow the spread of the invader. Alternatively, eradication of invaders at locations where they have already established is another possibility.

Our models for control of aquatic invader spread in a lake network start with a metapopulation-type model to track the proportion of lakes that are invaded in an ecosystem (see "Optimal Strategies," below). The model is extended to heterogeneous networks. Here, population dynamics within lakes can have an Allee effect (see "Preventing Invader Spread," below). The interaction between Allee effect and control may be sufficient to spatially *pin*, or halt, an invasion process, preventing further spread. In this case, we consider the optimal allocation of control effort for stopping an invasion, as well as the optimal location for invasion pinning.

MODELS OF BEHAVIORAL CHOICE

Since Alfred Marshall in the late nineteenth century, microeconomics has been focused on using the power of the abstract to mathematically describe the behavioral choices of humans, in their everyday life (as consumers), in the business place (as managers of firms), and in the public policy arena (as government policy makers). In this endeavor, economists have tried to reflect common sense, based on general rules and theories of human tendencies under certain conditions. While human behavior is, of course, complex, with subjective notions such as tastes at its roots, motives can be measured through real measures, such as prices and quantities, making it quite amenable to mathematical and statistical analyses.

These choices can be boiled down to how humans assess and make trade-offs. To analyze choices, economists have made liberal use of both static and dynamic optimization techniques (introduced in the next two subsections). Optimization can generate numerous testable hypotheses and predictions of behavior. Optimization is an appropriate method if behavior is purposeful, efficient, and consistent. Because humans definitely behave with purpose, usually make choices to efficiently use limiting resources, and are on average consistent in their behavior (i.e., rational), the use of these methods has become widespread.

In the economic framework, the role of the individual, and the trade-offs they face, frames the optimization problem. The optimization problem is mathematically defined in terms of objective function, which is to be maximized, subject to related state variables, constraints, and control variables. The objective function of an individual consumer is, in turn, summarized by a so-called ordinal utility function that defines the individual's preferences or tastes. Consumer tastes are defined over variables that they have control over, such as how much they purchase of commodities, and over state variables not under their immediate control, such as environmental quality, which might nonetheless be influenced by their choices. Consumers are constrained in their consumption choices by their budget, requiring that they make the best choices they can afford.

The objective function of a firm is assumed to be their economic profits. Profits are revenues from sales net of the costs of production. A firm is constrained by its production technology, represented by a production function that relates inputs to outputs, where some inputs are under the control of the firm, such as the capital and labor they hire, and some taken as given, such as environmental state variables beyond their control. The firm's choices are also influenced by the price it receives for its production and the costs of the inputs hired.

The government's objective function depends on the scale of the analysis. It may be a resource manager acting as a firm, or, at the economy level, it is usually taken to be the sum of consumer welfare (as given by consumer surplus) and producer welfare (as given by producer surplus). Both surplus measures relate to market outcomes, where consumer welfare is measured by the excess in the maximum that consumers would be willing to pay for a commodity (its value) over what is paid, and where producer welfare is measured by the excess in what firms receive for their product (the price) over the minimum they would be willing to accept for it (their marginal cost).

For the applications in this book, the behavioral choices of interest focus on the choices of individuals, firms, or resource managers who are influenced by biological invaders. Human choices, or control variables in dynamic contexts, can influence or control the state of the system such that there is a two-way feedback between humans and the invading organisms.

Behavioral Choices in a Static System

For an illustrative example, consider the static profit maximization problem of a firm faced with some impact of a generalized invasive species, for example, a power plant owner whose plant is adversely affected by invasive zebra mussels. In a static setting (i.e., the invasion state is evaluated at a fixed point in time), the firm produces its output y using inputs of capital k and labor l. The production process is influenced by the invasive species, whose abundance is u. The production function is defined mathematically by a nonlinear function:

$$y = F(k, l, u) . \tag{1}$$

where y represents production and $F(k, l, u)$ represents nonlinear dependence on capital k, labor l, and invader level u. The abundance of the invasive species, u, is taken to define the state facing the firm, so u is the relevant state variable. The production function has marginal products of capital and labor that are positive and decreasing, $\partial F/\partial i > 0$, $\partial F^2/\partial i^2 < 0$, $i = \{k, l\}$, and the invasive species affects production. These impacts may be through direct damages to the production process (e.g., preventing the power plant from operating at full capacity due to fouling of the cooling system, such that $\partial F/\partial u < 0$) or by making production more difficult such that the producer is forced to hire more inputs of capital or labor to produce the same output, effectively reducing the productivity of labor and capital (e.g., requiring additional labor and capital for the cleaning of zebra mussels from the cooling system, such that $\partial^2 F/\partial i \partial u < 0$).

The above case of zebra mussel invasion of power plants and water firms was modeled by Leung et al. (2002). Here, a modified Cobb-Douglas form was used for the production function

$$y = F(l, k, u) = \alpha l^a k^b G(u)^c \quad \text{where } G(u) = 1 - \exp(-\lambda/u).$$

The positive constants α, a, and b were estimated by linear regression of the log-transformed equation for production in the absence of zebra mussel infection ($u = 0$ so that $G = 1$ and hence $y = \alpha l^a k^b$):

$$\log(y) = \log(\alpha) + a \log(l) + b \log(k),$$

where y represents production, l represents labor, and k represents capital. The coefficients λ and c, associated with zebra mussel damages, were calculated in a detailed submodel (Leung et al. 2002).

The producer's profits π were constrained by the production function and given by revenues net of costs:

$$\pi = Py - rk - wl, \tag{2}$$

where Py indicates revenues; rk, capital costs; and wl, labor costs. Or, equivalently,

$$\pi = PF(k, l, u) - rk - wl, \tag{3}$$

where the output y is sold at a market price P, units of capital k cost r, and the wage rate of labor l is w. The power plant owner's optimal choices are then given by the capital and labor investments, k^* and l^*, that satisfy the first-order necessary conditions:

$$\frac{\partial \pi}{\partial k} = P\frac{\partial F(k,l,u)}{\partial k} - r = 0, \quad \frac{\partial \pi}{\partial l} = P\frac{\partial F(k,l,u)}{\partial l} - w = 0,$$

where the term on the left is the critical point with respect to capital investment, and the term on the right is the critical point with respect to labor investment. Each necessary condition requires that the relevant input be employed until its unit cost equals the value of its marginal product. The profits gathered by the firm will be affected by the invasive species whenever $\partial F/\partial u \neq 0$, and profit-maximizing choices of the firm will be affected by the invasive species whenever $\partial^2 F/\partial i \partial u \neq 0$. (Technically, for choices to be optimal, the second-order sufficient conditions must be checked. These require that the Hessian matrix of second derivative be negative definite.)

In the above example of zebra mussel, the first-order necessary conditions can be explicitly calculated to yield optimal capital and labor investments, k^* and l^*, as a function of invasion level, u:

$$k^* = \left(\frac{\alpha P b l^{*a} G(u)^c}{r}\right)^{1/(1-b)}, \quad l^* = \left(\frac{\alpha P a k^{*b} G'(u)^c}{r}\right)^{1/(1-a)} \tag{4}$$

This approach for calculating optimal capital and labor investments is applied in a stochastic dynamical setting (see "Behavioral Choices in a Stochastic Dynamic Setting," below).

Of course, this example presumes that all the producer can do is simply adapt to the invasive species. But what if the producer can take explicit action against the invader, such as prevention or eradication? In addition to simple adaptation by humans to invasive species, other costly strategies available to decision makers lay with efforts focused on preventing an invasion in the first place and in controlling the invader once established. However, both of these strategies require an intertemporal perspective, at the very least, and may also require an explicit incorporation of uncertainty given the inherent stochasticity of an invasion. Of course, the simple model laid out above becomes rapidly more complex when taking into account behavioral choices as circumstances change over time and when components of the problem are uncertain. Further, on top of the rapid escalation in analytical complexity, these additions require an incorporation of human preferences over time and human preferences toward risk.

Formulating Behavioral Choices in a Deterministic Dynamic Setting

A dynamic perspective on the choice problem starts with the state variable, in this case the invader abundance, having a dependence on time t, or $u(t)$. Strategies other than adaptation allow the firm a means of controlling the state variable, where these strategies are given in aggregate by the control variable $x(t)$. The evolution of the state variable, du/dt, then depends on the current value of the state and control variables at t and some given initial value u_0:

$$\frac{du(t)}{dt} = f(u(t), x(t)), \quad u(0) = u_0, \tag{5}$$

where the left-hand term is rate of change in the level of invasive species, and the right-hand terms are, respectively, nonlinear dependence on current invader levels and current control, and specified initial level of invader. Profits to the firm are now also time and state dependent:

$$\pi(u(t), x(t), t) = Pf(k(t), l(t), u(t)) - rk(t) - wl(t) - cx(t) \tag{6}$$

where all prices and unit costs, including that of the control variable c, are time invariant.

The firm's optimization problem is to choose a control trajectory that, together with the trajectory of the state variable, maximizes the firm's profits into perpetuity:

$$\max_{\substack{x(t) \\ 0 < t < \infty}} \int_0^\infty e^{-rt} \pi(u(t), x(t)) dt, \tag{7}$$

where the maximum is taken over all possible control strategies into the future, and the current value of profits is found by integrating discounted future profits into perpetuity: e^{-rt} is the discount on future profit, and the remaining term is the profit t time units into the future. If the agency controlling the state variable is the government or society, a function of the form π is typically referred to as the *welfare* function associated with the invasion problem, rather than the *profit* function.

Temporal attitudes are added by incorporating compound discounting of the future, where the relative importance of future costs and benefits is reflected by the discount rate r. Discounting captures the idea that current costs and benefits are given greater weight in any decision than future costs and benefits. Discounting arises because there is an opportunity cost for time—a dollar received today is worth more than a dollar received tomorrow or the day after tomorrow, or next year, and so on. The standard economic model assumes people have a constant discount rate, or marginal rate of time preference, when determining the current/present value of future streams of benefits (Strotz 1956). While the correct magnitude of the discount rate has been debated at length in the literature, the form of the discount rate is assumed constant under each of these scenarios (Stiglitz 1994).

It is worth noting there is some debate over aspects of the modeling presented in this section. Perhaps the most notable is the assumption of a constant discount rate. This assumption, common throughout economics, quickly extinguishes the importance of future states in calculating optimal choices over time. This feature may be quite appropriate when applied to monetary flows, but when applied toward future environmental and ecological states, its use becomes debatable, given the intergenerational social importance of these resources. In such instances, hyperbolic discounting has been proposed where discount rates decrease over time (see Weitzman 2001). With this decline in the discount rate over time, even those states in the distant future are given meaningful weight in the present value calculation (in relation to the higher discounted states in the near future). But hyperbolic discounting also allows for the possibility of time-inconsistent behavior (see Settle and Shogren 2006), where choices of optimal paths would differ according to the point in time in which the optimization is made. This has substantially limited use of the technique in practice.

There is also significant debate over the structure of the expected utility model. Human behavior can violate the assumptions and implications of welfare functions, leading to numerous proposals of alternative behavioral models of choice under uncertainty (see Machina 1987). But, much as with the discount rate debate, these issues have not been resolved, and none of the proposed theories has been uniformly accepted by the discipline as a general improvement over the status quo welfare function.

Finally, some object to the use of optimization, claiming humans seen as simply rational calculating machines neglects too much of human nature being based on instincts and habits. The debate is not new: in the early twentieth century, Thorstein Veblen decried orthodox economics for not taking an evolutionary view toward human behavior and society. Recently, the debate has been enumerated by some economists, with the employment of evolutionary game theory to explain human behavior (see Gintis et al. 2003). This type of game theory modeling is in its infancy and appears to be fruitful but generally has not yet outperformed current optimization techniques.

Calculating Optimal Choices in a Deterministic Dynamic Setting

The problem of modeling integrated decision-making responses to invasive species is often dynamic, given the evolution of the biological states over time. To understand, predict, and prescribe behavior in these situations requires calculating optimal time paths of variables, which may react to one another over time and may respond to one another. This is not a trivial task. In fact, until the development of optimal control theory and dynamic programming in the late 1960s, these types of questions were largely ignored by the economic profession. Since then, there has been an explosion of effort into methodologies for approaching these types of problems, the basics of which are summarized in this section.

The core problem is that an invasive species may affect human welfare in some way or another, and these impacts may occur over time (chapter 2). Furthermore, policies to respond to the invasive species tend to be costly. There are then trade-offs that must be assessed in describing behavioral responses in these settings. Dynamic optimization allows a consideration of these trade-offs.

To demonstrate optimal control theory, consider the system in equations 5–7. Our earlier static optimization problem used the fact that a necessary condition for a local maximum is that it is a critical point (has zero partial derivatives with respect to the control variables). The Maximum Principle (Pontryagin et al. 1962) derives an analogous set of conditions for maximizing a dynamical system that changes over time. The conditions given by the Maximum Principle are given in appendix 9.1. The Maximum Principle gives us a set of tools for deterministic dynamic optimization.

Behavioral Choices in a Stochastic Dynamic Setting

The above formulation is well suited for the problem of controlling invasive species when uncertainty as to outcomes does not dominate. For example, if we are interested in the proportion of lakes invaded over time in a very large network of lakes, uncertainty as to whether any particular given lake is invaded is unlikely to dominate the dynamics for the relevant state variable, which is the proportion of lakes invaded (see "Optimal Strategies," below). However, suppose we are interested in the dynamics of a single lake that starts off as being uninvaded. At any point in time, there may be a high level of uncertainty as to whether the lake is invaded yet (Jerde and Lewis 2007; see also chapters 5 and 7). Here a stochastic model is needed.

One approach is to assume that the state of the system at any time t can be adequately described by a random variable $U(t)$ that can take any number of discrete values u. For a single lake, these could be the level of invasion. For this stochastic process, we are given the state of the system at time t by $U(t) = u$ and can specify the probability that the state will transition to state v at the next time step $t+1$, given control effort $X(t) = x$:

$$\Pr(U(t+1) = v | U(t) = u, X(t) = x)$$

Suppose that we have a certain control policy, that is, the rule for defining control intensity at the lake at the given system state (e.g., the rule given equation 4). For the given initial state $U(0) = u_0$, a stochastic model with the above Markov transition probabilities simultaneously describes all possible scenarios of subsequent dynamics of the system, with probabilities associated with each of them. For each scenario, it is possible to calculate the welfare of the system. For a given state u at time t, we can define an optimization problem with respect to the control x that relates to the possible control problems associated with being at every possible state v at time $t+1$:

$$\begin{aligned}\max_x \; W(u,t) &= \pi(u(t), x(t), t) + \sum_v \Pr(U(t+1) \\ &= v | U(t) = u, X(t) = x) W(v, t+1)\end{aligned}$$

Here, $W(u, t)$ is referred to the cumulative welfare of the system from an end time horizon T to the current time t. For an infinite time horizon problem (T is infinity at which point $W = 0$ for any state), this is the so-called Bellman equation (Puterman 1994). This needs to be solved iteratively.

However, for biological invasion problems, a finite time horizon is often used, at which point there may be some residual welfare defined for the system being in one state as opposed to another. To find the optimal control, we need to find a policy that provides a maximum for W at all states u at each time step t. Stochastic dynamic programming solves this problem. For a finite time horizon, the procedure evaluates the optimal control and the maximum for W sequentially, moving backward in time from the end to the beginning. An excellent introduction to stochastic dynamic programming in ecology is given in Clark and Mangel (2000).

Leung et al. (2002) used the stochastic dynamic programming approach to calculate optimal strategies for combating zebra mussel spread into a previously uninvaded lake. They first made a hypothetical calculation to illustrate the generality of the framework to invasive species. Here, optimal capital and labor were calculated from the modified Cobb-Douglas functional form, using equation 4 (figure 9.1). The time frame was 25 years, after which the system was given no residual value. The time frame was a major determinant of the optimal strategy. As the terminal time was approached, sustained prevention effort could not be economically justified, because there was no incentive for preserving an uninvaded state of the system beyond the

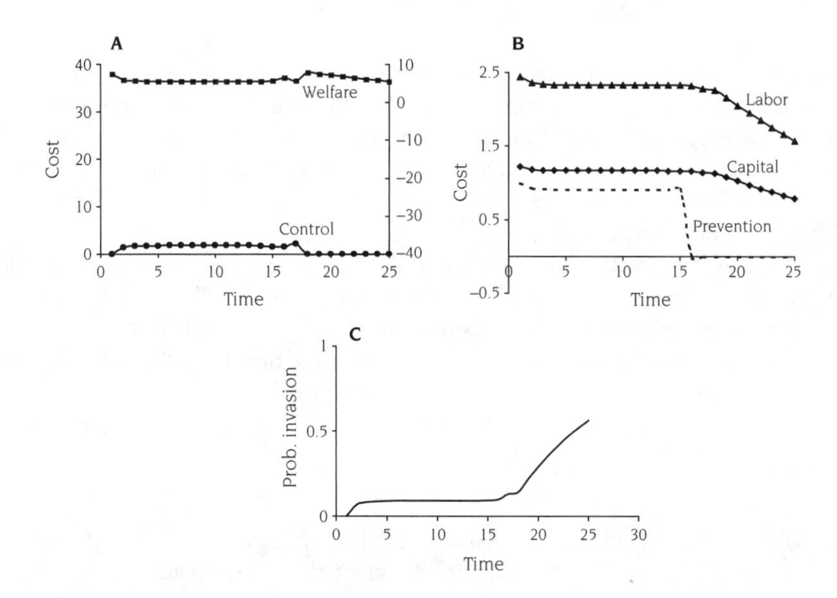

FIGURE 9.1.
Stochastic dynamic programming generates a probability distribution of states (i.e., population sizes) per time interval. The curves shown are based on the values (e.g., welfare, labor cost) of a state weighted by the probability of being in that state, summed across all states (i.e., Σ probability*value). Thus, each plotted datum is a weighted projection of costs and benefits for each time interval. Panels A–C show an initially uninvaded lake over a time horizon of 25 years. Modified from Leung et al. (2002), which gives full details of the model and simulations.

terminal time. Additional simulations over short (i.e., typical political) time scales of 5 years showed that little prevention effort could be justified, although it could be over longer time scales.

Leung et al. (2002) extended their model to a more detailed tactical model, specifically tailored to assess the optimal response by power plant owners and government to the threat of zebra mussel invasion. They focused on the impact to industry caused by reduced water intake due to pipe fouling by zebra mussels. Although lakewide eradication of established zebra mussels is not possible, industry can apply toxins to pipes or mechanical removal of zebra mussels. Because prevention of spread is possible via control and treatment of boat traffic as well as public education, the model asked how much should be spent on prevention to maximize net benefits. Specific details can be found in the article itself. The results were striking. To reduce the probability of invasion by 10%, it would be beneficial to pay up to $324,000 per year for a single lake containing a power plant. By way of contrast, the U.S., Fish and Wildlife Service in 2002 spent a total of $825,000 for prevention and control efforts for all aquatic invaders in all U.S. lakes. This indicates a dramatic underfunding of prevention and control.

OPTIMAL STRATEGIES FOR PREVENTING INVADER SPREAD IN A HOMOGENEOUS NETWORK

We now turn to the problem of controlling invaders in spatially distributed systems. If the species does establish, there is a numerical increase in the number of individuals over time as new locations in the landscape become invaded. In addition, the rate at which the species establishes to become invasive relies on propagule pressure, as well as establishment success, which is particular to both the species and the new landscape (see chapters 3–5).

Perhaps the simplest model to capture all three of these dynamical components describes the spread of an invader between connected patches. The total number of patches N is assumed to be sufficiently large to allow the invasion to be characterized by the proportion of invaded patches u (the number of invaded patches N_I divided by the total number of patches, $u = N_I/N$). The equation is

$$\frac{du}{dt} = A(u+b)(1-u), \tag{8}$$

with the initial proportion of infected lakes $u(0) = u_0$: du/dt is the rate of change of proportion of patches invaded, A is the maximum rate of establishment, $(u + b)$ is the invaded patches plus the external patches that provide new invaders, and $(1 - u)$ is the proportion of patches that can still be invaded. Here, the parameter A can be interpreted in terms of the average number of propagules transported from a source patch to any recipient patch per unit time (A_1) and establishment success (A_2) as

$$A = A_1 N \times A_2, \tag{9}$$

where $A_1 N$ is the maximum possible propagule pressure. This model of invasion *without background propagule pressure* ($b = 0$) can be interpreted as the Levins metapopulation model (Levins 1969) without extinction or simply as logistic growth (Clark 1990) in the proportion of invaded patches over time. A detailed derivation of the model is given in Potapov et al. (2007).

We now consider how prevention and control strategies can be incorporated into model represented by equations 8 and 9. Prevention is modeled as a reduction in propagule pressure A_1 or establishment success A_2. For example, if we consider u to represent the proportion of lakes invaded by zebra mussel in a large network of lakes, then prevention could be used to prevent zebra mussels from leaving a source lake (thereby reducing the propagule pressure A_1) or from establishing at an uninvaded lake (thereby reducing the establishment success A_2). In the case of zebra mussel spreading in a lake ecosystem, recreational boats are a primary infection vector. Boats, used in an infected lake, are transported to uninfected lakes, and zebra mussels on the boat exterior are transported by this means. Therefore, with zebra mussel, a useful way of reducing propagule pressure is by washing potentially contaminated boats as they leave invaded lakes. On the other hand, washing the boats at the other end of the transportation process, just prior to use in an uninfected lake, will reduce establishment success in the uninfected lake.

Prevention effort at a source lake $x(t)$ reduces the propagule pressure arising from a single source lake from A_1 to $A_1 a_1(x(t))$, $0 < a_1 \leq 1$. Here, $a_1(x(t))$ is the probability of a propagule escaping treatment, and $1 - a_1(x(t))$ is the proportion of propagules removed by the prevention strategy at source lakes. Similarly $s(t)$ and $a_2(s(t))$, $0 < b \leq 1$, are prevention effort and probability of escaping treatment at any recipient lake, respectively. Treatment at an uninvaded recipient lake would reduce establishment success from A_2 to $A_2\ a_2(s(t))$, $0 < a_2 \leq 1$. Potapov et al. (2007) derive reasonable forms for proportion of propagules escaping treatment at source and recipient lakes as

$$a_1[x(t)] = \exp[-k_1 x(t)], \quad a_2[s(t)] = \exp[-k_2 s(t)], \tag{10}$$

where k_1 and k_2 characterize heterogeneous efficiencies in boat processing. Under these assumptions, the spread of the invasion is described by equation 9 with $A = (A_1 N)\ A_2$ replaced by $A = (A_1 a_1 N)\ (A_2 a_2)$, so that equation 8 becomes

$$\frac{du}{dt} = Ae^{-k_1 x(t)} e^{-k_2 s(t)} (u + b)(1 - u). \tag{11}$$

Although the growth dynamics are a simple quadratic, the prevention efforts $x(t)$ and $s(t)$ enter the model in a highly nonlinear fashion as exponential terms.

Eradication effort $h(t)$ is used to return invaded patches to their earlier uninvaded state via invader eradication. While history has proved it difficult for such eradication ventures to be successful, they can nonetheless be included into our modeling framework. Unlike the highly nonlinear relationship between treatment effort and propagule reduction (equation 11), it is reasonable to expect a linear relationship between eradication effort and proportion of patches per unit time returning to the uninvaded state. Finnoff et al. (in press) argued that the rate of eradication of the invader from a proportion of lakes u should be $hu/(u + \alpha), \alpha << 1$. This modifies equation 11 to

$$\frac{du}{dt} = Ae^{-ks}e^{-kx}(u + b)(1 - u) - \frac{hu}{u + \alpha}, \tag{12}$$

where e^{-ks} and e^{-kx} are the invader flow reduction due to control at uninvaded and invaded lakes, respectively, and $u + b$ is the internal and external sources of the invader. In summary, there are three different means of control: (1) preventing invaders from leaving invaded lakes (with control effort xu), (2) preventing invaders from entering uninvaded lakes [with control effort $s(1 - u)$], and (3) eradication (with control effort h). In this case, it may be more natural to formulate a cost per unit time function for the invasion $C_I(t)$, rather than a productivity function $\pi(t)$:

$$C_I(t) = gu(t) + w_x x(t)u(t) + w_s s(t)[1 - u(t)] + C(h(t)),$$

where g describes losses per unit time for invasion of the lake ecosystem (see chapter 8; hence, gu represents the cost of invasion), w_x and w_s describe the cost per unit effort of preventing invaders from leaving invaded lakes and of preventing invaders from entering uninvaded lakes, respectively, so the corresponding terms in the equation represent cost of prevention at donor and recipient lakes, respectively, and $C(h)$ describes nonlinear cost dependence for eradication. This leads to a problem similar to the optimal control problem given in equation 7. However, the goal is to minimizing the total costs of an invasion, rather than maximizing the production function, and the minimization occurs with respect to three possible controls, $x(t)$, $s(t)$, and $h(t)$, rather than with respect to a single control, $x(t)$.

In the case where there is a finite time horizon, say, of length T, the optimal control problem must include the terminal value of the ecosystem, invaded to level $u(T)$ at time T. The optimal control problem consists in determining $x(t)$, $s(t)$, and $h(t)$ that correspond to minimum total discounted cost cost of invasion per lake J:

$$J[x, s, h] = \int_0^T e^{-rt} C_I(t)dt + e^{-rT} V_T(u(T)), \tag{13}$$

where $V_T(u(T))$ is terminal cost, related to future losses beyond the control horizon T.

This model can be solved numerically in the general case. However, the most instructive cases appear to be two versions of the model: (1) control via prevention without eradication and zero external propagules pressure ($h = b = 0$) and (2) control via eradication without prevention ($x = s = 0$). We briefly describe the most important findings for these two cases.

Control via Prevention

The model has been studied with the help of the Maximum Principle using the phase plane method (Potapov et al. 2007). The main findings are the following:

- There is a threshold value for the proportion of invaded lakes u_S (in typical cases $u_S = 0.5$) such that for $u < u_S$ it is optimal to control only invaded lakes; otherwise, only uninvaded.
- Control effort at invaded lakes $x(t)$ always decreases with t; hence, the most intensive control at the invaded lakes is to be applied in the beginning. The behavior of control effort at uninvaded lakes $s(t)$ depends on the discount rate r.
- There is a critical value of losses g below which control at uninvaded lakes is never optimal, so the control has to be limited only by the initial stage of the invasion.
- Without eradication, the invaded state of a lake is irreversible. This means that invasion can be optimally delayed, but not stopped.

As with the results from the Leung et al. (2002) model (figure 9.1), when the terminal cost is zero ($V_T(u(T)) = 0$), the time dependency of the control always has a distinguishing feature: at some moment $t_1 < T$, the control always becomes zero, so the invasion gets a "terminal boost." If T is small enough, it is optimal to have no control. The reason for this effect is quite clear: if we set $V_T(u(T)) = 0$, we actually do not care what happens to our lake system beyond the control horizon T: there will be no losses any longer. Hence, it is reasonable to save on control effort for times close to T.

If one does want to account for $V_T(u(T))$, it turns out that there is no established technique for its evaluation. The only reasonable approach is to continue the cost-benefit analysis beyond T and to consider an auxiliary infinite-horizon control problem. Its solution should bring $V_T(u(T))$. This technique has one important implication: it appears that the optimal control magnitude ceases to depend on T, and it always coincides with the solution of the control problem for $T = \infty$.

An example of the optimal control dependencies is shown in figure 9.2. It is interesting to note that, due to discounting, a problem with an infinite time horizon does actually have an effective equivalent of the finite time horizon: everything that occurs very far in the future (on time scales $t \gg 1/r$, where r is the discount rate) contributes little to the total cost. Mathematically, the infinite time horizon can be helpful. It allows use of analytical or numerical techniques that provide better insight and a more transparent management recommendation. For example, when the time horizon is infinite, the solution to the time-dependent control problem becomes autonomous; it depends only on the current invasion level u, not the remaining time horizon. This simplifies analysis and allows for the development of practical management policies. It is interesting to note that, qualitatively, the curves shown in figure 9.2 for the change in control effort and level of invasion over time have the same form as those in figure 9.1, even though the models giving rise to the curves are different.

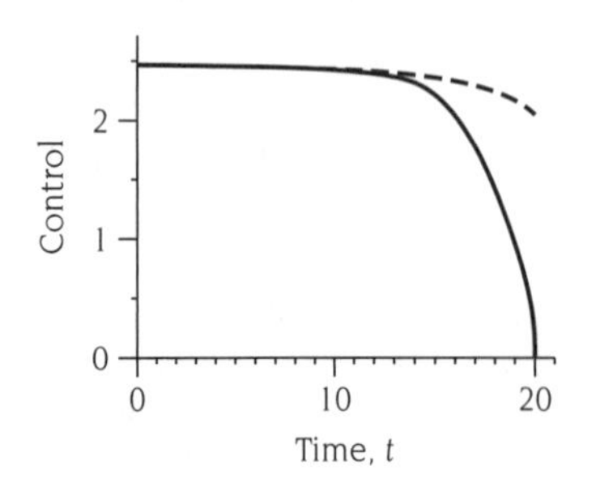

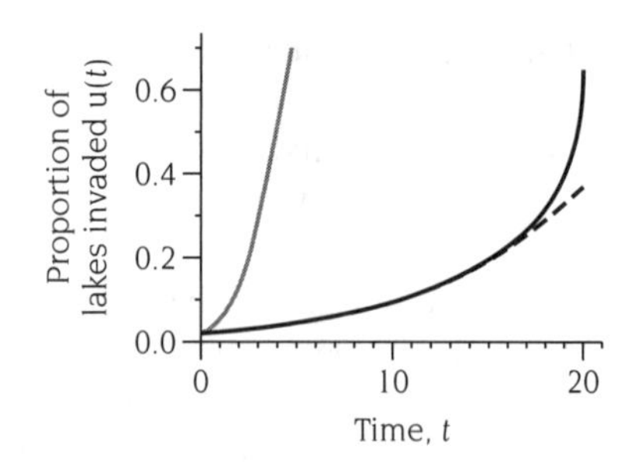

FIGURE 9.2.
An example of numerical solution of the optimal control problems 12 and 13 for $T = 20$, $r = 0.03$, $A = 1$, $w_X = w_S = 1$, $k = 2.5$, $g = 3$, $u(0) = 0.02$. Here, the focus is on prevention, so it is assumed that there is no eradication ($h = 0$ in equation 12). The left panel shows the control $x(t) + s(t)$ for zero terminal cost (solid line) and for infinite horizon estimate of the terminal cost (dashed line). The right panel shows the invasion progress $u(t)$ for both cases, as well as in case of no control (gray solid line).

Acceptance of problem statement with an infinite control horizon allows a more rigorous approach to the search for simpler management policies (Finnoff et al. in press). It is known that solutions of such problems tend to a steady state. All control problems may be roughly subdivided into two classes: those where the main goal is optimal transition to this steady state, and those where the main interest is in the type of the eventual steady management regime. In the latter case, the problem reduces to search for control-dependent steady states, and choosing the level of control that gives the best result.

Control via Eradication

For the optimal eradication problem ($x = s = 0$), it first appeared that standard Maximum Principle analysis led to analytically intractable results. However, using the above steady-state approach, we were able to find quite simple management strategies. The properties of the model depend on the value of eradication effort h. For large eradication effort [$h > (A+b)/4A$], there is only one slightly invaded state $u_S(h) \approx \alpha b/(h-b)$. For intermediate eradication effort values, the system is bistable: besides u_S, there is an essentially invaded state that depends on the eradication effort $u_F(h)$. As eradication effort h decreases to levels near the background propagule pressure, $h \approx b$, the state $u_S(h)$ disappears. As the eradication effort decreases further toward zero, the essentially invaded state approaches the fully invaded state $u_F(h) \to 1$. From a practical point of view, the two invasion states, u_S and u_F, correspond to two different outcomes from management policy: protection of the whole lake system from invasion, such that only a small proportion of lakes is invaded at every moment, or protection of a comparatively small proportion of lakes. Analyzing the benefits and costs gives the following results, which depend on eradication cost $C(h)$:

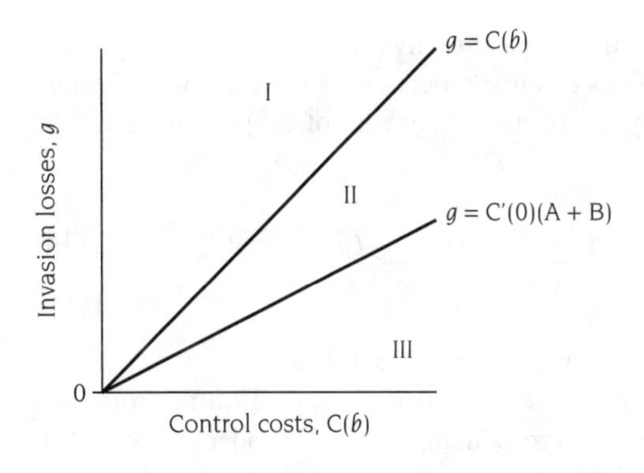

FIGURE 9.3.
Graphic representation of three possible control policies in the problem of control via eradication: I, protection of the whole system; II, protection of a small reserve; III, no invasion management.

- If invasion losses g are large relative to costs associated with eradicating at a rate equal to the background propagule pressure $b[C(b) < g]$, then it is optimal to protect the whole lake system.
- If invasion losses g are at an intermediate level $[C'(0)(A + b) < g < C(b)]$, then it is optimal to protect a reserve of the size $1 - u_F(h_F)$, where h_F is the solution of $C'(h_F) = g(A + b)$.
- If invasion losses g exceed a critical level, $C'(0)(A + b)$, then no control is optimal.

We have estimated model parameters for zebra mussel invasions in midwestern lakes (Finnoff et al. in press). Schematically, the optimal management policy is shown in figure 9.3. Thus, analysis of the model allowed us to suggest quite simple management rules, within the limitations of the aggregate description of the whole system.

PREVENTING INVADER SPREAD IN A HETEROGENEOUS NETWORK WITH AN ALLEE EFFECT

In the metapopulation model described above, the invasion cannot stop without eradication. This is the consequence of the model assumptions: the spread rate is proportional to mean invader flow. Invasion prevention cannot make the flow exactly zero, and any nonzero flow is enough for the invasion to progress. However, for many species, an Allee effect takes place: very small or rare populations cannot grow and go extinct (for a review, see Taylor and Hastings 2005). It is natural to suggest that, in the presence of an Allee effect, control of the invader spread can stop the invasion. However, the metapopulation model does not allow inclusion of an Allee effect. To include this, we considered a spatially explicit model of connected patches where the nonlinear population dynamics have a threshold a, below which the local population goes extinct, and above which the local population grows to carrying capacity (Potapov and Lewis 2008). This Allee effect appears to be present in a number of well-known invasive species, particularly those, such as zebra mussels, that reproduce sexually (Leung et al. 2004).

Rather than tracking the proportion of patches invaded over time $u(t)$, spatially explicit models for invasion dynamics can track the level of invader in each patch. When patches are coupled spatially, the result is a network of coupled models:

$$\frac{du_i}{dt} = f(u_i) + \eta \sum_{j=1}^{N} T_{ij}u_j - \eta \sum_{j=1}^{N} T_{ji}u_i, \tag{14}$$

where $u_i(t)$ denotes the level of invader in patch i, $f(u_i)$ denotes local population dynamics, η is the mean number of invades per boat, T_{ij} is boat traffic from lake j to lake i, hence $\eta T_{ij}u$ describes the corresponding flow of invaders from and to other patches. Potapov and Lewis (2008) explored the case where f describes non-linear population dynamics associated with an Allee effect, which is characterized by reduced per-capita growth rates at low population densities (see also Keitt et al. 2001).

In the model depicted by equation 14, there are $N(N-1)$ connections T_{ij} between the N patches. (Connections T_{ii} go from lake i to itself and are typically set to zero.) In practice, these connections are rarely measured directly (but see chapter 6). However, one approach that has been successfully applied to biological invasions is to approximate connection weights using gravity models. Details for formulating the weights for gravity models are given in appendix 9.2.

An Allee effect means that for populations below certain size a, the growth rate is negative (figure 9.4). Even if there is an incoming flow of the invader, the population can still decline, providing the flow does not exceed a critical value w_0, which is $|\min_{0<u<1} f(u)|$ (figure 9.4). This suggests the idea of invasion stopping: if one organizes control such that the incoming flow into each lake remains at all times below w_0, the invasion will not spread (figure 9.4).

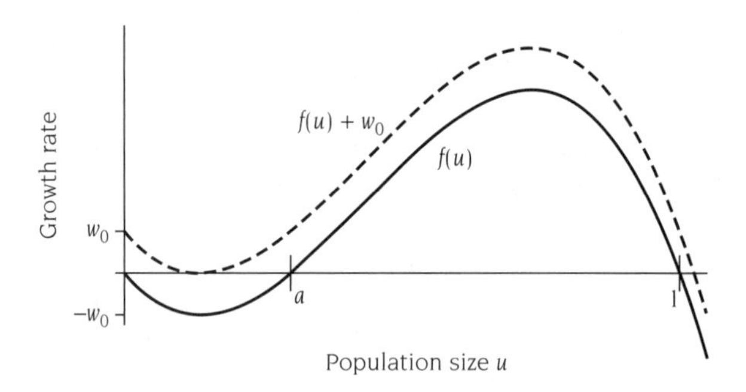

FIGURE 9.4.
Graphic representation of Allee dynamics. We shall characterize the population of the invader within the lake only by its proportion of carrying capacity u: from uninvaded $u = 0$ to fully invaded $u = 1$. In the absence of inflow or outflow in equation 14, a population that starts below the threshold a has negative growth and goes extinct. When the population starts above the threshold a, growth is positive and it goes to carrying capacity $u = 1$. If the net inflow exceeds w_0, then $f + w_0$ (dotted line) is positive for $0 < u < 1$, and the population can grow to the carrying capacity.

We assume that the boat traffic is the major source of the invader flow. The boat traffic is described by the number T_{ij} of boats per year moved from lake j to lake i. Since the invader traffic is proportional to T_{ij}, it is possible to introduce the critical boat traffic $T_0 = w_0/\eta$ instead of w_0. We also introduce the cost of a single boat treatment at invaded (x_i) and uninvaded (s_i) lakes with the same exponential model treatment efficiency: the number of invaders decreases by factor $\exp(-kx_i)$ or $\exp(-ks_i)$. When patches are considered to be lakes in a network, then the application of control via boat washing with efforts x_j at invaded lake j and s_i at uninvaded lake i yields a modification of the inflow term in equation 14 to

$$\eta e^{-ks_i} \sum_{j=1}^{N} T_{ij} e^{-kx_j} u_j, \tag{15}$$

Here each individual invader flow from lake j to lake i $\eta T_{ij} u_j$ is multiplied by treatment factor at lake j, then these flows are summed and multiplied by treatment factor at lake i. The outflow term is modified similarly.

The losses related to the invasion at each lake are denoted by g_i. Because we want to consider the problem of invasion stopping, both x_i and s_i do not depend on time; all lakes are either invaded ($u = 1$) or uninvaded ($u = 0$), and we can minimize costs per year instead of total costs. Then the problem of invasion stopping splits into two optimization problems: optimal allocation of control and optimal invasion stopping.

Optimal Allocation of Control

For the given configuration of invaded/uninvaded lakes $\{u_i\}$, find the optimal allocation of control at each lake x_i or s_i, which minimizes the total control cost E. This is a typical optimization problem: find the control cost per year, E_C, which is a sum of cost of treatment of all departing boats and all arriving boats at each lake,

$$E_C(u) = \min_{\{x\},\{s\}} \sum_{i=1}^{N} \left(x_i \sum_{j=1}^{N} T_{ji} + s_i \sum_{j=1}^{N} T_{ij} \right),$$

where x_i and s_i are cost of treatment one departing and arriving boat at lake i respectively. The minimum should be found under the following constraints: $x_i, s_i > 0$, and for each uninvaded lake i the incoming boat traffic from invaded lakes times the treatment factors should be less than critical boat traffic, T_0,

$$e^{-ks_i} \sum_{j=1}^{N} T_{ij} e^{-kx_j} u_j < T_0.$$

This problem can be efficiently solved numerically. Examples of spatial control distribution are given in Potapov and Lewis (2008) and in figure 9.5.

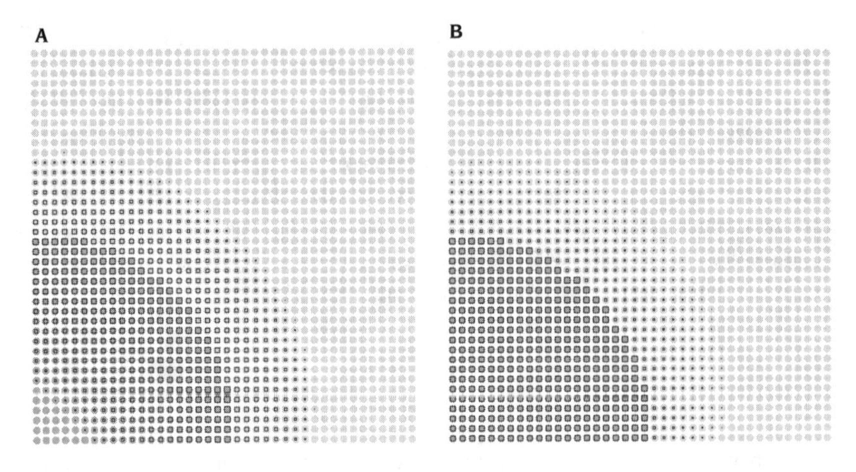

FIGURE 9.5.
Optimal allocation of control for a system of lakes on 40 × 40 grid. Uninvaded lakes are shown as light gray circles, invaded lakes as dark gray, and the intensity of control by the size of the square. The connections are generated according to gravity model, $T_{ij} = m_i m_j \varphi(d_{ij})$. (A) Localized connections $[\varphi(d) = \exp(-bd)]$. (B) Slowly decaying connections $[\varphi(d) = b(d/d_0)^{-\gamma}]$.

Optimal Invasion Stopping

Find the optimal configuration $\{u_i\}$ for which the sum of control costs E_C and invasion losses $g_i u_i$ is minimal. This problem is more complicated—it is related to optimization on a discrete set of $\sim 2^N$ possible combinations, to find one giving minimum total invasion cost per year

$$E(u) = E_C(u) + \sum_{i=1}^{N} g_i u_i.$$

There is no efficient way to solve this problem completely for N exceeding approximately 20, since one has to look though all possibilities. Practical methods of solving problems of this type use various heuristics to find a satisfactory solution in a reasonable time. In our case, simplification may be related with lake clustering. By a cluster we mean a group of lakes with between-group connections stronger than outside-group ones. The boundaries between clusters appear to be natural boundaries for invasion stopping. Intuitively it is clear, because control of weaker connections requires fewer resources. However, the optimal invasion-stopping policy essentially depends on the losses due to the invasion. To illustrate this dependence, we consider two systems of identical lakes with identical connection strengths. However, in one system each lake is connected with each other (figure 9.6a), while in the second one the lakes are organized in two clusters with a single connection between the clusters (figure 9.6b). In both cases, there is a critical value of losses per lake g^*, such that for $g > g^*$ the optimal policy is to stop the invader where it is now. For $g < g^*$,

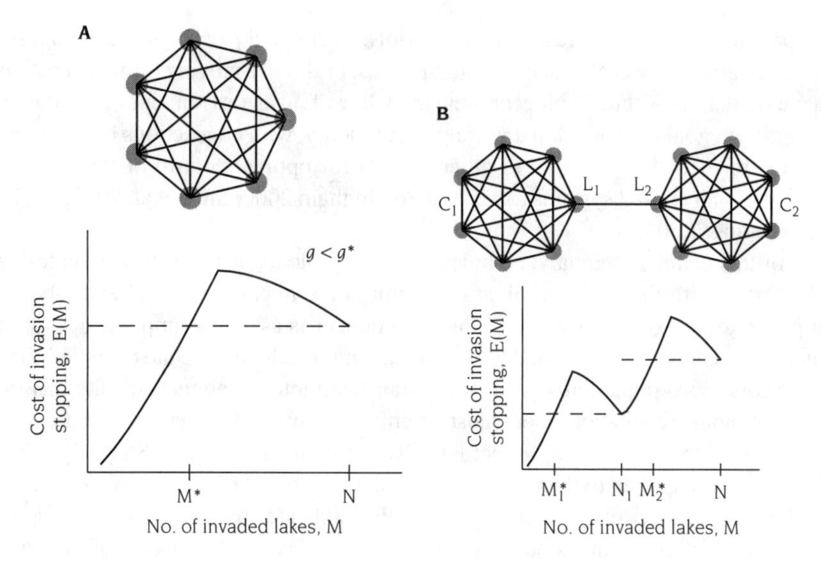

FIGURE 9.6.
Simplified model showing the effect of clustering on the cost of stopping the invasion after it has occupied M lakes out of N. All lakes and connections are identical. For g exceeding a critical value g^*, the total invasion cost always grows with the invasion progress. For $g < g^*$, there is a critical number of invaded lakes M*, such that for M > M* it becomes optimal not to control at all (A) or to abandon the invaded cluster and to protect uninvaded clusters (B).

there is a critical proportion of invaded lakes in a cluster, and when it is exceeded, it becomes optimal to abandon the cluster and to control the next one.

Model calculations show that, in more realistic model systems, clustering also influences the optimal control policy (Potapov and Lewis 2008). Clustering also may allow a multiscale approach for spatial control problems. For a large group of lakes, the problem can be formulated in terms of clusters instead of individual lakes. After the problem is solved on a bigger scale, it may be considered for each cluster separately.

Our models have incorporated the dynamical effects of control effort on species invasions. Mathematically, these have been depicted as the effects of s (prevention effort at uninvaded patches), x (prevention effort at invaded patches), and h (eradication effort at invaded patches) in equations 12–15. In other words, the proportion of lakes invaded (equations 12 and 13) or the number of lakes invaded (in systems 14 and 15) will be reduced in the presence of control effort. We expect that this effort will confer a benefit through reduced damages to the ecosystem.

DISCUSSION

In this chapter we have developed and applied bioeconomic theory for the control of invasive species in networks of patches. While the lessons and the applications have focused on the spread of aquatic invaders through networks of lakes, there is the potential to apply the results more broadly, to coupled patch systems, including archipelagos of physical islands, habitat islands, or disease moving through a patchy metapopulation. Although bioeconomic modeling of invasions remains in its infancy, a long-term goal is to develop a more general theory, with connections between our models for networks of lakes to other models for optimal control of invasions in terrestrial and coastal systems (Shea and Possingham 2000; Shea et al. 2002; Taylor and Hastings 2004).

In this chapter we analyzed bioeconomic problems of increasing complexity. We started with the static problem of optimal investments in capital and labor in response to an existing invader. Here the producer is asked to simply adapt to the invasive species. Once the model is defined, simple calculus dictates coupled first-order conditions (equation 4) that, when translated into economics, dictate that the relevant inputs (capital or labor) must be employed until their marginal cost equals the value of their marginal product (see "Behavioral Choices in a Static System," above). When applied to the problem of combating spread of an invasive species to a lake with a power plant on it, the calculation indicates reduced labor and capital investments after the invasion occurs (figure 9.1). The explanation is of the basic trade-off: after an invasion, the value of the marginal product of both labor and capital is reduced (by the damages caused by the invader) to a point where they are less than marginal (unit) costs. To maximize profit, producers necessarily reduce their input employment, increasing the value of the marginal product of the inputs and therefore bringing the trade-off back into balance. Of course, the profits the producer realizes are less than before the invasion because they produce less given fewer inputs. The producer is able to adapt to the invasion, but not perfectly, and the welfare falls.

Next, we moved to the problem of behavioral choices in dynamic settings. Here, the producer/and or government can respond to the invader by control (prevention or eradication). It can adjust its behavioral strategies, depending on the current state of the system. This leads to time-dependent optimal control problems. Depending on the application, a stochastic formulation or a deterministic formulation is the most appropriate. The stochastic formulation has the advantage of encapsulating uncertainties associated with vagaries of invasion outcomes; the deterministic model has the advantage of a straightforward mathematical formulation with an associated powerful method (Maximum Principle) for calculating optimal behavioral choices. For example, when considering the invasion status of a single lake, the level of uncertainty associated with when/if the invasion will occur suggests the stochastic formulation. By way of contrast, a model tracking the proportion of lakes invaded across a large network of connected lakes is possible with a deterministic formulation.

Despite complexities in formulating the models tailored to specific invasion biology questions, a number of principles have emerged:

First, if the system is initially uninvaded, or is invaded at a low level, no control is always the best policy response for an invasive species if one or both of the following are true: (1) a short (e.g., typical political) time horizon, with zero defined economic value for the ecosystem at the end of the time horizon, and (2) a sufficiently high discount rate. This can be understood in terms of the ecological time scale required for the invader to establish, grow, and inflict significant damages, relative to the short time horizons associated with the control problem or with significant returns on investments made elsewhere. In other words, strong human preferences for the present period (higher discount rates) and short decision-making time horizons provide disincentives for adequate control of invasive species. This may explain the relatively low investment evident in current control of species such as zebra mussels (see "Behavioral Choices in a Stochastic Dynamic Setting," above).

Second, strategies to control invasive species range from slowing spread by "buying some time" so as to delay an inevitable invasion process, to actually halting the invasion process. Here, it is the details of the nonlinear population dynamics that indicate which is possible. In the presence of an Allee effect, the models predict possible halting of the invasion process through prevention alone. In populations with no Allee effect, ongoing eradication effort is additionally needed to halt or reverse an invasion process. Even though these outcomes of delaying versus preventing an invasion are ecologically distinct, economically there may be little difference: if an invasion is delayed for sufficiently long, the economical discount applied to discount future damages may be sufficient to make the invasion economically unimportant from today's perspective.

In the model with eradication, three possible approaches could be optimal, depending upon the cost of eradicating new invaders $C(b)$ relative to the damages inflicted by invaders: protection of the whole landscape (low costs relative to damages inflicted), protection of a small reserve (intermediate costs relative to damages inflicted), and no protection (high costs relative to damages inflicted) (figure 9.3). Although they arise from a different mechanism, the same triage treatment was evident in heterogeneous networks of lakes with an Allee effect (figure 9.6). Depending on the costs and configurations of lakes, it may be optimal to preserve all the lakes, a cluster of lakes, or no lakes.

To date, the application of tactical models for lake networks has been constrained to the analysis of control of a single lake with a power plant. However, the development of large-scale tactical models is under way.

Strong human preference for the present period coupled to a constant discount rate explains how reduced control of invasive species may emerge as a rational decision from a bioeconomic perspective. However, this bioeconomic explanation also provides a challenge to ecologists and economists alike. Given the intergenerational social importance of the environment, how do we appropriately value our environment so as to ensure investments in prevention and eradication needed to accomplish goals on the time horizons that citizens care about? If we wish to preserve ecosystems for future generations through rational economic behavior, we must necessarily consider new economic incentives or revised methods for valuing ecosystems that can be used to achieve this goal.

APPENDIX 9.1. THE MAXIMUM PRINCIPLE

The procedure for applying the Maximum Principle to the optimization problems 5–7 follows from writing a current value Hamiltonian:

$$H = \pi(u,x) + \mu f(u,x),$$

where μ is the costate variable that represents the sensitivity of the maximized problem to changes in the state variable. In this setting, it is therefore the current marginal valuation of the state variable and known as shadow price. The maximum principle requires (1) that optimal controls x^* be such that H reaches its maximum at each t:

$$\frac{\partial H}{\partial x} = 0 = \frac{\partial \pi(u,x)}{\partial x} + \mu \frac{\partial f(u,x)}{\partial x};$$

(2) that the shadow price satisfy

$$\frac{d\mu}{dt} = r\mu - \frac{\partial H}{\partial u};$$

and (3) that the state constraint holds:

$$\frac{du}{dt} = f(u,x).$$

These conditions, in conjunction with any relevant transversality conditions (end-point conditions) are necessary for the solution to the problem. Sufficient conditions for optimality are met in the problem if the maximized Hamiltonian is concave in the state variable. This will be satisfied if the functions π and g are concave in the state and control variables (Kamien and Schwartz 1991).

Thus, for any deterministic dynamic optimization problem, one simply needs to construct the problem with the appropriate structure and apply the maximum principle. The approach outlined here can be generalized to multiple controls x and multiple state variables u. In simple problems, the resulting differential equations describing the time paths of the control variable and shadow price can be analytically solved, but in many cases auxiliary techniques are required. In any case, the maximum principle determines the optimal control $x^*(t)$, optimal evolution of the state $u^*(t)$, and current marginal valuation of the state variable, $\mu^*(t)$. Given the renewable nature of invasive species, an infinite horizon perspective is appropriate (for more comments on control time horizon, see "Optimal Strategies," above). Thus, not only are the time paths of these variables of interest, but also the steady state to which they (may) be on the path toward, where the solution becomes invariant over time. Again, the steady state may not be analytically solvable, and again, one may have to rely upon a diagrammatic approach or an approximation.

APPENDIX 9.2. GRAVITY MODEL WEIGHTS

For gravity models, it is assumed that that the weights in equation 14 are given by

$$T_{ij} = m_i m_j \varphi(d_{ij}), \quad i \neq j, \quad T_{ij} = 0,$$

where m_i characterizes "attractiveness" of a network node, and d_{ij} is the distance between nodes i and j. Then, it is necessary to determine only N parameters m_i, and parameters related to the distance weighting $\phi(d)$. In most practical cases, $\phi(d)$ is chosen to be either exponential,

$$T_{ij} = m_i m_j \exp(-\beta d_{ij}),$$

or power law,

$$T_{ij} = m_i m_j (d_{ij}/d_0)^{-\gamma}.$$

Although methods for estimation of gravity model coefficients originally were developed outside the area of biological invasions in the context of transportation networks (Sen and Smith 1995), they have been developed for and successfully applied to invasive pathways between lakes (Buchan and Padilla 1999; Bossenbroek et al. 2001; MacIsaac et al. 2004).

References

Bossenbroek, J. M., C. E. Kraft, and J. C. Nekola. 2001. Prediction of long-distance dispersal using gravity models: zebra mussel invasion of inland lakes. Ecological Applications 11:1778–1788.

Buchan, L. A. J., and D. K. Padilla. 1999. Estimating the probability of long-distance overland dispersal of invading aquatic species. Ecological Applications 9:254–265.

Clark, C. W. 1990. Mathematical bioeconomics: the optimal management of renewable resources. Wiley, New York.

Clark, C. W., and M. Mangel. 2000. Dynamic state variable models in ecology. Oxford University Press, Oxford.

Finnoff, D., A. Potapov, and M. A. Lewis. In press. The costs of rules of thumb in invasive species management. Forthcoming *in* C. Perrings, H. Mooney, and M. Williamson, editors. Bioinvasions and globalization. Oxford University Press, Oxford.

Gintis, H., S. Bowles, R. Boyd, and E. Fehr. 2003. Explaining altruistic behavior in humans. Evolution and Human Behavior 24:153–172.

Jerde, C. J., and M. A. Lewis. 2007. Waiting for invasions: a framework for the arrival of non-indigenous species. American Naturalist 170:1–9.

Johnson, L. E., and J. T. Carlton. 1996. Post-establishment spread in large-scale invasions: dispersal mechanisms of the zebra mussel *Dreissena polymorpha*. Ecology 77:1686–1690.

Johnson, L. E., A. Riccardi, and J. T. Carlton. 2001. Overland dispersal of aquatic invasive species: a risk assessment of transient recreational boating. Ecological Applications 11:1789–1799.

Kamien, M. I., and N. L. Schwartz. 1991. Dynamic optimization: the calculus of variations and optimal control in economics and management. North-Holland, Amsterdam.

Keitt, T. H., M. A. Lewis, and R. D. Holt. 2001. Allee effect, invasion pinning, and species' borders. American Naturalist 157:203–216.

Leung, B., J. M. Drake, and D. M. Lodge. 2004. Predicting invasions: propagule pressure and the gravity of Allee effects. Ecology 85:1651–1660.

Leung, B., D. M. Lodge, D. Finnoff, J. F. Shogren, M. A. Lewis, and G. Lamberti. 2002. An ounce of prevention or a pound of cure: bioeconomic risk analysis of invasive species. Proceedings of the Royal Society of London Series B Biological Sciences 269:2407–2413.

Levins, R. 1969. Some demographic and genetic consequences of environmental heterogeneity for biological control. Bulletin of the Entomological Society of America 15:237–240.

Machina, M. J. 1987. Choice under uncertainty: problems solved and unsolved. Journal of Economic Perspectives 1:121–154.

MacIsaac, H. J., J. V. M. Borbely, J. R. Muirhead, and P. A. Graniero. 2004. Backcasting and forecasting biological invasions of inland lakes. Ecological Applications 14:773–783.

Pontryagin, L. S., V. G. Boltyanskii, R. V. Gamkrelize, and E. F. Mishchenko. 1962. The mathematical theory of optimal processes. Wiley, New York.

Potapov, A., and M. A. Lewis. 2008. Allee effect and control of lake system invasion. Bulletin of Mathematical Biology 70:1371–1397.

Potapov, A., M. A. Lewis, and D. Finnoff. 2007. Optimal control of biological invasions in lake networks. Natural Resource Modeling 20:351–379.

Puterman, M. L. 1994. Markov decision processes. Wiley, New York.

Rejmanek, M., and M. J. Pitcairn. 2002. When is eradiation of exotic pest plants a realistic goal? Pages 249–253 *in* C. R. Veitch and M. N. Clout, editors. Turning the tide: the eradication of invasive species. Proceedings of the International Conference on Eradication of Island Invasives. IUCN Species Survival Commission No. 27 International Union for Conservation of Nature, Gland, Switzerland.

Sen, A., and T. E. Smith. 1995. Gravity models of spatial interaction behavior. Springer, Berlin.

Settle, C., and J. F. Shogren. 2006. Does integrating economic and biological systems matter for public policy? The case of Yellowstone Lake. Topics in Economic Analysis and Policy 6(1) (www.bepress.com/bejeap/topics/vol6/iss1/art9/).

Shea, K., and H. Possingham. 2000. Optimal release strategies for biological control agents: an application of stochastic dynamic programming to population management. Journal of Applied Ecology 37:77–86.

Shea, K., H. Possingham, W. W. Murdoch, and R. Roush. 2002. Active adaptive management in insect pest and weed control: intervention with a plan for learning. Ecological Applications 12:927–936.

Stiglitz, J. E. 1994. Discount rates: the rate of discount for benefit-cost analysis and the theory of the second best. Pages 116–159 *in* R. Layard and S. Glaister, editors. Cost-benefit analysis, 2nd edition. Cambridge University Press, Cambridge, UK.

Strotz, R. 1956. Myopia and inconsistency in dynamic utility maximization. Review of Economic Studies 23:165–180.

Taylor, C. M., and A. Hastings. 2004. Finding optimal control strategies for invasive species: a density-structured model for *Spartina alterniflora*. Journal of Ecology 41:1045–1057.

Taylor, C. M., and A. Hastings. 2005. Allee effects in biological invasions. Ecology Letters 8:895–908.

Weitzman, M. L. 2001. Gamma discounting. American Economic Review 91:260–271.

10

The Laurentian Great Lakes as a Case Study of Biological Invasion

David W. Kelly, Gary A. Lamberti, and Hugh J. MacIsaac

In a Clamshell

The Laurentian Great Lakes have an extensive history of human-mediated biological invasions, beginning at least 150 years ago. During this interval, a number of transitions have occurred with respect to both the types of non-indigenous species (NIS) that established and the mechanisms that vectored them to the lakes. Fish and plants were the most common NIS prior to 1900, with most introductions resulting from deliberate human releases. Algae and invertebrate establishments became more common after transoceanic shipping converted to the use of liquid ballast around 1890. The ship vector was dominant during much of the twentieth century. Since the expanded St. Lawrence Seaway opened in 1959, ships' ballast water has been the leading vector for approximately 55% of new established species. Eurasia was the source of 68% of NIS that have established in this period, followed by North America (14%), and palearctic/nearctic (7%). We review select NIS that have caused ecological and economic harm in the Great Lakes and, in some cases, spread to inland lakes. We close with a discussion of adaptive management for NIS on the Great Lakes.

Economic and financial factors strongly influence many human decisions and activities, which in turn can affect ecological systems and their associated services. Expansion of global trading networks and trade liberalization policies (e.g., NAFTA, GATT) may inadvertently expose ecosystems to new nonindigenous species (NIS) (Tatum et al. 2006; Tatum and Hay 2007). Levine and D'Antonio (2003) determined that accumulation of mollusk, plant pathogen, and insect NIS were all positively associated with the cumulative value of imported products to the United States. Similarly, Dehnen-Schmutz et al. (2007) noted that frequency of sale and seed price were significant predictors of invasion success for ornamental plants sold in Britain. Once established, NIS may adversely affect a variety of ecosystem services (see, e.g., Cook et al. 2007) and motivate significant expenditure on control activities (e.g., Pimentel et al. 2005; Colautti et al. 2006; Xu et al. 2007; see also chapter 8). Although commercial vectors (e.g., shipping, horticulture, pet trade) are responsible for most NIS introductions, the costs caused by invasive species are generally borne by government, private individuals, or commercial sectors (e.g., agriculture) other than those that introduced the species.

Forty-one percent of humans live in coastal habitats worldwide, and 21 of the world's largest 33 cities are located within 100 km of the coast (Martinez et al. 2007). Cities are located near coastal areas for a number of reasons, of which ready access to sea ports and marine food sources are particularly important. Costanza et al. (1997) estimated that fully 63% of total global ecosystem services were derived from oceanic ecosystems, principally within coastal zones. Martinez et al. (2007) increased this value to 77%, owing to their inclusion of benefits derived from terrestrial habitats located less than 100 km from the coast.

Cities and coastal ports are often located at the mouth of major river systems. For example, cities and major ports in Europe are located at the outflow of the Rhine (Rotterdam), Danube (Constanta), Schelde (Antwerp), Vistula (Gdansk), Neva (St. Petersburg), Bug (Mykolaiv), and Dnieper (Kherson) rivers. These ports are often linked to other regions by networks of constructed canals. One of the most important of these, the Rhine-Main-Danube canal, was opened in 1992 to link the Black Sea basin with the Rhine River basin. Development of canals facilitates enhanced trade and recreational travel. However, these canals also have been instrumental in the spread of NIS both regionally (Bij de Vaate et al. 2002) and internationally (Ricciardi and MacIsaac 2000), including to the Laurentian Great Lakes. Aquatic ecosystems, in general, are among the most vulnerable to invasion by NIS and attendant externalities (Sala et al. 2000; Connelly et al. 2007). Once invaded, these ecosystems may serve as NIS "hubs" and, by interacting with a global transport network, may facilitate the worldwide spread of invasive species (e.g., Drake and Lodge 2004; Muirhead and MacIsaac 2005; Tatum et al. 2006; Tatum and Hay 2007).

Impacts of invasion of aquatic ecosystems by NIS are numerous and include reduced or enhanced native species diversity (e.g., see Ward and Ricciardi 2007), disease transmission (Indiana Department of Natural Resources 2005), altered nutrient cycling patterns (Mellina et al. 1995), hybridization with native species (see Roman and Darling 2007), and biofouling of industrial and municipal water intake structures (Connelly et al. 2007). Strong adverse impacts associated with NIS establishment and spread have lead to global recognition of the problem, but responses by governments vary widely. Some countries, such as Australia and New Zealand, have implemented programs designed to assess and reduce the likelihood of future invasions and to eliminate or control NIS established within their countries (e.g., Cook et al. 2007; Keller et al. 2007). These efforts can effectively slow invasion rates but come with significant costs for quarantine, risk assessment, and restricted trade patterns.

The Laurentian Great Lakes of North America share many of the attributes described above for oceanic coastal regions, and thus are an ideal model system to examine the history, ecological impacts, economic effects, and societal responses to biological invasions. We review the invasion history of the Great Lakes, identify some major ecological and economic impacts of such invasions, and show how, through risk prevention and control, the biodiversity impacts of invasive species are inextricably linked to the Great Lakes economy.

LAURENTIAN GREAT LAKES

The Laurentian Great Lakes border Canada and the United States and are among the most heavily utilized water bodies in the world. Containing an estimated 20% of the planet's surface freshwater, the lakes provide more than 40 million coastal residents with access to drinking and industrial water, hydroelectric supplies, recreation, food, and transportation. Major metropolitan areas and their economies depend on the Great Lakes and the St. Lawrence Seaway, including Chicago, Detroit, Toronto, and Montreal, with more than $200 billion of economic activity conducted annually within the basin (U.S. Policy Committee for the Great Lakes 2002). For the United States alone, the region generates more than 50% of total manufacturing output. Fifty million metric tons of cargo annually passes through the Great Lakes in the international shipping trade, with the main commodities being grain, steel and iron ore, coal, coke, and petroleum products. About one-half of this cargo travels to and from overseas ports, mainly in Europe, the Middle East, and Africa.

In 1829, humans facilitated international commerce in the Great Lakes via construction of the Welland Canal, linking Lakes Erie and Ontario to provide navigable waterways between those water bodies. Subsequent development of the lock system between Lake Ontario and the Saint Lawrence River in 1847, and between Lakes Superior and Huron in 1855, allowed uninterrupted passage from the Atlantic Ocean to Lake Superior (Mills et al. 1993). However, after the Second World War, growth in international trade created the need to deepen the St. Lawrence waterway and allow larger ships to enter the Great Lakes. Canals were later expanded and the modern Saint Lawrence Seaway was opened in 1959. The development of a navigable water network has opened the Great Lakes region both to international trade and to the introduction of NIS. Attendant with the expansion of navigable waters was a major change in the risk of invasion from ships, as ballast utilized to stabilize incoming vessels changed from solid materials (e.g., stone, sand, soil, cobble) to liquid around 1890 (Mills et al. 1993).

In the ground-breaking retrospective study by Mills et al. (1993) of the invasion history of the Great Lakes, clear patterns emerged with respect to the nature of NIS that invaded during different time periods. The initial phase of invasion began in the early 1830s with the introduction of sea lamprey (*Petromyzon marinus*) to Lake Ontario via the Erie Canal. The Erie Canal was constructed between Lake Erie and the Hudson River to reduce the cost of transporting produce from the East Coast to growing human settlements in the Great Lakes basin and those farther west. Sea lamprey had profound ecological and economic impacts, causing severe declines in whitefish (*Coregonus clupeaformis*) and lake trout (*Salvelinus namaycush*) and negatively affecting commercial catches. Eight other NIS became established prior to 1850, all marsh-dwelling plants likely released through the nursery trade or from food cultivars (Mills et al. 1993). Between 1850 and 1900, the rate at which NIS were discovered increased and included a wider variety of species. Although marsh plants continued to be the dominant NIS during this period, shoreline trees, invertebrates, and fishes were also introduced. The dominant vectors of introduction were ships' solid ballast and accidental releases or food cultivar escapees. Solid ballast used to maintain ship stability during long ocean voyages was often discarded at Great Lakes destination ports, thus enabling the concomitant release of NIS (Mills et al. 1993).

Deliberate release (mainly by government agencies) also was an important vector of introduction (Ricciardi 2006), particularly for chinook salmon (*Oncorhynchus tshawytscha*), rainbow trout (*O. mykiss*), and brown trout (*Salmo trutta*), which were introduced to develop recreational and commercial fisheries. Several aquatic invertebrate NIS were apparently released to increase biological diversity (Mills et al. 1993). Among the "invasive" (i.e., harmful) species discovered between 1850 and 1900 were the submerged macrophytes curly pondweed (*Potamogeton crispus*) and spiny naiad (*Najas marina*), the wetland species purple loosestrife (*Lythrum salicaria*) and narrow-leaved cattail (*Typha angustifolia*), and the fishes alewife (*Alosa pseudoharengus*) and common carp (*Cyprinus carpio*).

Between 1900 and 1958, an additional 54 species were added to the NIS inventory of the Great Lakes (Mills et al. 1993). A recent analysis (see Kelly 2008) indicates that a further five species became established during this interval, for a total of 59 NIS. The taxonomic composition of NIS changed dramatically during the 1900–1958 period, with the first reports (7 species) of algae and a growing importance of invertebrates (17 species) and fishes (11 species). This NIS shift coincided with the gradual replacement of solid by liquid ballast in ships. Water was advantageous as ballast since it was readily available in foreign source ports and its volume could be easily adjusted to maintain ship draft. Thus, it is not surprising that ballast water release was responsible for the first appearance of planktonic invertebrates, including the copepod *Eurytemora affinis* and the water fleas *Daphnia galeata galeata* and *Eubosmina coregoni*. Notable "invasive" species that established in the Great Lakes during this period include the plants fanwort (*Cabomba caroliniana*), Eurasian watermilfoil (*Myriophyllum spicatum*), and water chestnut (*Trapa natans*), all of which outcompete native plants. Mechanisms of introduction varied widely across taxa. Because a particular species may have been introduced by more than one mechanism, in the following section we focus on the dominant vector but recognize that some assignments could be incorrect. Fishes were most often released deliberately (up to five species) or accidentally (up to four species) from fish hatcheries or the aquarium trade. Dominant introduction mechanisms for invertebrates were unintentional release (up to five species) associated with imported ornamental plants or the aquarium trade, while two species each were likely introduced with solid ballast or as cultivation escapees, and finally one by canal. Algae were almost exclusively released from ballast water (six species). Cultivation escapees (up to five species), ships' solid ballast (four species), and accidental releases (up to three species) were dominant introduction mechanisms of submerged plants, although two species may have entered attached to ship hulls. In summary, the early phase of invasion in the

Great Lakes was dominated by wetland plants, while algae, invertebrates, and fishes became far more common additions after 1900. Until 1959, accidental introductions associated with shipping, canals, and other economically motivated activities accounted for up to 37 NIS introductions (16 animals, 21 plants), whereas intentional introductions associated with stocking programs and other activities accounted for up to 26 NIS introductions (19 animals, 6 plants, 1 pathogen) (modified from Mills et al. 1993; Ricciardi 2006).

INVASIONS IN MODERN TIME

Determination of the vector (i.e., mode of introduction), pathway (i.e., source), and timing of species invasion is a valuable tool for management and control efforts (Ruiz and Carlton 2003). Invasion histories and vulnerability of ecosystems to future invasions can be re-created and predicted by examining putative vectors (Mills et al. 1993), including international shipping (e.g., Colautti et al. 2003; Drake and Lodge 2004), or by phylogeographic assessments of genetic structure of native and introduced populations (e.g., Cristescu et al. 2001; Kelly et al. 2006). These approaches are particularly relevant to the Laurentian Great Lakes, where vectors, pathways, and the composition of NIS changed markedly after 1959. For instance, a number of studies have explored invasion patterns in the lakes following the opening of the expanded, modern St. Lawrence Seaway in 1959, an event that permitted larger foreign vessels access to the entire Great Lakes system. Grigorovich et al. (2003) and Ricciardi (2006) attributed 67% and 65% of post-seaway introductions, respectively, to ballast water. Ballast water that is loaded in regions outside of the Great Lakes may entrain large numbers of viable freshwater species that are discharged during cargo loading in Great Lakes ports. BOB ("ballast water on board") ships were recognized as a major vector of NIS introduction to the Great Lakes, and between 1989 and 1993 the United States and Canada introduced ballast water control policies aimed at reducing further introductions (Locke et al. 1991; U.S. Coast Guard 1993). Since 1993, BOB ships have been required to conduct ballast water exchange (BWE) in open-ocean marine waters to purge freshwater NIS from tanks and kill those that remain by exposure to saltwater. Despite these management efforts, 19 new NIS have been reported in the Great Lakes system since 1993, nine of which were most likely introduced by the ship ballast vector (see table 10.1).

Several factors could account for continuing invasions of the Great Lakes. Up until the early 1980s, a large number of Russian ships entered the Great Lakes to collect grain (I. Lantz, Shipping Federation of Canada, personal communication). Most of these vessels are believed to have carried ballast water of Baltic Sea origin, the native or introduced range of many Great Lakes invaders, and these ships may have introduced many NIS. The decline in Russian grain ship visits toward the end of the 1980s was likely due to a need to reduce debts to Western creditors, improved home grain yields, and the imposition of a U.S. embargo on grain sales to the former USSR. During the 1980s, the characteristics of transoceanic ships also changed. An increasing proportion of ships declared no-ballast-on-board status for some or

TABLE 10.1. Vectors and origins of NIS reported in the Great Lakes since 1959 (modified from Kelly 2008).

Species	Common name	Year discovered	Geographic origin	Primary vector	Secondary vector
Pisidium supinum	Humpback pea clam	1959	Atlantic North America	SB	SB
Trapa natans	Water chestnut	1959	Europe	DR	UI
Persicaria longiseta	Lady's thumb	1960	East Asia	U	U
Glugea hertwigi	Protozoan	1960	Eurasia	DR	DR
Lepisosteus platostomus	Shortnose gar	1962	Mississippi	RE	RE
Bangia atropurpurea	Red alga	1964	Atlantic Europe	SB/SF	SB/SF
Epilobium parviflorum	Hairy willow herb	1966	Eurasia	U	U
Dugesia lugubris	Flatworm	1968	Europe	SB	SB
Myxobolus cerebralis	Whirling disease	1968	Unknown	A	A
Solidago sempervirens	Seaside goldenrod	1969	Atlantic North America	U	U
Enneacanthus gloriosus	Bluespotted sunfish	1971	Atlantic North America	UI	RE
Cyclops strenuous	Copepod	1972	Europe	SB	RE
Nitocra hibernica	Copepod	1973	Eurasia	SB	SB
Lotus corniculatus	Birdsfoot trefoil	1975	Eurasia	DR	DR
Renibacterium salmoninarum	Bacterium	1975	Unknown	A	A
Nitellopsis obtuse	Green alga	1978	Eurasia	SB	SB
Biddulphia laevis	Diatom	1978	Widespread	U	U
Enteromorpha prolifera	Green alga	1979	Widespread	U	U
Corbicula fluminea	Asiatic clam	1980	East Asia	RB	RB
Ripistes parasita	Oligochaete	1980	Eurasia	SB	SB
Lupinus polyphyllus	Lupine	1982	Western North America	DR	ND
Phallodrilus aquaedulcis	Oligochaete	1983	Europe	SB	SB
Bythotrephes longimanus	Spiny water flea	1984	Eurasia	SB	SB
Gymnocephalus cernuus	Eurasian ruffe	1986	Europe	SB	SB
Apeltes quadracus	Fourspine stickleback	1986	Atlantic North America	SB	UI
Thalassiosira baltica	Diatom	1988	Baltic Sea	SB	SB
Argulus japonicus	Copepod	1988	Asia	UI	UI
Dreissena polymorpha	Zebra mussel	1988	Ponto-Caspian	SB	SF
Bosmina maritima	Water flea	1988	Eurasia	SB	SB
Scardinius erythrophthalmus	Rudd	1989	Eurasia	UI	RE

Dreissena rostriformis bugensis	Quagga mussel	1989	Ponto-Caspian	SB	SF
Apollonia melanostoma	Round goby	1990	Ponto-Caspian	SB	SB
Proterorhinus marmoratus	Tubenose goby	1990	Ponto-Caspian	SB	SB
Potamopyrgus antipodarum	New Zealand mud snail	1991	Australasia	SB	A
Neascus brevicaudatus	Digenean fluke	1992	Eurasia	SB	SB
Dactylogyrus amphibothrium	Monogenetic fluke	1992	Eurasia	SB	SB
Acanthostomum sp.	Digenean fluke	1992	Eurasia	SB	SB
Trypanosoma acerinae	Flagellate	1992	Ponto-Caspian	SB	SB
Dactylogyrus hemiamphibothrium	Monogenetic fluke	1992	Eurasia	SB	SB
Ichthyocotylurus pileatus	Digenean fluke	1992	Ponto-Caspian	SB	SB
Scolex pleuronectis	Cestode	1994	Ponto-Caspian	SB	SB
Neoergasilus japonicus	Copepod	1994	East Asia	A	A
Megacyclops viridis	Copepod	1994	Europe	SB	SB
Sphaeromyxa sevastopoli	Mixosporidian	1994	Ponto-Caspian	SB	SB
Echinogammarus ischnus	Amphipod	1995	Ponto-Caspian	SB	SB
Alosa aestivalis	Blueback herring	1995	Atlantic North America	RE	RE
Heteropsyllus nr. Nunni	Harpacticoid copepod	1996	Unknown	U	U
Cercopagis pengoi	Fishhook water flea	1998	Ponto-Caspian	SB	SB
Schizopera borutzkyi	Copepod	1998	Ponto-Caspian	SB	SB
Nitocra incerta	Copepod	1999	Eurasia	SB	SB
Daphnia lumholtzi	Water flea	1999	Africa	RB	RB
Heterosporis sp.	Microsporidian	2000	Unknown	UI	UI
Gammarus tigrinus	Amphipod	2001	Atlantic North America	SB	SB
Piscirickettsia cf. salmonis	Muskie pox	2002	Unknown	A	A
Largemouth bass virus	Iridovirus	2003	Unknown	U	U
Enteromorpha flexuosa	Green alga	2003	Widespread	SF	SB
Viral hemorrhagic septicemia (VHS)	Fish virus	2003	Atlantic North America	U	U
Rhabdovirus carpio	Carp viremia	2003	Eurasia	RE	RE
Hemimysis anomala	Opossum shrimp	2006	Ponto-Caspian	SB	SB

Vectors: SB, ships' ballast; SF, ship fouling; DR, deliberate release; UI, unauthorized intentional; U, unknown; RE, range extension; A, aquaculture; RB, recreational boating; ND, natural dispersal. Primary vector is the most likely vector based on available information; secondary vector is the next most likely vector.

all of their ballast tanks; these "NOBOBs" carried cargo and were exempt from ballast control. However, their tanks still carried large volumes of residual water and sediment. Economically, these vessels experience higher operational efficiency since they backhaul cargoes from the Great Lakes after delivering steel, petrochemicals, or other products to Great Lakes ports (Colautti et al. 2003). Studies have shown that residual ballast harbors large numbers of viable NIS, which pose a risk of discharge to the Great Lakes during multiport operations (Colautti et al. 2003; Bailey et al. 2005). Regulations introduced in Canada in 2006 and the United States in 2008 require that NOBOB tanks be flushed at sea to eliminate freshwater residue (unpublished data; Canada Shipping Act 2006). Since these rules augment those for BOB vessels, and should affect transport of many of the same species, it will take some time before the effectiveness of these policies can be assessed.

A further problem in determining BWE efficiency for ballasted vessels is time lags (Costello et al. 2007), of which two may occur. First, species may not be detected in the lakes until well after they were introduced. Second, a gap may exist between when species are first detected and when they are first reported (e.g., time to positively identify a species). Time lags almost certainly vary in length, depending on the conspicuousness and invasiveness of the species and the habitats that they colonize.

The high profile of ballast-mediated invasions may also have distracted researchers from consideration of alternative vectors, leading to uncertainty in evaluations. For example, large numbers of fouling organisms—species that usually have sessile adults capable of attaching to the hull, anchor chain, or other external surfaces—have been found attached to ships, a subvector that is dominant in marine environments (Ruiz et al. 2000; Gollasch 2002; Drake and Lodge 2007). Also, the live fish market, aquarium, and aquaculture industries have received less attention, but recent research has indicated that these vectors pose a significant risk to the Great Lakes (Kolar and Lodge 2002; Rixon et al. 2004; Cohen et al. 2007; Keller and Lodge 2007).

In a recent study, Kelly (2008) prioritized alternative vectors to assess the most likely origin and pathway of introduction of NIS since 1959. Eurasia has been the dominant donor region, accounting for 67% of NIS in the Great Lakes since 1959 (figure 10.1, table 10.1). Within Eurasia, Europe and the Black and Caspian Sea basins (Ponto-Caspian) contributed most of the species, with Southeast Asia and the Baltic Sea being of lesser importance. North America has been the next most important donor region, accounting for approximately 14% of NIS, most of which originate in the Northwest Atlantic coastal region. Ship ballast was the strongest vector and accounted for up to 55% of all NIS primary vector assignments (table 10.1). It is interesting that 94% (31 of 33 species) of ship-ballast invasions originated in Eurasia (figure 10.1). Overall, Europe and the Black Sea basin were the main donor regions, which is consistent with coarse measures of propagule pressure—a product of the number of introduced individuals or infective stages and their frequency of introduction—in the Great Lakes. Very little vessel traffic to the Great Lakes originates in Black Sea ports (Colautti et al. 2003), and thus few species—with the exception of quagga mussels (*Dreissena bugensis* (= *D. rostriformis bugensis* [Andrusov

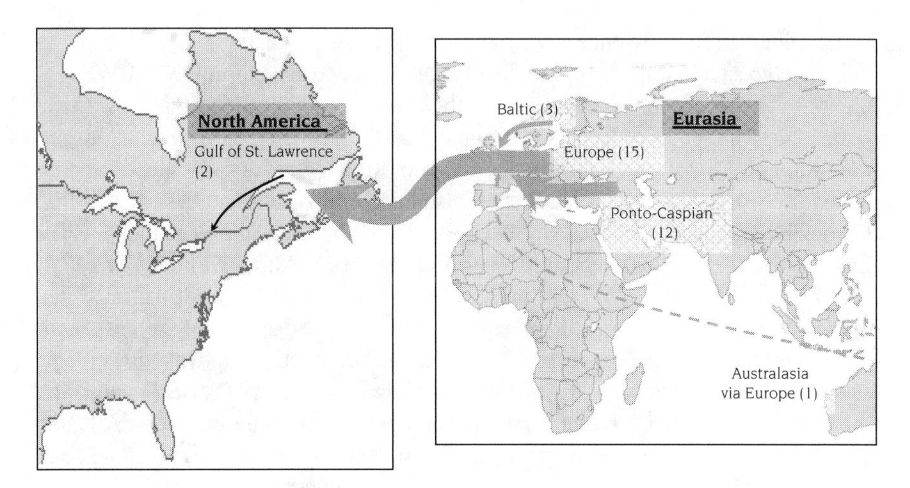

FIGURE 10.1.
Pathways of ship-ballast–vectored invasions to the Great Lakes since 1959. Arrow width is proportional to the strength of each donor subregion. Numbers in parentheses are the number of species originating in each region. Reprinted from Kelly (2008), with permission of the Transportation Research Board of The National Academy of Sciences of the USA.

(1897)]))—were likely brought in directly from this region. Most of the Black Sea species that have established in the Great Lakes also have an invasion history in western European ports, from which most inbound vessel traffic to the Great Lakes does originate (Ricciardi and MacIsaac 2000). Indeed, the opening of the Main Canal in 1992, which connects the Danube with the Rhine system, provided a major westward colonization pathway from the Black, Azov, and Caspian Sea basins (Bij de Vaate et al. 2002). The Main Canal was completed despite concerns regarding possible spread of NIS. It should be noted, however, that concern in Europe regarding NIS has increased dramatically in the past decade, in part due to the spread of Ponto-Caspian species. This invasion pathway, coupled with the impacts of Ponto-Caspian species in western Europe, provided a basis for several key studies that warned of future potential invaders from this region (e.g., Ricciardi and Rasmussen 1998; Grigorovich et al. 2003). However, despite the implementation of ballast control policies between 1989 and 1993, Ponto-Caspian species have continued to invade the Great Lakes (table 10.1). It is possible that such recent invasions were due to a lack of awareness of risks due to residual sediment and water in NOBOB vessels, a ship subvector that has only recently received the attention of policy makers, or to time lags in discovery or reporting.

The only non-Eurasian species likely introduced to the Great Lakes in ship ballast were of Northwest Atlantic origin. Both *G. tigrinus* and *A. quadracus* were likely introduced from the Gulf of St. Lawrence. It is difficult to discern the type of vessel that may have brought these species into the Great Lakes, because it could have been accomplished by "salties" (ships transiting the seaway to conduct international trade), by "lakers" (ships that trade mainly within the Great Lakes but venture into the St. Lawrence River), or by coastal vessels that operate within North America (coastal marine areas and the Great Lakes) and are exempt from BWE regulations. For example, lakers occasionally offload cargo and load ballast water at ports in the St. Lawrence River and estuary (e.g., Quebec City, Sorel, Baie-Comeau) downstream of Montreal (Eakins 1999, 2000; M. Rup, University of Windsor, unpublished data). Vessels with NOBOB tanks also may offload cargo at ports such as Port Cartier, downstream of the seaway, before taking on ballast and proceeding to Great Lakes ports for cargo collection (Colautti et al. 2003). Although both *G. tigrinus* and *A. quadracus* are seemingly innocuous NIS, transfer of coastal ballast water, or of infected fishes contained therein, is a possible mechanism (along with naturally migrating fishes) responsible for the recent introduction of viral hemorrhagic septicemia (VHS) into the Great Lakes. Although the origin of VHS is uncertain, molecular studies indicate that it may have originated from the northeast Atlantic. The economic consequences of this disease introduction are not yet known, although it is likely to be profound, because VHS affects more than 40 species of fish.

Since 1959, ship fouling has been implicated in only two species introductions (the algae *E. flexuosa* and *B. atropurpurea*), while deliberate release, aquaculture, range extensions, and unauthorized intentional vectors were individually of minor importance, but collectively represented 26–32% of all NIS introductions. Both recreational boating and natural dispersal were of minor importance. Only three introductions have occurred via canals and one via recreational boating (table 10.1). The small proportional contribution of canal-mediated introductions to the total introductions in the Great Lakes is a finding consistent with previous works (see Mills et al. 1993; Ricciardi 2006). However, the unimpeded canal pathways from the Mississippi and Hudson river basins continue to pose a serious risk of future introductions (figure 10.2). This risk is highlighted by the recent construction of an electric fish barrier on the Chicago Sanitary and Ship Canal (CSSC) to prevent spread of round gobies from the Great Lakes to the Mississippi River, but which is now being used to prevent entry by Asian silver and bighead carp into the Great Lakes (Stokstad 2003). These Asian carp, which dispersed via the Mississippi River into the lower Des Plaines River, immediately downstream of the CSSC, pose an additional risk to the Great Lakes since some are infected with carp viremia (*Rhabdovirus carpio*), a virus of Eurasian origin (Nelson 2003). If Asian carp overcome the electrical barrier, there is a risk that fish in the Great Lakes could become infected with the virus. Herborg et al. (2007) utilized an environmental niche model to identify the Great Lakes as vulnerable to invasion by both carp species, whereas Kolar and Lodge (2002) predicted the opposite based upon life history characteristics of the carp.

Prior to 1959, the Erie Canal was of relatively minor importance for Great Lakes NIS, but it has allowed invasion by several North Atlantic species that have had substantial impacts (Mills et al. 1993). Although only a single species, the blueback herring, invaded via this pathway since 1959, the canal could be an entrance mechanism in the future.

Thus, consideration of all possible vectors since 1959 indicates that ship ballast has been responsible for the greatest numbers of NIS introduced to the Great

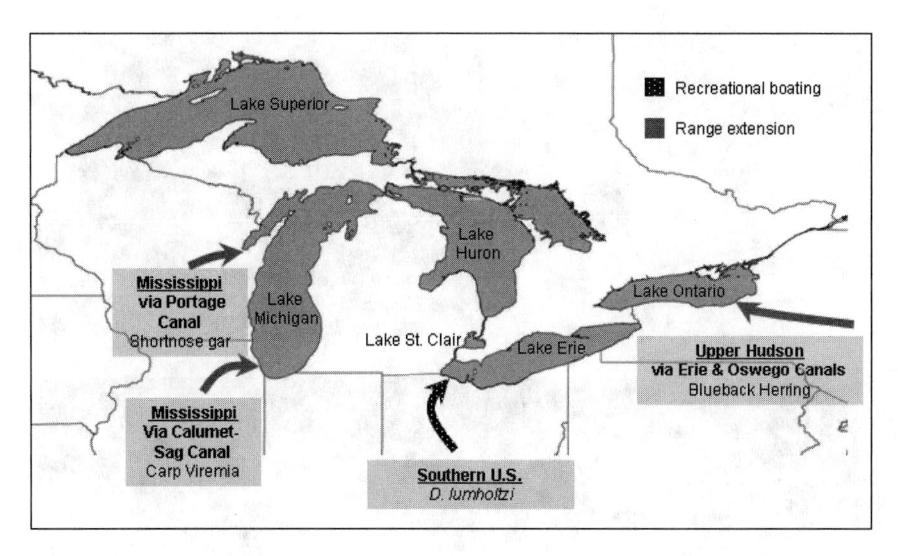

FIGURE 10.2.
Canal and recreational boating pathways of invasion to the Great Lakes since 1959. Reproduced from Kelly (2008), with permission of the Transportation Research Board of the National Academy of Science of the USA.

Lakes. As mentioned above, this continuing risk has focused attention of U.S. and Canadian governments on ballast management policies whose effectiveness is difficult to determine. What is clear is that NIS continue to be discovered in the Great Lakes, the majority of which have a European or Black Sea origin, with ship ballast as the likely vector. Other species are colonizing key port areas in Europe, and so the Great Lakes remain at risk of ballast-mediated introductions from this region. This continuing risk is illustrated by the most recent NIS, the mysid shrimp *Hemimysis anomala*, which was predicted to pose a risk in recent assessment models (Ricciardi and Rasmussen 1998; Grigorovich et al. 2003). Ballast water of European origin was the most likely vector of this species.

In the following sections, we highlight case studies of NIS that have had significant ecological and economic effects in the Great Lakes basin. Species considered include the water fleas *Bythotrephes longimanus* and *Cercopagis pengoi*, dreissenid mussels *D. polymorpha* and *D. bugensis*, and the round goby *Apollonia melanostoma* (= *Neogobius melanostomus*). In each case, we consider the vector of invasion, distribution, and consequences in the Great Lakes and secondary spread to other systems.

Predatory Water Fleas

The water fleas *Bythotrephes longimanus* and *Cercopagis pengoi* (figure 10.3) are predators of other zooplankton species and share similar life histories. Both are

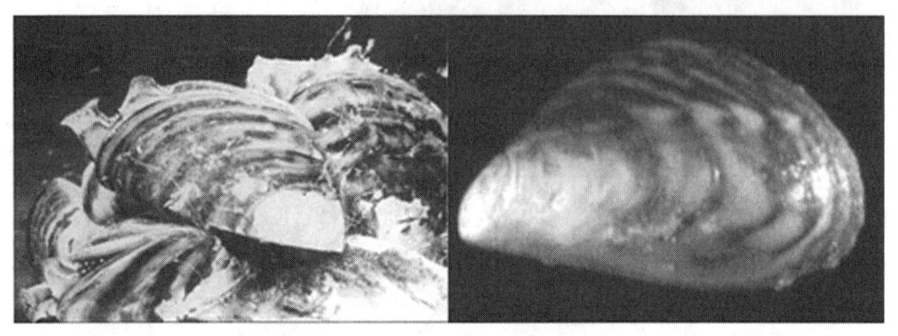

FIGURE 10.3.
Representative introduced species that have caused significant ecological and economic harm to the Great Lakes include water fleas *Cercopagis pengoi* and *Bythotrephes longimanus* (upper and lower, respectively, in the top left image), zebra and quagga mussels (left and right, respectively, in middle row), and round gobies and sea lamprey (left and right, respectively, in lower row). Upper right image highlights water flea fouling (*Cercopagis*) of commercial gill nets on Lake Erie (Ontario Ministry of Natural Resources). *Cercopagis* and *Bythotrephes* image courtesy of Dr. H. Vanderploeg (with permission of the Canadian Journal of Fisheries and Aquatic Sciences, CRC Press); round goby courtesy of Shedd Aquarium.

members of the crustacean family Cercopagidae, both have alternating modes of sexual and asexual reproduction, and both have invasion histories in Europe and the Great Lakes. *Bythotrephes* has a Eurasian native distribution, whereas *Cercopagis* is native to the Caspian, Black, Azov, and Aral seas. The first record of *Bythotrephes* in North America is from Lake Huron (1984), followed by Lakes Ontario and Erie (1985), Lake Michigan (1986), and Lake Superior (1987). *Cercopagis* was also discovered first in Lake Ontario (1998) and subsequently spread to Lake Michigan (1999) and Lake Erie (2001). *Cercopagis* has not yet been reported from Lake Superior or Huron. Both species were almost certainly transferred to the Great Lakes in contaminated ballast water from Europe (table 10.1) and subsequently moved within the Great Lakes by internal ballast water transfers by salties or lakers. The latter vessels load and discharge disproportionately more water within the Great Lakes system (M. Rup, unpublished data) and thus are the more likely regional vector.

Bythotrephes has spread rapidly to inland lakes, beginning with Lake Muskoka in central Ontario in 1989. The species has continued to spread, with 108 lakes now reported invaded in the province, including 18 new reports in 2006 (A. Cairns, N. Yan, and J. Muirhead, unpublished data; figure 10.4). Inland lakes have also been invaded

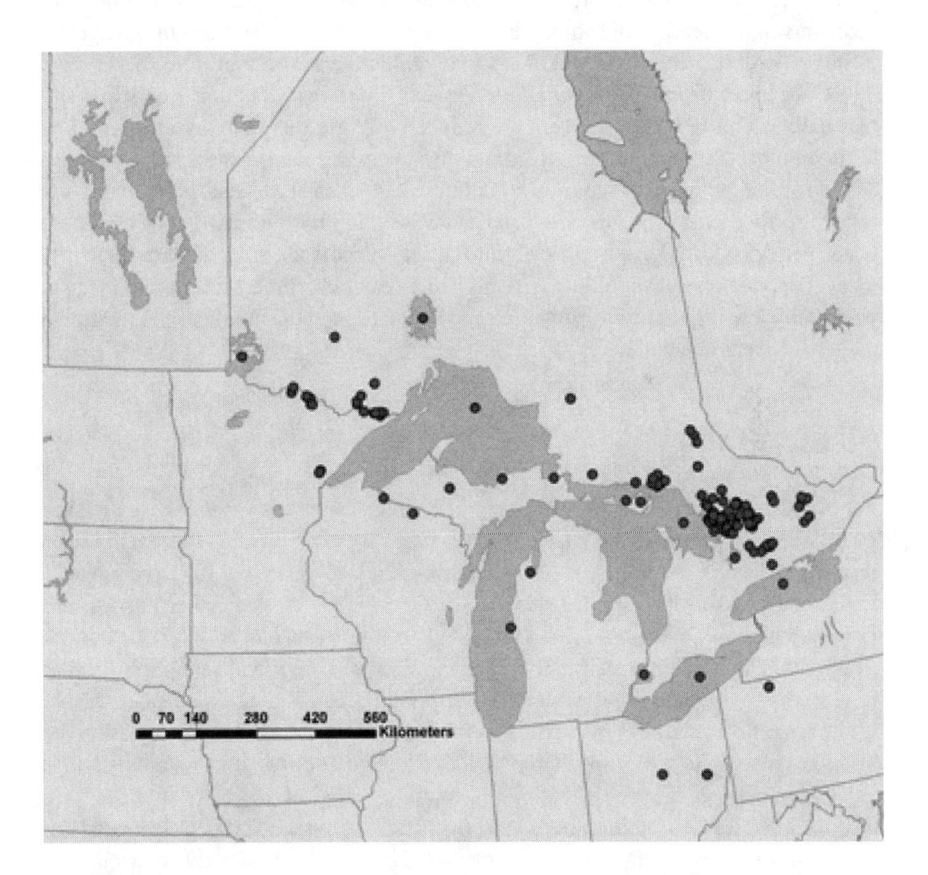

FIGURE 10.4.
Distribution of the spiny water flea *Bythotrephes longimanus* in Canada and United States (2007). The species colonized the Great Lakes in the early 1980s and spread to inland lakes in the United States and Canada beginning around 1989. Data kindly provided by the Canadian Aquatic Invasive Species Network and Dr. Jim Muirhead.

in the United States, although reports are seemingly an order of magnitude fewer than in Canada (figure 10.4). In Minnesota, *Bythotrephes* has been confirmed in 16 inland lakes, mainly on the border with Ontario (U.S. Geological Service Invasive Species Database, unpublished data). *Bythotrephes* has also been confirmed in at least four lakes in Michigan, two reservoirs in Ohio, and two lakes in Wisconsin (figure 10.4). It is not clear whether the differential occurrence on Canadian and U.S. sides of the Great Lakes is real or the result of differential sampling and reporting. Interestingly, recreational activities and numerous hydraulic connections west of Lake Superior may allow the species to move between the two countries outside of the Great Lakes.

Bythotrephes has spread much faster to inland systems than has *Cercopagis*, which, other than the Finger Lakes in New York, has failed to colonize inland systems. The differential occurrence of *Bythotrephes* in Canada and the United States and the differential rate of spread of the two species are puzzling since both species produce resting stages that foul fishing lines and are believed to be the primary mechanism of spread to and among inland systems (figure 10.3). In addition, boaters in states bordering the Great Lakes are likely as active as those in Canada.

Once lakes are invaded by *Bythotrephes* or *Cercopagis*, abundance and diversity of small and midsize zooplankton are reduced . While this nonmarket effect may seem unimportant, it could lead to competition between the invaders and larval fish for food. Future work is required to determine if changes in fish populations are related to food web changes associated with invasion by these two species. These water fleas may also have a more direct impact on sport fisheries and consequently local economies because fouling of fishing lines can hinder recovery of fishing gear, potentially leading to angler frustration and a reduction in the number of recreational anglers visiting invaded lakes.

Dreissenid Mussels

Zebra mussels (*Dreissena polymorpha*) and quagga mussels (*D. bugensis*) (figure 10.3) are mollusks from the Black Sea basin that were reported in the Great Lakes in 1988 and 1989, respectively. *D. polymorpha* has an extensive invasive range in Europe and was probably introduced to the Great Lakes from northern or western Europe, whereas *D. bugensis* has a limited distribution and has only recently begun to spread. The species are virtually identical in morphology and seemingly similar in reproductive biology and other traits, although quagga mussels occur in deeper water not inhabited by zebra mussels. Both species were likely introduced as larvae in ballast water from Europe (table 10.1), although fouling of structures such as anchor chains, sea chests, or floating macrophytes by adult mussels may have been responsible (Horvath and Lamberti 1997). It is this fouling ability, coupled with high dispersal ability by natural and human-mediated means, that has likely contributed to the regional spread of zebra and quagga mussels within the Great Lakes and beyond (see chapter 12 for a more comprehensive treatment of establishment and dispersal of zebra mussels). Both species now have extensive histories of spread in temperate eastern North America with ecological effects that are more profound than any other aquatic NIS (for reviews, see MacIsaac 1996; Ward and Ricciardi 2007). The recent discovery of zebra mussels in San Justo Reservoir in San Benito County, California, in January 2008 and quagga mussels in Lake Mead in Nevada and lakes in California highlights the importance of the Great Lakes as a hub for regional spread (see chapter 12).

Both mussels are capable of having significant adverse economic impacts, mainly through fouling of water intake facilities, including hydroelectric units, other power plants, and municipal water supply plants. However, even for these important species, economic data are incomplete; the best evidence comes from hydro plants, which individually spend between $400,000 and $1,500,000 Canadian per year to prevent colonization (Colautti et al. 2006). Connelly et al. (2007) placed the total expenditure of water treatment plants and hydro installations at $267 million U.S. between 1989 and 2004, or about $44,000 per facility per year. Estimates by both Colautti et al. (2006) and Connelly et al. (2007) are far lower than the projections of the U.S. Fish and Wildlife Service's estimate of $5 billion over 10 years. Colautti et al.'s (2006) value was based on tractable direct expenses, whereas the latter was based upon an extrapolation for all manufacturers and municipalities using raw water from the system.

Damage and control costs associated with the quagga mussel invasion of Lake Mead could be much higher owing to the presence of the massive Hoover Dam hydro works on that system. Leung et al. (2004) estimated that it was cost-effective to spend up to $324,000 U.S. annually to prevent colonization of a single midwestern lake with a large power plant located on it. Many other economic changes wrought by zebra and quagga mussels are nonmarket (e.g., extirpation of native unionid mussels) and not well studied (see chapter 12).

The Round Goby

Round gobies [*Apollonia melanostoma* (=*Neogobius melanostomus*)] (figure 10.3) were first reported in North America from Lake St. Clair (see figure 10.2) in 1990 (Jude et al. 1992). Another Black Sea native, this species was likely introduced via ballast water (table 10.1). Round gobies have spread to each of the Great Lakes, often forming very large populations. The species has also been found in the lower Gulf of St. Lawrence, in the CSSC, in at least two inland lakes in Michigan, and in Pefferlaw Brook, a tributary of Lake Simcoe, a large inland lake in Ontario. Gobies may disperse naturally or in ballast water within the Great Lakes, or as unrecognized baitfish contaminants to inland systems. A population in Pefferlaw Brook, Canada, was subjected to a $250,000 Canadian eradication campaign during 2005 to protect a recreational fishery valued at approximately $200 million Canadian per year. While the eradication effort seemed successful initially, gobies were found in the same system a year later. Round gobies may have a wide variety of trophic effects, including adverse effects on recruitment of native fishes via predation on their eggs, but also possibly beneficial consumption of smaller size classes of zebra mussels—a preferred prey item (Bauer et al. 2007). Ominously, infected round gobies may have contributed to the spread of VHS (table 10.1) in the Great Lakes owing to the large biomass "reservoir" they represent. If this is correct, it represents a form of "invasional meltdown," where the dispersal and impact of one invasive species is facilitated by another (Ricciardi 2001).

Round gobies also have been implicated in die-offs of diving waterfowl (e.g., common loons, mergansers), involving a chain of events beginning with growth and consumption of *Clostridium botulinum* (type E) bacteria by zebra mussels, which in turn are eaten by round gobies, which are consumed by waterfowl, which then fall critically ill (Yule et al. 2006). The economic impact of both VHS and botulism poisoning is not known, though it could be enormous, particularly for VHS because it affects more than 40 species of sport and commercial fishes in the basin.

Largely in response to invasion of the Great Lakes by round gobies, an electrical barrier was constructed in the CSSC to prevent spread of the fish to the Mississippi drainage via the Illinois River (figure 10.2). By the time the $1.3 million U.S. barrier was constructed and operational, round gobies had already passed downstream. This barrier does, however, provide a serendipitous defense against movement of bighead carp (*Hypophthalmichthys nobilis*) and silver carp (*H. molitrix*) into the Great Lakes from the Mississippi River. A second barrier, valued at $9.1 million U.S., has been constructed as a backup defense but is not yet operational. While spread of either of these fishes to the Great Lakes is highly undesirable, they are cultured and sold as food elsewhere, and bighead carp are harvested from the Upper Mississippi River for food. There is a high risk that Asian carp could spread via the CSSC to the Great Lakes, although it is uncertain whether the species would thrive in the Great Lakes (Kolar and Lodge 2002; Herborg et al. 2007). Because the species are also sold in Asian food stores in both the United States and Canada, these stores could provide a secondary mechanism of introduction of the species to the Great Lakes. This possibility seems remote, however, because all Great Lakes states and the province of Ontario have implemented bans on live sale, possession, or transport of these fishes.

MANAGEMENT OF INVASIVE SPECIES IN THE GREAT LAKES

Biological invasions have clearly wrought irrevocable changes to the nature of food webs in the Great Lakes and how humans interact with those resources. In the United States alone, the total economic loss due to invasive species is estimated to exceed $120 billion annually (Pimentel et al. 2005). No clear estimates exist for the monetary costs of invasive species in the Great Lakes, but the annual total must certainly be in the billions of dollars. Furthermore, the impact of invasions on Great Lakes ecosystems and society clearly have ecological and nonmarket costs, and the latter are difficult to quantify. For example, the extirpation by zebra mussels of native unionid mussels (some of which are of conservation concern) in many inland lakes has left "graveyards" of unionid shells in place of once-thriving native mussel beds; such ecological costs could be quantified using the frameworks outlined in chapter 8, although this has not yet been done.

With the attendant ecological and economic impacts of NIS in the Great Lakes, the system serves as a useful model to illustrate how NIS management can benefit from an adaptive tiered approach. Virtually all experts recognize the inherent value of prevention of invasions (see Ruiz and Carlton 2003; Lodge et al. 2006), and in the current chapter, the focus on vectors, timing, and identity of species invading the Great Lakes has a number of strengths. First, explicit prioritization of vectors allows funds and efforts to be focused on mechanisms most important in transmitting NIS to the lakes. The value of this approach is that we can prevent invasions both by the many species we are aware of and by others not yet identified but that may use a particular vector. Second, the number of species invasions prevented is most likely to be maximized by prioritizing and eliminating the strongest vectors to the lakes. Thus, recent patterns of invasion to the Great Lakes indicate that management of ballast of ships arriving from Europe should reduce the risk of future invasions (e.g., see figure 10.1, table 10.1). Analyses of invasion timelines and the identity of particular NIS can help inform the efficacy of current management programs as well as direct future programs. For example, although midocean BWE policies for transoceanic ships reduces the risk of introducing species intolerant of high salinity, these strategies appear to have been less effective for sediment-dwelling species or those capable of producing resistant resting stages. For example, nine NIS were likely introduced in ballast sediment since ballast water control policies were implemented in 1993 (see table 10.1). This information underpinned recent programs aimed at the management of NOBOB residuals in the Great Lakes (U.S. Coast Guard 2005; Canada Shipping Act 2006). As of 2008, all vessels from non-North American source ports must flush ballast water and/or ballast residuals before entering the Great Lakes if they intend to perform any ballast discharges while operating on the Great Lakes. This policy should effectively eliminate new introductions of European or Asian species via the ballast vector. Because of time lags, however, it might be some time before the efficacy of this policy can be assessed.

Modeling efforts may be useful to identify whether specific organisms pose an invasion risk based upon assessments of life history attributes, propagule pressure, environmental suitability, or a combination of these approaches (e.g., Kolar and Lodge 2002; Muirhead and MacIsaac 2005; Herborg et al. 2007). These approaches will likely be limited to only those species for which excellent background information exists and that are perceived as potentially problematic (e.g., Chinese mitten crabs *Eriocheir sinensis*, Asian carps). However, managers can utilize the output of these models to discriminate between NIS that may or may not establish and spread, and those likely to have large versus small impacts. Management efforts could be tailored accordingly to guard against introduction of those NIS most likely to survive and become problematic in the Great Lakes.

A focus on prevention cannot be expected to prevent all invasions. In such cases, early detection is desirable, although often difficult. Once new NIS incursions are detected, scientific risk assessments are required to determine the appropriate management response. Some NIS may be perceived as having little potential for establishment, spread, or impact following establishment. These assessments can often be made by examining the life history attributes of the species (e.g., Kolar and

Lodge 2002) and interspecific interactions and economic impacts in regions where the species is established. If the risk of establishment and/or the risk of adverse impacts is deemed to be low, then it might be appropriate to take no further action other than managing the vector that was responsible for the introduction. If the species is deemed a moderate to high risk, then additional actions may be warranted. These actions consist of eradication or, failing that, a control-the-spread strategy. The number of invertebrate, aquatic NIS that have been successfully eradicated is quite small (e.g., black striped mussel *Mytilopsis* in Australia and green alga *Caulerpa* in California; Bax et al. 2002; Williams and Schroeder 2004). A central problem is detection of the incursion at a sufficiently early stage that the population size and range of an NIS are very small and relatively easy to manage. Nevertheless, cases may occur where it is economically advantageous to establish monitoring programs to facilitate early detection of nascent invasions, particularly where the potential for biofouling is large or the threat to native biodiversity is great. The 100th Meridian Project was designed with this in mind (see chapter 12), although the recent establishment of quagga mussels in the western United States highlights the difficulty in completely eliminating vector activity. Creation of barriers to dispersal, including the electrical field barrier in the CSSC, is an example of a control-the-spread strategy that may be effective not only for target species (e.g., silver and bighead carp) but other NIS, as well.

When prevention and eradication are ineffective, managers and society must adapt to life with the established NIS. At this point, managers are essentially helpless with respect to distribution of the NIS, as for the case with dreissenid mussels in the Great Lakes. Here, management efforts may consist of limiting damage associated with the NIS by controlling its local or regional abundance, as is done on the Great Lakes through chlorination of water intake pipelines to reduce mussel biofouling, and application of biocides to specific streams to reduce recruitment of sea lamprey. In a limited number of cases, new markets may be created to exploit the NIS, thereby reducing abundance and economic or ecological impact, as has been done by instituting a bighead carp fishery on the Upper Mississippi River.

In summary, the introduction of NIS has emerged as a critically important form of human-mediated global change. The Great Lakes have been highly receptive to NIS and are now greatly disturbed by them, with society bearing the economic impacts of those invasions. Most evidence points to a small number of vectors, especially ballast contents, as the predominant source of new NIS to the Great Lakes. Development of appropriate strategies to manage NIS in the Great Lakes is clearly a work in progress, but much can be learned from previous invasions both within and outside of the basin to shape management programs of the future.

Acknowledgments We are grateful for funding from the National Science Foundation, the Natural Sciences and Engineering Research Council, and the Canadian Aquatic Invasive Species Network.

References

Bailey, S. A., I. C. Duggan, P. T. Jenkins, and H. J. MacIsaac. 2005. Invertebrate resting stages in residual ballast sediment of transoceanic ships. Canadian Journal of Fisheries and Aquatic Sciences 62:1090–1103.

Bauer, C. R., A. M. Bobeldyk, and G. A. Lamberti. 2007. Predicting habitat use and trophic interactions of Eurasian ruffe, round gobies, and zebra mussels, in nearshore areas of the Great Lakes. Biological Invasions 9:667–678.

Bax, N., K. Hayes, A. Marshall, D. Parry, and R. Thresher. 2002. Man-made marinas as sheltered islands for alien marine organisms: establishment and eradication of an alien invasive marine species. Pages 26–39 *in* C. R. Veitch and M. N. Clout, editors. Turning the tide: the eradication of invasive species. Proceedings of the International Conference on Eradication of Island Invasives. IUCN Species Survival Commission No. 27. International Union for Conservation of Nature, Gland, Switzerland.

Bij de Vaate, A., K. Jażdżewski, H. A. M. Ketelaars, S. Gollasch, and G Van der Velde. 2002. Geographical patterns in the range extension of Ponto-Caspian macroinvertebrate species in Europe. Canadian Journal of Fisheries and Aquatic Sciences 59:1159–1174.

Canada Shipping Act. 2006. Ballast water control and management regulations. SOR/2006-129.

Cohen, J., N. Mirotchnick, and B. Leung. 2007. Thousands introduced annually: the aquarium pathway for non-indigenous plants to the St Lawrence Seaway. Frontiers in Ecology and the Environment 5:528–532.

Colautti, R. I., S. A. Bailey, C. D. A. van Overdijk, K. Admunsen, and H. J. MacIsaac. 2006. Characterized and projected costs of nonindigenous species in Canada. Biological Invasions 8:45–59.

Colautti, R. I., A. J. Niimi, C. D. A. van Overdijk, E. L. Mills, K. Holeck, and H. J. MacIsaac. 2003. Spatial and temporal analysis of transoceanic shipping vectors to the Great Lakes. Pages 227–246 *in* G. M. Ruiz and J. T. Carlton, editors. Invasive species: vectors and management strategies. Island Press, Washington, DC.

Connelly, N. A., C. R. O'Neill Jr., B. A. Knuth, and T. L. Brown. 2007. Economic impacts of zebra mussels on drinking water treatment and electric power generation facilities. Environmental Management 40:105–112.

Cook, D. C., M. B. Thomas, S. A. Cunningham, D. L. Anderson, and P. J. De Barro. 2007. Predicting the economic impact of an invasive species on an economic system. Ecological Applications 17:1832–1840.

Costanza, R., R. d'Arge, R. de Groot, S. Farber, M. Grasso, B. Hannon, S. Naeem, K. Limburg, J. Paruelo, R. V. O'Neill, R. G. Raskin, P. Sutton, and M. van den Belt. 1997. The value of the world's ecosystem services and natural capital. Nature 387:253–260.

Costello, C., J. M. Drake, and D. M. Lodge. 2007. Evaluating an invasive species policy: ballast water exchange in the Great Lakes. Ecological Applications 17:655–662.

Cristescu, M. E. A., P. D. N. Hebert, J. D. S. Witt, H. J. MacIsaac, and I. A. Grigorovich. 2001. An invasion history for *Cercopagis pengoi* based on mitochondrial gene sequences. Limnology and Oceanography 46:224–229.

Dehnen-Schmutz, K., J. Touza, C. Perrings, and M. Williamson. 2007. A century of plant trade and its impact on invasion success. Diversity and Distributions 13:527–534.

Drake, J. M., and D. M. Lodge. 2004. Global hot spots of biological invasions: evaluating options for ballast-water management. Proceedings of the Royal Society of London Series B 271:575–580.

Drake, J. M., and D. M. Lodge. 2007. Hull fouling is a risk factor for intercontinental species exchange in aquatic ecosystems. Aquatic Invasions 2:121–131.

Eakins, N. 1999. Lakers and salties 1998–1999. Canadian Coast Guard, Point Edward, Ontario.

Eakins, N. 2000. Lakers and salties 1999–2000. Canadian Coast Guard, Point Edward, Ontario.

Gollasch, S. 2002. The importance of ship hull fouling as a vector of species introductions into the North Sea. Biofouling 18:105–121.

Grigorovich, I. A., R. I. Colautti, E. L. Mills, K. Holeck, A. Ballert, and H. J. MacIsaac. 2003. Ballast-mediated animal introductions in the Laurentian Great Lakes: retrospective and prospective analyses. Canadian Journal of Fisheries and Aquatic Sciences 60:740–756.

Herborg, L.-M., N. E. Mandrak, B. C. Cudmore, and H. J. MacIsaac. 2007. Comparative distribution and invasion risk of snakehead and Asian carp species in North America. Canadian Journal of Fisheries and Aquatic Sciences 64:1723–1735.

Horvath, T. G., and G. A. Lamberti. 1997. Drifting macrophytes as a mechanism for zebra mussel (*Dreissena polymorpha*) invasion of lake-outlet streams. American Midland Naturalist 138:29–36.

Indiana Department of Natural Resources 2005. Largemouth bass virus. www.in.gov/dnr/invasivespecies/LMBV.

Jude, D. J., R., H. Reider, and G. R. Smith. 1992. Establishment of Gobiidae in the Great Lakes basin. Canadian Journal of Fisheries and Aquatic Sciences 49:416–421.Keller, R. P., and D. M. Lodge. 2007. Species invasions and commerce in living aquatic organisms: problems and possible solutions. BioScience 57:428–436.

Keller, R. P., D. M. Lodge, and D. C. Finnoff. 2007. Risk assessment for invasive species produces net economic benefit. Proceedings of the National Academy of Sciences of the United States of America 104:203–207.

Kelly, D. W. 2008. Vectors and pathways for nonindigenous aquatic species in the Great Lakes. Committee on the St. Lawrence Seaway: Options to eliminate introduction of nonindigenous species into the Great Lakes, phase 2. In: Great Lakes Shipping, Trade, and Aquatic Invasive Species, Transportation Research Board Special Report 291, National Research Council of the National Academies, Washington, DC.

Kelly, D. W., J. R. Muirhead, D. D. Heath, and H. J. MacIsaac. 2006. Contrasting patterns in genetic diversity following multiple invasions of fresh and brackish waters. Molecular Ecology 15: 3641–3653.

Kolar, C. S., and D. M. Lodge. 2002. Ecological predictions and risk assessment for alien fishes in North America. Science 298:1233–1236.

Levine, J. M., and C. M. D'Antonio. 2003. Forecasting biological invasions with increasing international trade. Conservation Biology 17:322–326.

Leung, B., J. M. Drake, and D. M. Lodge. 2004. Predicting invasions: propagule pressure and the gravity of Allee effects. Ecology 85:1651–1660.

Locke, A., D. M. Reid, W. G. Sprules, J. T. Carlton, and H. C. van Leeuwen. 1991. Effectiveness of mid-ocean exchange in controlling freshwater and coastal zooplankton in ballast water. Canadian Technical Report of Fisheries and Aquatic Sciences No. 1822, Great Lakes Laboratory for Fisheries and Aquatic Sciences, Burlington, Canada.

Lodge, D. M., S. Williams, H. J. MacIsaac, K. R. Hayes, B. Leung, S. Reichard, R. N. Mack, P. B. Moyle, M. Smith, D. A. Andow, J. T. Carlton, and A. McMichael. 2006. Biological invasions: recommendations for U.S. policy and management. Ecological Applications 16:2035–2054.

MacIsaac, H. J. 1996. Potential abiotic and biotic impacts of zebra mussels on the inland waters of North America. American Zoologist 36:287–299.

Martinez, M. L., A. Intralawan, G. Vazquez, O. Perez-Maqueo, P. Sutton, and R. Landgrave. 2007. The coasts of our world: ecological, economic and social importance. Ecological Economics 63:254–272.

Mellina, E., J. B. Rasmussen, E. L., and Mills. 1995. Impact of mussel (*Dreissena polymorpha*) on phosphorus cycling and chlorophyll in lakes. Canadian Journal of Fisheries and Aquatic Sciences 52:2553–2573.

Mills, E. L., J. H. Leach, J. T. Carlton, and C. L. Secor. 1993. Exotic species in the Great Lakes: a history of biotic crises and anthropogenic introductions. Journal of Great Lakes Research 19:1–54.

Muirhead, J. R., and H. J. MacIsaac. 2005. Development of inland lakes as hubs in an invasion network. Journal of Applied Ecology 42:80–90.

Nelson, R. 2003. Exotic spring viremia of carp virus confirmed in common carp taken from the Calumet-Sag channel near Chicago, Illinois. U.S. Fish and Wildlife Service. http://news.fws.gov/newsrelease/r3/E5DE11CB-EF51-49E8-A147A9955185C7C9.html. Pimentel, D., R. Zuniga, and D. Morrison. 2005. Update on the environmental and economic costs associated with alien-invasive species in the United States. Ecological Economics 52:273–288.

Ricciardi, A. 2001. Facilitative interactions among aquatic invaders: is an "invasional meltdown" occurring in the Great Lakes? Canadian Journal of Fisheries and Aquatic Sciences 58:2513–2525.

Ricciardi, A. 2006. Patterns of invasion in the Laurentian Great Lakes in relation to changes in vector activity. Diversity and Distributions 12:425–433.

Ricciardi, A., and H. J. MacIsaac. 2000. Recent mass invasion of the North American Great Lakes by Ponto-Caspian species. Trends in Ecology and Evolution 15:62–65.

Ricciardi, A., and J. B. Rasmussen, 1998. Predicting the identity and impact of future biological invaders: a priority for aquatic resource management. Canadian Journal of Fisheries and Aquatic Sciences 55:1759–1765.

Rixon, C. A. M., I. C. Duggan, N. M. N. Bergeron, A. Ricciardi, and H. J MacIsaac. 2004. Invasion risks posed by the aquarium trade and live fish markets on the Laurentian Great Lakes. Biodiversity and Conservation 14:1365–1381.

Roman, J., and J. A. Darling. 2007. Paradox lost: genetic diversity and the success of aquatic invasions. Trends in Ecology and Evolution 22:454–464.

Ruiz, G. M., and J. T. Carlton. 2003. Invasion vectors: a conceptual framework for management. Pages 459–504 *in* G. M. Ruiz and J. T. Carlton, editors. Invasive species: vectors and management strategies. Island Press, Washington, DC.

Ruiz, G. M., P. W. Fofonoff, J. T. Carlton, M. J. Wonham, and A. H. Hines. 2000. Invasion of coastal marine communities in North America: apparent patterns, processes, and biases. Annual Review of Ecology and Systematics 31:481–531.

Sala, O. E., F. S. Chapin, J. J. Armesto, E. Berlow, J. Bloomfield, R. Dirzo, E. Huber-Sanwald, L. F. Huenneke, R. B. Jackson, A. Kinzig, R. Leemans, D. M. Lodge, H. A. Mooney, M. Oesterheld, N. L. Poff, M. T. Sykes, B. H. Walker, M. Walker, and D H. Wall. 2000. Global biodiversity scenarios for the year 2100. Science 287:1770–1774.

Stokstad, E. 2003. Can well-timed jolts keep out unwanted exotic fish? Science 301:157–158.

Tatum, A. J., and S. I. Hay. 2007. Climatic similarity and biological exchange in the worldwide airline transportation network. Proceedings of the Royal Society Series B 274:1489–1496.

Tatum, A. J., D. J. Rogers, and S. I. Hay. 2006. Global transport networks and infectious disease spread. Advances in Parasitology 62:293–343.

U.S. Coast Guard. 1993. Ballast water management for vessels entering the Great Lakes. 33 CFR 151.1510.

U.S. Coast Guard. 2005. Ballast water management practices for NOBOB vessels. 33 CFR Part 151.2035 Fact Sheet. uscg.mil/hq/gm/mso/docs/fact_sheet_nobobbmps.pdf.

U.S. Policy Committee for the Great Lakes. 2002. Great Lakes Strategy 2002: a plan for the new millennium. U.S. Policy Committee for the Great Lakes. http://www.epa.gov/glnpo/gls/gls2002.pdf.

Ward, J. M., and A. Ricciardi. 2007. Impact of *Dreissena* invasions on benthic macroinvertebrate communities: a meta-analysis. Diversity and Distributions 13:155–165.

Williams, S. L., and S. L. Schroeder. 2004. Eradication of the invasive seaweed *Caulerpa taxifolia* by chlorine bleach. Marine Ecology Progress Series 272:69–76.

Xu, H., H. Ding, M. Li, S. Qiang, J. Guo, Z. Han, Z. Huang, H. Sun, S. He, H. Wu, and F. Wan. 2007. The distribution and economic losses of alien species invasion to China. Biological Invasions 8:1495–1500.

Yule, A. M., I. K. Barker, J. W. Austin, and R. D. Moccia. 2006. Toxicity of *Costridium botulinum* type E neurotoxin to Great Lakes fish: implications for avian botulism. Journal of Wildlife Diseases 42:479–493.

11

A Case Study on Rusty Crayfish: Interactions between Empiricists and Theoreticians

Caroline J. Bampfylde, Angela M. Bobeldyk, Jody A. Peters, Reuben P. Keller, and Christopher R. McIntosh

In a Clamshell

Interactions among empirical ecologists, theoretical ecologists, and economists are often difficult because each discipline has different values, backgrounds, perspectives, and languages. Despite the difficulties, multidisciplinary collaborations are essential to answer many questions important to science, management, and policy. We describe three recent collaborative research projects that address slowing the spread, and controlling established populations, of the invasive rusty crayfish (*Orconectes rusticus*). Rusty crayfish are an aggressive invader across a number of North American freshwater ecosystems. They displace resident crayfish species, interfere with fish reproduction, overgraze macrophytes and invertebrates, and affect lake substrates. They are a useful study system because their invasion in the U.S. Upper Midwest has been well studied by empirical ecologists for more than 30 years, providing a wealth of data. Additionally, there is widespread concern about rusty crayfish impacts on sport fish populations and the economic consequences of their loss from recreational lakes. The projects presented here draw from the long history of ecological research, and additionally incorporate other relevant data (e.g., economic values), to parameterize models that could be used to guide management efforts. Each model framework demonstrates the integration of empirical, theoretical, and economic data and can be used to determine if management is feasible or cost-effective.

The last several decades have seen rapid growth in the number of studies investigating biological invasions (Lodge 1993; Sax et al. 2005; Callaway and Maron 2006; Strayer et al. 2006; Bampfylde et al. unpublished manuscript). This growth has been led by empirical ecologists, and the work produced demonstrates the range of negative ecological impacts of nonindigenous species. This research approach generates crucial information regarding the impacts and life history of invasive species and the invasion process as an integral component of the decision-making process. However, we argue that to effectively address the impacts of invasive species, collaborative efforts among empiricists, theoreticians, and economists are required. Such an approach has been effective in other areas of ecology (Carpenter 1998; Belovsky et al. 2004; Dobson et al. 2007). We believe that justification for prevention and control efforts could be enhanced by greater development and use of predictive models in which economic and ecological systems are integrated, and where management efforts can be empirically tested.

The research framework supported by this book requires the collaboration of ecologists and economists, and of empiricists and theoreticians. The goal of this chapter is to demonstrate how these collaborations can occur, using many of the methods described in preceding chapters. We use the history of research into the rusty crayfish invasion of the U.S. Upper Midwest to illustrate a collaborative approach. In the following, we briefly describe the history of ecological research on the rusty crayfish invasion and then describe three recent efforts that use theoretical and bioeconomic frameworks to assess current and potential future management scenarios. Although there are examples of integrated research approaches to investigate management of freshwater invasive species (e.g., Leung et al. 2002; Settle and Shogren 2002; Settle et al. 2002; Horan and Lupi 2005), there is overall a lack of multidisciplinary work on these issues (Bampfylde et al. unpublished manuscript).

For the purposes of this chapter, we define *empirical* research to be any projects that attempt to describe the functioning of a system (in this case, an ecosystem containing one or more invasive species). *Theoretical* work is defined as research that aims to formalize ecosystem dynamics into mathematical relations that can be used to predict future states of the system. Finally, we define *bioeconomic* research as any work that combines ecological and economic data into theoretical models.

RUSTY CRAYFISH INVASION OF THE UPPER MIDWEST

Rusty crayfish are native to the Ohio River basin but have been introduced by anglers, who use them as bait, to many lakes and streams throughout the northern United States and Canada (Capelli and Magnuson 1983; Hobbs and Jass 1988; figure 11.1). Researchers from the University of Notre Dame and the University of Wisconsin have been studying the spread and impacts of rusty crayfish in northern Wisconsin and Michigan's Upper Peninsula for the last 30 years (Capelli 1982; Lodge et al. 1994; Hein et al. 2006; Rosenthal et al. 2006). The majority of this research has been empirical, with researchers utilizing long-term field surveys and short-term experiments to demonstrate the negative impacts of rusty crayfish on native crayfish, macroinvertebrates, aquatic plants, and fish (Momot et al. 1978; Lodge et al. 1986; Olsen et al. 1991; Luttenton et al. 1998; Garvey et al. 2003; Wilson et al. 2004; Olden et al. 2006; Rosenthal et al. 2006). Examples of data collected include crayfish density, sex ratios, size classes, habitat use, habitat characteristics, and lake water

FIGURE 11.1.
Rusty crayfish *Orconectes rusticus*. Photo by Angela Bobeldyk.

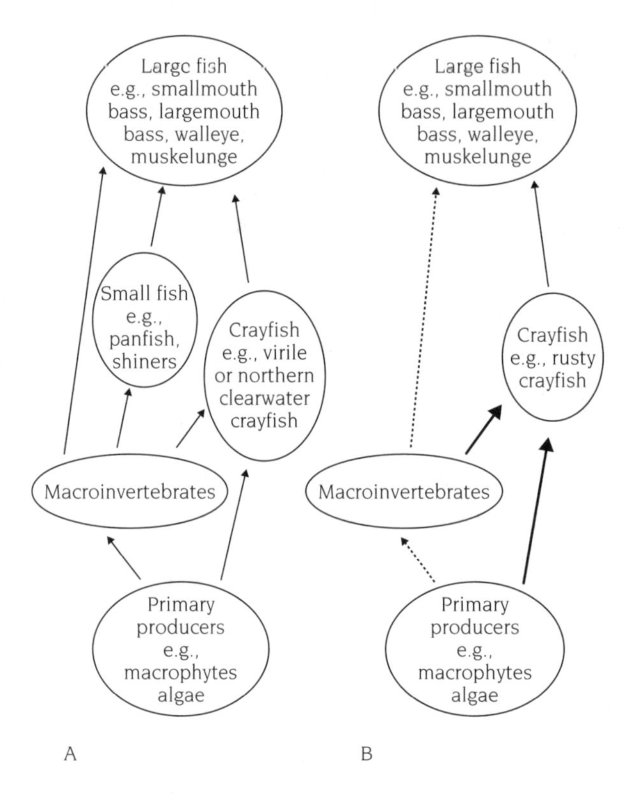

FIGURE 11.2.
Lake food webs in the U.S. Upper Midwest before (A) and after (B) invasion by rusty crayfish. Arrow thickness indicates extent of energy flow between groups. Dashed arrows indicate very little movement. Invasions have economic consequences including a reduction in, or complete loss of, some species of sport fish (e.g., panfish). Ecological impacts of rusty crayfish include a reduction in vegetation and invertebrate density and diversity and the extirpation of native crayfish (O. *virilis* and O. *propinquus*). Invertebrate density reduction leads to less energy transport through the food web. Destruction of macrophyte beds removes shelter necessary for small and juvenile fish.

chemistry. From this work, the historical food web of Upper Midwest lakes is now well known (figure 11.2A), in addition to the changes that can result after rusty crayfish invade (figure 11.2B).

Although empirical ecologists have clearly demonstrated the negative ecological impacts of rusty crayfish, little has been done (apart from poorly enforced laws banning the use of rusty crayfish as bait, and signs at lakes [Gunderson 1995]) to prevent the further spread of rusty crayfish or to control established populations. Rusty crayfish continue to spread, and their total impacts continue to increase. We believe the lack of prevention and management effort has occurred at least partly because research has not answered the questions of most relevance to managers and policy makers. For example, we are aware of only one study (Hein et al. 2006, 2007) that focused on developing methods for controlling established rusty crayfish populations. Additionally, questions about accurately predicting rusty crayfish invasions and the economic losses associated with such invasions have only recently received attention.

Over the last few years, we have attempted to redress this shortfall with multidisciplinary analyses of rusty crayfish management options. Here, we describe three projects that have combined empirical data with bioeconomic and theoretical tools to address (1) the economic value of preventing further rusty crayfish spread, (2) biological control of established populations, and (3) control at the population growth stage.

INTEGRATED APPROACHES

A Bioeconomic Model to Determine the Benefits of Preventing the Spread of Rusty Crayfish

Rusty crayfish were first sampled in Vilas County, Wisconsin, in the mid-1970s (Capelli and Magnuson 1983). Of the 64 lakes originally sampled, rusty crayfish were found in 14. In this project, we evaluated how effective management efforts based on these data would have been at preventing future invasions. We then built an economic model to test whether such efforts would have returned economic dividends to Vilas County (for more detail, see Keller et al. 2007).

We took Capelli and Magnuson's (1983) set of 64 lakes and began by removing the 13 lakes that are unsuitable for crayfish (calcium levels < 2.5 mg/l in Vilas County; Capelli and Magnuson 1983), assuming that managers would not expend resources to protect these. The remaining 51 lakes form the "historical data set." Collecting only data that were available in the mid-1970s, we assembled information on nine lake characteristics (calcium level, lake size, number of lakeshore cabins, and the densities of six fish species: muskellunge [*Esox masquinongy*], northern pike [*E. lucius*], walleye [*Sander vitreus*], smallmouth bass [*Micropterus dolomieu*], largemouth bass [*M. salmoides*], and panfish [*Centrarchidae*]) that might reasonably be related to invasion risk for each lake in the historical data set.

Multiple logistic regression was used to parameterize a model relating these variables to invasion status for the historical data set lakes. This "predictive occurrence model" contained the variables walleye, largemouth bass, calcium, cabins, and lake size. Model diagnostics (area under receiver operator characteristic curve = 0.847) showed that if this model had been created in the mid-1970s, there would have been justification to use it as a tool for determining the likelihood of other lakes becoming invaded in the future.

Next, we assembled the "modern data set," consisting of the 48 additional lakes for which we had data and that were sampled after 1975. Thirty-two of these were invaded by rusty crayfish (assumed to be at some date post-1975). Each of these lakes was assessed with the predictive occurrence model and given a score between 0 and 1 (a metric of the probability in 1975 that the lake would become invaded sometime in the future).

Although predictive model scores were strongly related to the invasion status of the lakes in the modern data set, there was overlap in scores of lakes that did and did not become invaded. Prevention efforts based on this model would thus have protected some lakes that ultimately did not become invaded, and would have assumed to be safe some lakes that did become invaded. We constructed a simple cost–benefit projection model to determine whether the use of these predictions would have produced economic benefit over the last 30 years.

The cost in our economic model was the money spent by managers to put personnel on boat ramps. Managers decide which lakes to protect by selecting a threshold score from the predictive occurrence model and placing personnel at lakes with scores higher than this threshold. Annual cost for this protection was estimated to be $6,897 per lake based on the cost of a similar program in New Hampshire (New Hampshire Lakes Association 2005). An assumption of our model is that all rusty crayfish lake invasions have been caused by anthropogenic transport. To test the possibility that some lakes became invaded through connecting streams, and thus that our modeled prevention efforts would not have been effective at those lakes, we also ran the model under the assumption that prevention efficacy was 75%.

Although rusty crayfish have a variety of impacts on lake ecosystems (Lodge et al. 1986; Hill and Lodge 1999; Rosenthal et al. 2006), the best-resolved impact with economic implications is the decline in panfish populations (Wilson et al. 2004; but also see chapter 8, for impacts of rusty crayfish [table 8.1] and an application of nonmarket valuation [table 8.2]). The benefits in the economic model were the continuing value of panfish populations in uninvaded lakes, whereas the invaded lakes lost their value 2 years after invasion, at which point it was assumed that the lake would no longer attract panfish anglers. The U.S. Fish and Wildlife Service Survey of Fishing, Hunting, and Wildlife-Associated Recreation (U.S. Department of the Interior 2002) was used to determine the annual value of a hectare of lake in Wisconsin to be $232.16 (hereafter referred to as the "panfish value"). This is likely a low estimate because rusty crayfish impacts extend beyond declines in panfish (Lodge et al. 1985). To test for the benefits of protection guided by the predictive occurrence model when the value of lakes is higher, we also ran the model under the assumption that the value of lakes is double ("panfish × 2") and triple ("panfish × 3") the value calculated from money spent on panfish angling.

Other methods are available for assessing the economic benefits from a fishery. Our use of the money spent by anglers is relevant to local management agencies for determining how investing in lake protection would maintain the flow of money from anglers. As an alternative, however, if we wanted to determine the change in total angler welfare caused by invasions, the relevant value for the lakes would be the difference between what anglers actually pay to go fishing (i.e., the panfish value) and their total willingness to pay for their angling experience. This could be determined through a nonmarket survey approach (McIntosh et al. 2007; see also chapter 8), or with the travel-cost method (e.g., Hanley et al. 2007).

The total expenditure of anglers on lakes in the modern data set are the sum of the expenditures on lakes that do and do not become invaded, minus the money expended to protect a subset of those lakes. Assuming that a dollar of angler spending is exactly offset by a dollar of cost for lake protection, the economic benefit from lakes that do not become invaded (B_U) are equal to

$$B_U = \sum_{l=1}^{n_u} \left[\left(H_l V - C_l\right) \sum_{t=1}^{30} \frac{1}{(1+r)^t} \right], \tag{1}$$

where n_u is the number of lakes that do not become invaded in the simulation, t is the number of years after 1975, r is the annual discount rate, H_l is the size (in hectares) of lake l, and V is the value per hectare of uninvaded lakes in that simulation (i.e., panfish value, panfish × 2, or panfish × 3). C_l is a binary variable for the cost of protection. Depending on the predictive occurrence model score for lake l and the threshold chosen, the value of C_l is either 0 (i.e., threshold greater than model score) or \$6,897 (i.e., threshold lower than model score).

The expenditures arising from lakes that do become invaded (B_I) are equal to

$$B_I = \sum_{l=1}^{n_i} \left[\left(H_l V - C_l\right) \sum_{t=1}^{D+2} \frac{1}{(1+r)^t} \right], \tag{2}$$

where D is the number of years after 1975 that lake l becomes invaded, and other variables are as for equation 11.1. There are 48 lakes in the modern data set, so $n_i + n_u = 48$.

We projected this system of benefits and costs over 30 years (1975–2005) to test whether the expected net economic benefit that would have been gained from implementing protection at thresholds of 0.0, 0.1, 0.2, . . . , 0.9 is greater than the economic benefit that has been gained from the actual policy followed over the last 30 years (i.e., threshold = 1.0, no protection).

Our results show that the optimum protection threshold depended on the value of lakes (figure 11.3). This occurs because the cost of protection is fixed but angler expenditures vary across the three value assumptions. Lower expenditures lead to lower net present values from protection efforts. For the lowest value of lakes ("Panfish"), it is economically rational to protect some but not all lakes (optimal threshold = 0.2). For higher lake values, it becomes most rational to protect all lakes (panfish × 2

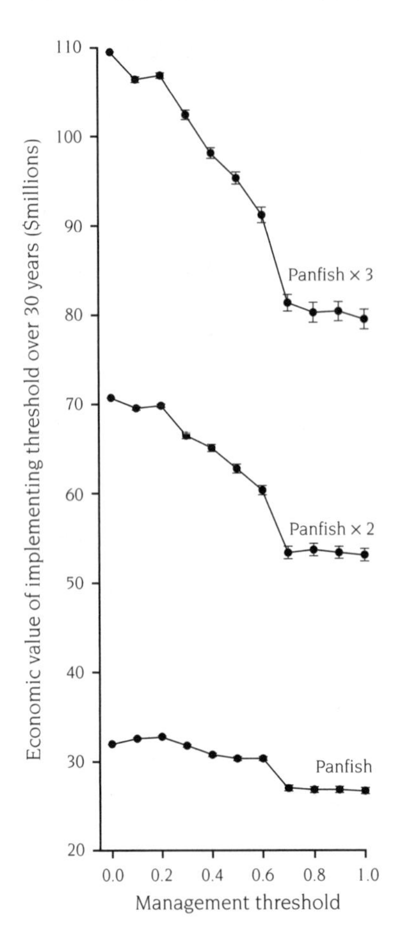

FIGURE 11.3.
Total 30-year discounted value ± 95% confidence intervals (in 2005 $US) derived from lakes in the modern data set when different thresholds for management are applied. Results are from the cost-benefit model described in text, with each line coming from a different assumption of the value of a lake: panfish is the actual amount of money spent annually by anglers when targeting panfish in Wisconsin; panfish × 2 and panfish × 3 assume that the impacts of rusty crayfish go beyond declines in panfish populations and are double and triple, respectively, the base value.

and panfish × 3, optimum thresholds = 0.0). Modeling prevention to be only 75% successful did not change the optimum threshold for management for any of the three expenditure scenarios.

Overall, this work demonstrates a number of the benefits from combining ecological and economic models to support invasion management. First, it shows that data available early in an invasion may provide sufficient justification for large-scale and costly efforts to prevent future spread. Managers and ecologists are often conservative when few data are available and prefer to wait until more is known about the invasion dynamics before acting. Second, our results indicate that, although expensive, invasion prevention based on ecological models may produce significant economic benefit over reasonable time horizons. Indeed, we note here that our model both underestimates the benefits (e.g., other invaders would have been prevented by personnel on boat ramps) and overestimates the costs (e.g., placing personnel at road junctions would reduce the number needed but effectively protect many more lakes) of protection in this system.

A Theoretical Model for Biological Control of Rusty Crayfish

The model presented in the preceding section addresses the potential benefit of preventing lakes from being invaded by rusty crayfish. For lakes where the invader has already established, control becomes the only option for reducing impacts. The spatial scale of control is different from that for prevention: most likely, individual lakes rather than multiple lakes at the landscape level.

Several options have been suggested for controlling established populations of rusty crayfish in the U.S. Upper Midwest. One of these is to use smallmouth bass (*Micropterus dolomieu*), a native species and natural predator of rusty crayfish (Stein 1977; Momot 1984; Didonato and Lodge 1993; Garvey et al. 2003), as a biological control agent (Lodge and Hill 1994; Roth and Kitchell 2005; Hein et al. 2006). Smallmouth bass are found across the Upper Midwest, and lakes with this species are generally of high recreational value (Department of Fisheries and Oceans Canada 2004; Leonard 2005). However, in many lakes invaded by rusty crayfish, the density of smallmouth bass has been observed to decline (Wilson et al. 2004). Juvenile smallmouth bass compete with many life stages of rusty crayfish for food and shelter (Hamr 2001), but undergo an ontogenetic niche shift in their second year when crayfish become their preferred diet (Olson and Young 2003). The combination of interspecific competition and predation between the crayfish and bass (i.e., intraguild predation) may result in a "competitive bottleneck." Therefore, it may be possible to perturb the lake system to switch the bottleneck to benefit smallmouth bass and allow bass to control the crayfish. We developed a theoretical model to determine under what circumstances biological control of rusty crayfish is possible and identified key parameters necessary for model structure and prediction (more detail can be found in Bampfylde and Lewis 2007).

The model we developed considers the population dynamics of crayfish and bass and their shared resources within an invaded lake (figure 11.2). Intraspecific competition occurs between individual conspecifics and between individuals of different species. Smallmouth bass predation occurs on rusty crayfish. The growth of each population in the absence of the other is logistic. Since both species interact in the littoral zone, we assumed that species disperse randomly throughout a one-dimensional region. If we denote $c(x,t)$ and $b(x,t)$ as the densities of rusty crayfish and smallmouth bass, respectively, at each location x and time t, then we have the following dynamics for both species:

$$\frac{\partial c}{\partial t} = \left[1 - c - \left(\alpha_{bc} + \frac{a}{1+hc}\right) b\right] c + \frac{\partial^2 c}{\partial x^2},$$
$$\frac{\partial b}{\partial t} = r\left[1 - b - \left(\alpha_{cb} - \frac{\beta/r}{1+hc}\right) c\right] b + \varepsilon \frac{\partial^2 b}{\partial x^2}, \qquad (3)$$

where $r = r_b/r_c$ is the ratio of intrinsic growth rates of bass to crayfish, α_{bc} is the competition coefficient measuring the effect of bass on crayfish, α_{cb} is the competition coefficient measuring the effect of crayfish on bass, a is the encounter rate (or capture efficiency), h is the handling time, β is the conversion efficiency from crayfish to bass biomass, and $\varepsilon = D_b/D_c$ is the ratio of diffusion coefficients of bass to crayfish. This is a dimensionless model, and the dependent variables c and b are scaled by their carrying capacities (i.e., $c = 1$ is equivalent to crayfish at carrying capacity).

We use the seven dimensionless system parameters to investigate whether biological control of rusty crayfish by smallmouth bass is possible. First, we investigate the spatially uniform (time-dependent only) steady-state solutions (figure 11.4). The ratio of intrinsic growth rates ($r = r_b/r_c$) was used as a control parameter because it can be manipulated either by stocking bass (increasing r_b increases r), fishing for

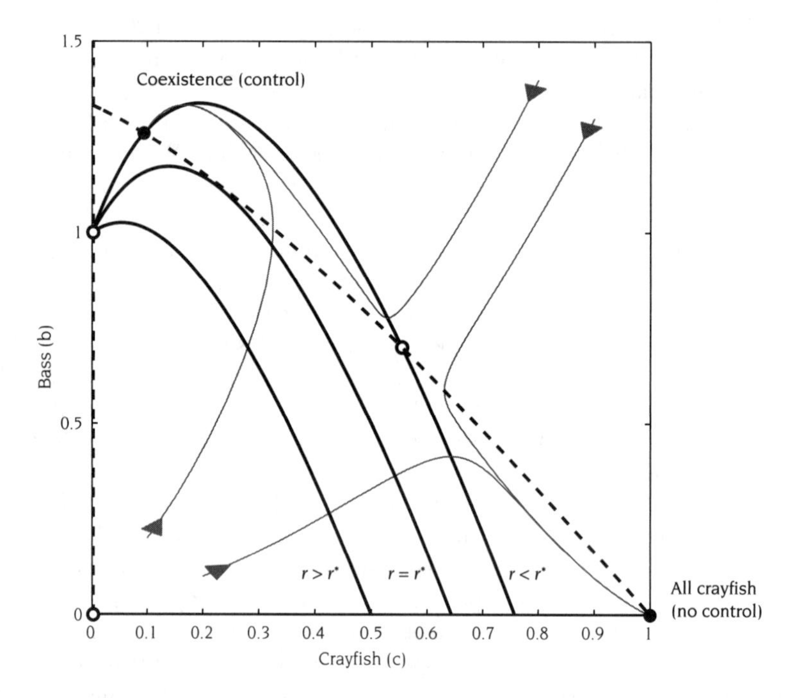

FIGURE 11.4.
Phase-plane diagram of equations 11.3 for the cases $r = 1 > r^*$, $r = r^* = 0.8$, and $r = 0.7 < r^*$, with $h = 1.5$, $\alpha_{bc} = 0.5$, $a = 0.25$, $\alpha_{cb} = 6$, and $\beta = 7$. r^* is the critical value of r below which alternative stable states are observed. Solid lines indicate bass zero-growth isoclines, and dashed lines crayfish zero-growth isoclines, of equations 11.3. The intersections of the zero-growth isoclines give the steady states (circles). The bass zero-growth isocline and the number of steady states (intersections) are dependent on the value of r. Closed circles indicate stable steady states; open circles, unstable. For $r < r^*$, there are two stable steady states (coexistence and all crayfish) and two unstable steady states (all bass and coexistence). For $r > r^*$, there is only one stable steady state (all crayfish) and one unstable steady state (all bass). As r increases through r^*, the system changes from bistability to monostability. Example time courses of the system are shown by gray arrows for $r = 0.7 < r^*$. For each trajectory, the initial condition is behind the arrow, and as time progresses the solution moves toward the other end of the gray line, toward a steady state. When $r > r^*$, all trajectories tend to the all-crayfish steady state.

bass (decreasing r_b decreases r), or removing crayfish (increasing r_c decreases r). A critical value of $r = r^*$ exists below which alternative stable states are observed (a coexistence steady state and a crayfish monopoly state) and above which the only feasible steady state is a crayfish monopoly. The coexistence steady state can be interpreted as bass controlling crayfish, while the crayfish monopoly steady state can be interpreted as bass being unable to control crayfish. In the bistable case ($r < r^*$ line in figure 11.4), it may be possible to perturb the population densities to allow the system to follow a trajectory to a different, more desirable, steady state.

The model analysis provides counterintuitive results. As the control parameter $r = r_b/r_c$ increases, the ratio of bass fecundity to crayfish fecundity increases, and it might be expected that the bass would fare better. However, from the invasibility criteria ($\alpha_{cp} + a < 1$ for crayfish persistence, and $\alpha_{cp} - \beta(1 + h)^{-1}/r < 1$ for bass persistence), we see that as r increases, the bass loses persistence in the system due to the loss of the coexistence steady state.

The model also exhibits spatiotemporally varying solutions. We found traveling-wave solutions joining the spatially uniform steady states, the coexistence state and the crayfish-only state, within a lake's littoral zone. We investigated the speed (and direction) of the traveling waves. There is another critical value of r, $r_0 (= 0.67$ for the example parameter set), which is less than r^*. When $r_0 < r < r^*$, the wave retreats and there is a wave of extinction of bass. When $r < r_0$, the wave advances and there is a wave of control of crayfish by bass.

We present some numerical simulations (figure 11.5) for different scenarios in a model lake with low bass and high crayfish density on one side of the lake and high bass and low crayfish density on the other side of the lake. In figure 11.5A–C ($r > r_0$), a rightward-moving wave is observed. The wave profile is fixed and moving at constant speed. As the wave moves, areas of low crayfish density are replaced by areas of high crayfish density, while areas of high bass density are replaced by areas of low bass density. In figure 11.4D–F ($r < r_0$), a leftward-moving wave is observed. In this case, as the wave moves, the bass are able to control the crayfish population and the bass-excluded area is replaced by a coexistence zone.

The results of this model show under which circumstances biological control of rusty crayfish by smallmouth bass is likely to be successful. If a large enough perturbation is generated to create low crayfish density locally, then a wave of coexistence can move throughout the lake, enabling the bass to control the crayfish.

In order for the model to be most useful to managers, parameter estimation needs to be completed to determine which scenario a particular lake matches. Future empirical work should focus on collecting relevant field data to estimate the model parameters. An example of a fully parameterized dynamical model is given in the following section.

Efficient Mechanical Control of Rusty Crayfish

The preceding section considered biological control of crayfish within an invaded lake. The model's generality makes it strategic: useful for understanding general

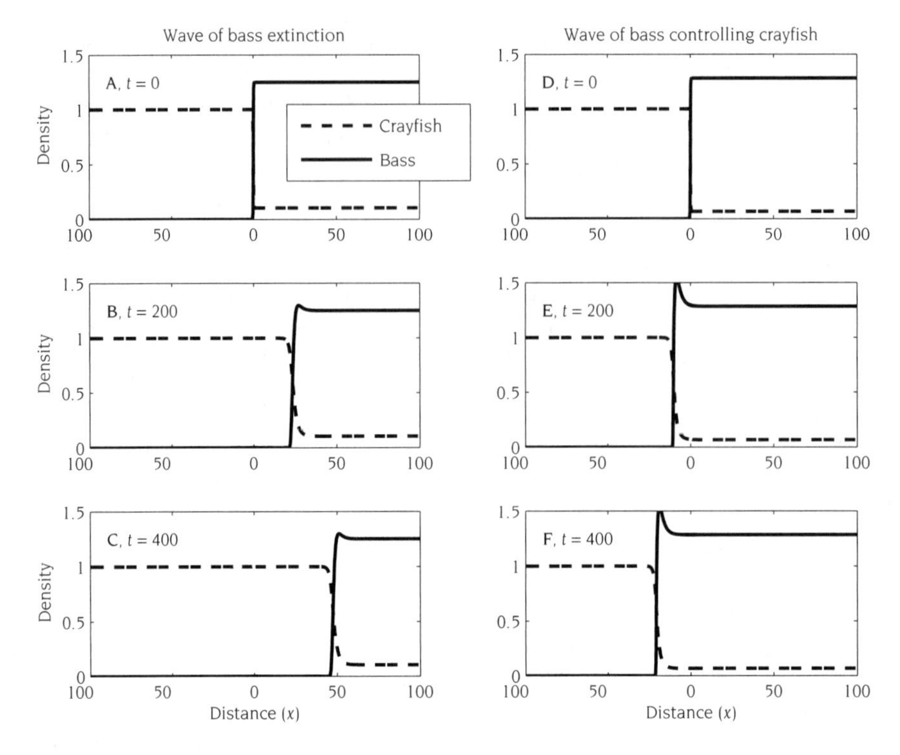

FIGURE 11.5.
Traveling-wave solutions of equations 11.3. Initial conditions: low bass and high crayfish density on the left-hand side of the lake and high bass and low crayfish density on the right-hand side of the lake. (A–C). A rightward-moving wave of bass extinction. Areas of high bass and low crayfish are replaced by areas of low bass and high crayfish. Solutions as a function of space throughout the lake are shown after 0 (A), 200 (B), and 400 (C) time intervals. (D–F). A leftward-moving wave of bass control of crayfish. Areas of low bass and high crayfish are replaced by areas of high bass and low crayfish. Solutions after 0 (D), 200 (E), and 400 (F) time intervals are shown. Parameter values used in A–F are $(\alpha_{bc}, a, h, \alpha_{cb}, \beta, \varepsilon) = (0.50, 0.25, 1.5, 6, 7, 1 \times 10^{-8})$. In A–C, $r = 0.62$; in D–F, $r = 0.72$.

ecological interactions while allowing mathematical analysis. However, a tactical model specific to the system of interest, and parameterized with field data, would be more suited to making predictions useful for management purposes. In this section, we describe a tactical model integrating empirical data and theory into a framework to find the most efficient combination of control strategies for rusty crayfish in an invaded lake.

A variety of mechanical control measures have been used in the attempt to reduce populations of rusty crayfish in lakes. In one example, an ongoing experiment has intensively trapped rusty crayfish in Sparkling Lake, Wisconsin, over the last 5 years (Hein et al. 2006, 2007; Carpenter et al. 2007; Kratz et al. 2007). While the rusty crayfish has not been extirpated, it has been driven to low population abundance (Hein et al. 2007; Kratz et al. 2007). In the framework, we consider two strategies:

trapping and trawling. Trapping involves placing baited traps throughout the lake and collecting the crayfish inside at a later date. Trapping is selective for larger crayfish (Hein et al. 2006, 2007; Kratz et al. 2007). Trawling removes a high density of crayfish in a local area by a technique similar to dredging the lake. Trawling is selective for smaller crayfish (B. Peters, personal communication). For the purposes of our model, we made the simplifying assumption that the lake in question was appropriate for trawling, but recognize that submerged logs and other substrate features will make some lakes, or areas of some lakes, inappropriate for this approach.

This model framework considers the dynamics of rusty crayfish within an invaded lake. We use a discrete-time model because individual crayfish reproduce annually. The model is not explicitly spatial. Crayfish are divided into two stage classes: small crayfish (vulnerable to trawling and predation) and large crayfish (vulnerable to trapping and invulnerable to predation). We assume that small and large crayfish compete for resources. The ecological system is described by a system of difference equations. Population model parameters were estimated from field data and the literature, while the person-hour costs of alternative strategies were estimated from field trials. Model formulation, detailed analysis, and full parameterization of this system can be found elsewhere (Bampfylde and Lewis unpublished manuscript). The model objective was to determine what the best combination of removal strategies would be to reduce crayfish levels to sustainable preinvasion (native) crayfish densities within a fixed number of years.

The best combination of approaches is the one that minimizes cost; the costs of controlling crayfish are person-hours required to complete each removal (C. Hein, personal communication; B. Peters, personal communication). The cost of control over 5 years is given by the sum of costs in all years, subject to the conditions that the total crayfish population is below the threshold by year 5 and that trawling or trapping in each year can only remove up to 90% of the population. Once low densities are reached, the efficacy of removal decreases (Hein et al. 2007). Results are shown in figure 11.6. A more desirable outcome might be to eliminate rusty crayfish from Sparkling Lake within the duration of the removal efforts. However, while low abundances have been reached, total elimination does not seem possible in practice, or in our model framework, although this has been proposed elsewhere (Hein et al. 2006; Carpenter et al. 2007).

The optimal control strategy given by our model is a combination of trawling and trapping. A combination of approaches is most effective because each approach targets different sizes of crayfish. The optimal number of trap days is much less than was used in Sparkling Lake and decreases over time. The total number of trawls is small, but each trawl unit removes a considerable number of crayfish. Even though trapping targets large crayfish that have a higher fecundity than small crayfish, survival of this size class is low. Trawling can remove a much larger number of small crayfish that have higher survival and are able to reproduce over a number of years. The effort required for the optimal strategy is approximately 930 person-hours, while the effort spent in Sparkling Lake (Hein et al. 2007) was four times this amount. We reiterate, however, that lake characteristics not included in this model may affect trawling efficiency. The model could, however, easily be modified to take account of these factors.

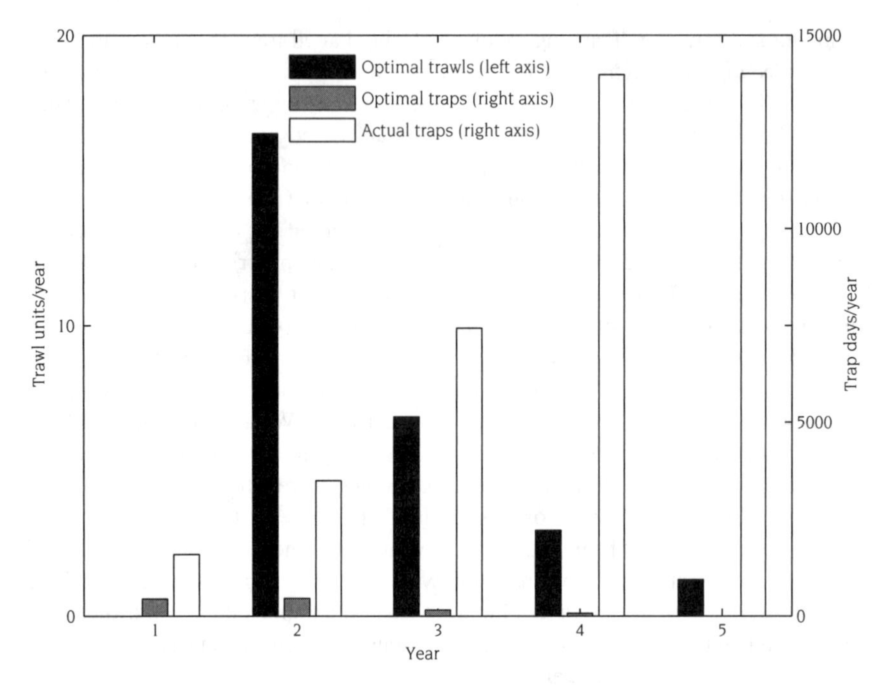

FIGURE 11.6.
The optimal control strategy for reducing crayfish below threshold population within 5 years. The optimal control is given by the combination of optimal trawls and optimal traps. The actual control (actual traps) carried out took four times more effort than the most time-effective solution which is 439, 453, 154, 69, and 0 traps and 0, 17, 7, 3, and 1 trawl units in years 1–5, respectively. The units for control are a "trap day" (which is a single trap fished for 24 hr) and a trawl unit is the area trawled within 10 hours.

The advantage of this model's framework is the ability to extend it to include monetary costs and benefits of control (similar to the bioeconomic model for prevention above). The benefits could be obtained from market and nonmarket impact surveys. In addition, the cost of stocking bass (and the impact on crayfish population dynamics) could also be included in a modified model. Including stocking would potentially increase the control costs. Different control strategies can also be evaluated for their effectiveness.

DISCUSSION

The majority of studies into freshwater invasions have been empirical, with only a small proportion having either theoretical or bioeconomic components (Bampfylde et al. unpublished manuscript). Empirical studies have produced a large amount of information about the impacts of invasive species, and this has supported an accounting of, for example, the relative strength of different threats to freshwater biodiversity

(Sala et al. 2000). In contrast, the economic factors that lead to invasions, and how those invasions then manifest economic consequences, are less well understood. Our thesis in this chapter, and indeed, the thesis of this book, is that biological invasions occur because of interactions among economic and ecological factors, and that they have impacts at each of these levels. Thus, empirical ecological research is, while necessary, not sufficient to fully understand or respond to freshwater invasions.

The ecological impacts of rusty crayfish have been described since their introduction in the 1970s (Capelli and Magnuson 1983) by empirical ecologists from several universities (Capelli and Munjal 1982; Lodge et al. 1994; Charlebois and Lamberti 1996; Wilson et al. 2004). Although few invaders are better understood than rusty crayfish (Olden et al. 2006), little has been done proactively to prevent new invasions or manage established populations (e.g., reduce population densities). We believe that this is because the perspective of managers and policy makers encompasses factors that cannot be addressed through empirical research alone. Before expending resources on control or prevention, managers should reasonably ask whether those expenditures will be justified by the ecological and economic returns. The three projects described above are, we believe, the first that allow these types of questions to be answered for rusty crayfish management.

In combination, the three projects show that expensive efforts to prevent the spread of rusty crayfish would have been economically beneficial if they had been applied between 1975 and 2005. In the absence of prevention efforts, rusty crayfish have spread to many additional lakes, and their impacts have subsequently expanded (Puth and Allen 2005; Olden et al. 2006). Depending on the value assigned to the lost resources after crayfish invasion, it may have been worthwhile to protect all 48 lakes in the modern data set. Once a lake has been invaded by rusty crayfish, the second and third projects offer guidance for management. The second project uses a theoretical approach to show that, under some circumstances, it may be possible to establish a wave of bass dominance (resulting in fewer crayfish) that could eventually reduce the impacts of crayfish throughout a lake. The third project offers trapping and trawling advice when mechanical removal is the available management option.

Although sufficient data were available for each model, data quality needs to be considered. In many cases, it was difficult or impossible to obtain or estimate confidence intervals for each parameter, so point estimates were more realistic. Additionally, often data were available only for congeners (e.g., Rabeni 1992; Roth et al. 2006), or data were available for the correct species but in a different region. This was the case in the third project, which estimated the inter- and intrastage competition coefficients for rusty crayfish using data that described its congener *O. virilis* (Momot and Gowing 1974). Hence, although the models produce very precise results, these need to be seen in the context of input data that may not be of high accuracy (see chapter 7 for a discussion of data quality and uncertainty). Bayesian methods are available to determine the degree of exchangeability of data from related species or locations and to assess the accuracy and precision of these transfers (Finetti 1974; DuMouchel and Harris 1983).

In many cases, the lack of bioeconomic data was even more acute than the lack of ecological data. For example, in the first project, we assumed that the average amount spent per hectare on panfishing in the whole of Wisconsin was the same as the amount of money spent on panfishing per hectare in Vilas County, Wisconsin. In reality, however, fish population densities may be different, or anglers in Vilas County may represent a specific subpopulation of anglers. The value we were trying to model may be more or less than the value we used. Ideally, the lost economic value from panfish fisheries would have been determined from surveys of Vilas County anglers. Surveys could be used to quantify market and nonmarket values (e.g., the value of a family's history of fishing in Vilas County, which may be lost as fish populations decline), and we acknowledge that estimates based on actual dollars spent can rarely capture nonmarket values. However, the surveys required to quantify the total costs and benefits of an invasion are costly (McIntosh et al. 2007; see also chapter 8) and beyond the scope of most research projects.

Each assumption about data will take the final result further from answering the original question asked. Thus, it is necessary to evaluate not only each assumption, but also the cumulative effects of many assumptions. If the data used are too distant from the system they are being applied to, it may not be possible to produce a defensible analysis (however, there are methods to address such issues, e.g., Finetti 1974; DuMouchel and Harris 1983). This issue is alleviated when data are available for exactly the parameter of interest, or for data that are closely related to that parameter. Because early management of invaders is desirable, data about the exact system of interest may not be available when management has the greatest chance of success. In these cases, which are very common, judgments must be made about whether data available for other species and regions can give results accurate enough for management guidance.

Although the three models presented in this chapter represent steps toward the type of integrated models advocated by this book, none reaches the larger goal of including feedbacks among economic and ecological systems. For example, the first model assumes that placing personnel on boat ramps to check boats and bait buckets for rusty crayfish would not affect angler behavior. In reality, such personnel may lead to anglers avoiding those lakes, and this effect may be particularly strong for anglers who prefer to use rusty crayfish as bait. If this were the case, then prevention would be effective at those lakes where personnel are placed, but lakes initially identified as having a low risk of invasion would experience a higher risk as they receive more angler visits. Although the dynamics of rusty crayfish invasions in Wisconsin are relatively well known, the type of data that would allow feedbacks such as this to be included in the models have never been collected. Gathering and analyzing relevant data often require a combination of economic, ecological, theoretical, and empirical expertise. This, in turn, requires multidisciplinary collaborations. Future research that generates models of appropriate complexity and accuracy early in the invasion sequence could allow a more proactive approach to the management of species invasions.

Acknowledgments We thank the University of Alberta biological invasions working group, which is supported by the Integrated Systems for Invasive Species project (www.math.ualberta.ca/~mathbio/ISIS), funded by the National Science Foundation Integrated Research Challenge in Environmental Biology (DEB 02-13698). We are pleased to acknowledge additional financial support from a Wain

International Travel Fellowship from the Biotechnology and Biological Sciences Research Council, a National Sciences and Engineering Research Council of Canada Collaborative Opportunity Grant, and Mathematics of Information Technology and Complex Systems to C.J.B.; the Center for Aquatic Conservation, University of Notre Dame to A.M.B.; and a National Science Foundation graduate fellowship to J.A.P. Data sources: North Temperate Lakes Long Term Ecological Research program (lter.limnology.wisc.edu), National Science Foundation; James A. Rusak, Trout Lake Station, University of Wisconsin–Madison; and Lake Ottawa Trawl programme, Brett W. Peters, Department of Biological Sciences, University of Notre Dame.

References

Bampfylde, C. J., and M. A. Lewis. 2007. Biological control through intraguild predation: case studies in pest control, invasive species and range expansion. Bulletin of Mathematical Biology 69: 1031–1066.

Belovsky, G. E., D. B. Botkin, T. A. Crowl, K. W. Cummins, J. F. Franklin, M. L. Hunter, A. Joern, D. B. Lindenmayer, J. A. MacMahon, C. R. Margules, and J. M. Scott. 2004. Ten suggestions to strengthen the science of ecology. BioScience 54:345–351.

Callaway, R. M., and J. L. Maron. 2006. What have exotic plant invasions taught us over the past 20 years? Trends in Ecology and Evolution 21:369–374.

Capelli, G. M. 1982. Displacement of northern Wisconsin crayfish by *Orconectes rusticus* (Girard). Limnology and Oceanography 27:741–745.

Capelli, G. M., and J. J. Magnuson. 1983. Morphoedaphic and biogeographic analysis of crayfish distribution in northern Wisconsin. Journal of Crustacean Biology 3:548–564.

Capelli, G. M., and B. L. Munjal. 1982. Aggressive interactions and resource competition in relation to species displacement among crayfish of the genus Orconectes. Journal of Crustacean Biology 2:486–492.

Carpenter, S. R. 1998. The need for large-scale experiments to assess and predict the response of ecosystems to perturbation. Pages 287–312 *in* M. L. Pace and P. Groffman, editors. Successes, limitations, and frontiers in ecosystem science. Springer, New York.

Carpenter, S. R., B. J. Benson, R. Biggs, J. W. Chipman, J. A. Foley, S. A. Golding, R. B. Hammer, P. C. Hanson, P. T. J. Johnson, A. M. Kamarainen, T. K. Kratz, R. C. Lathrop, K. D. McMahon, B. Provencher, J. A. Rusak, C. T. Solomon, E. H. Stanley, M. G. Turner, M. J. Vander Zanden, C. H. Wu, and H. L. Yuan. 2007. Understanding regional change: a comparison of two lake districts. BioScience 57:323–335.

Charlebois, P. M., and G. A. Lamberti. 1996. Invading crayfish in a Michigan stream: direct and indirect effects on periphyton and macroinvertebrates. Journal of the North American Benthological Society 15:551–563.

Department of Fisheries and Oceans Canada. 2004. 2000 survey of recreational fishing in Canada. Department of Fisheries and Oceans, Ottawa, Ontario.

Didonato, G. T., and D. M. Lodge. 1993. Species replacements among Orconectes crayfishes in Wisconsin lakes—the role of predation by fish. Canadian Journal of Fisheries and Aquatic Sciences 50:1484–1488.

Dobson, A. P., W. R. Turner, and D. S. Wilcove. 2007. Conservation biology: unsolved problems and their policy implications. Pages 172–189 *in* R. M. May and A. R. McLean, editors. Theoretical ecology: principles and applications. Oxford University Press, New York.

DuMouchel, W. H., and J. E. Harris. 1983. Bayes methods for combining the results of cancer studies in humans and other species. Journal of the American Statistical Association 78:293–308.

Finetti, B. D. 1974. Bayesianism—its unifying role for both foundations and applications of statistics. International Statistical Review 42:117–130.

Garvey, J. E., J. E. Rettig, R. A. Stein, D. M. Lodge, and S. P. Klosiewski. 2003. Scale-dependent associations among fish predation, littoral habitat, and distributions of crayfish species. Ecology 84:3339–3348.

Gunderson, J. 1995. Three-state exotic species boater survey: what do boaters know and do they care? Aquatic Nuisance Species Digest 9:8–10.

Hamr, P. 2001. Orconectes. Pages 585–608 *in* D. M. Holdich, editor. Biology of freshwater crayfish. Blackwell, Oxford.

Hanley, N., J. F. Shogren, and B. White. 2007. Environmental economics in theory and practice, 2nd edition. Palgrave Macmillan, Basingstoke, UK.

Hein, C. L., B. M. Roth, A. R. Ives, and M. J. Vander Zanden. 2006. Fish predation and trapping for rusty crayfish (*Orconectes rusticus*) control: a whole-lake experiment. Canadian Journal of Fisheries and Aquatic Science 63:383–393.

Hein, C. L., M. J. Vander Zanden, and J. J. Magnuson. 2007. Intensive trapping and increased fish predation cause massive population decline of an invasive crayfish. Freshwater Biology 52:1134–1146.

Hill, A. M., and D. M. Lodge. 1999. Replacement of resident crayfishes by an exotic crayfish: the roles of competition and predation. Ecological Applications 9:678–690.

Hobbs, H. H., and J. P. Jass. 1988. The crayfishes and shrimp of Wisconsin (Cambaridae, Palaemonidae). Milwaukee Public Museum, Milwaukee.

Horan, R. D., and F. Lupi. 2005. Tradeable risk permits to prevent future introductions of invasive alien species into the Great Lakes. Ecological Economics 52:289–304.

Keller, R. P., D. M. Lodge, and K. Frang. 2007. Prevention guided by ecological predictions produces environmental and economic benefits for lakes at risk from biological invasion. Conservation Biology 22:80–88.

Kratz, T. K., J. J. Magnuson, and J. A. Rusak. 2007. North Temperate Lakes LTER: crayfish abundance. North Temperate Lakes–Long Term Ecological Research program, Center for Limnology, University of Wisconsin-Madison, Madison.

Leonard, J. 2005. Black bass and trout fishing in the United States. Report 2001-10. Division of Federal Assistance U.S. Fish and Wildlife Service, Arlington, VA.

Leung, B., D. M. Lodge, D. Finnoff, J. F. Shogren, M. A. Lewis, and G. Lamberti. 2002. An ounce of prevention or a pound of cure: bioeconomic risk analysis of invasive species. Proceedings of the Royal Society of London Series B-Biological Sciences 269:2407–2413.

Lodge, D. M. 1993. Biological invasions—lessons for ecology. Trends in Ecology and Evolution 8:133–137.

Lodge, D. M., A. L. Beckel, and J. J. Magnuson. 1985. Lake-bottom tyrant. Natural History 94:32–36.

Lodge, D. M., and A. M. Hill. 1994. Factors governing species composition, population size, and productivity of cool-water crayfishes. Nordic Journal of Freshwater Research 69:111–136.

Lodge, D. M., M. W. Kershner, J. E. Aloi, and A. P. Covich. 1994. Effects of an omnivorous crayfish (*Orconectes rusticus*) on a fresh-water littoral food-web. Ecology 75:1265–1281.

Lodge, D. M., T. K. Kratz, and G. M. Capelli. 1986. Long-term dynamics of 3 crayfish species in Trout Lake, Wisconsin. Canadian Journal of Fisheries and Aquatic Sciences 43:993–998.

Luttenton, M. R., M. J. Horgan, and D. M. Lodge. 1998. Effects of three *Orconectes* crayfishes on epilithic microalgae: a laboratory experiment. Crustaceana 71:845–855.

McIntosh, C. R., J. F. Shogren, and D. Finnoff. 2007. Valuing delaying the inevitable: evidence from a national survey. Working Paper, University of Minnesota, Duluth.

Momot, W. T. 1984. Crayfish production—a reflection of community energetics. Journal of Crustacean Biology 4:35–54.

Momot, W. T., and H. Gowing. 1974. The cohort production and life cycle turnover ratio of the crayfish *Orconectes virilis*, in three Michigan lakes. Freshwater Crayfish 2:489–511.

Momot, W. T., H. Gowing, and P. D. Jones. 1978. Dynamics of crayfish and their role in ecosystems. American Midland Naturalist 99:10–35.

New Hampshire Lakes Association. 2005. Lake host program summary and statistics. New Hampshire Lakes Association, Concord.

Olden, J. D., J. M. McCarthy, J. T. Maxted, W. W. Fetzer, and M. J. Vander Zanden. 2006. The rapid spread of rusty crayfish (*Orconectes rusticus*) with observations on native crayfish declines in Wisconsin (USA) over the past 130 years. Biological Invasions 8:1621–1628.

Olsen, T. M., D. M. Lodge, G. M. Capelli, and R. J. Houlihan. 1991. Mechanisms of impact of an introduced crayfish (*Orconectes rusticus*) on littoral congeners, snails, and macrophytes. Canadian Journal of Fisheries and Aquatic Sciences 48:1853–1861.

Olson, M. H., and B. P. Young. 2003. Patterns of diet and growth in co-occurring populations of largemouth bass and smallmouth bass. Transactions of the American Fisheries Society 132: 1207–1213.

Puth, L. M., and T. F. H. Allen. 2005. Potential corridors for the rusty crayfish, *Orconectes rusticus*, in northern Wisconsin (USA) lakes: lessons for exotic invasions. Landscape Ecology 20:567–577.

Rabeni, C. F. 1992. Trophic linkage between stream centrarchids and their crayfish prey. Canadian Journal of Fisheries and Aquatic Sciences 49:1714–1721.

Rosenthal, S. K., S. S. Stevens, and D. M. Lodge. 2006. Whole-lake effects of invasive crayfish (*Orconectes* spp.) and the potential for restoration. Canadian Journal of Fisheries and Aquatic Sciences 63:1276–1285.

Roth, B. M., C. L. Hein, and M. J. Vander Zanden. 2006. Using bioenergetics and stable isotopes to assess the trophic role of rusty crayfish (*Orconectes rusticus*) in lake littoral zones. Canadian Journal of Fisheries and Aquatic Science 63:335–344.

Roth, B. M., and J. F. Kitchell. 2005. The role of size-selective predation in the displacement of Orconectes crayfishes following rusty crayfish invasion. Crustaceana 78:297–310.

Sala, O. E., F. S. Chapin, J. J. Armesto, E. Berlow, J. Bloomfield, R. Dirzo, E. Huber-Sanwald, L. F. Huenneke, R. B. Jackson, A. Kinzig, R. Leemans, D. M. Lodge, H. A. Mooney, M. Oesterheld, N. L. Poff, M. T. Sykes, B. H. Walker, M. Walker, and D. H. Wall. 2000. Biodiversity—global biodiversity scenarios for the year 2100. Science 287:1770–1774.

Sax, D. F., S. D. Gaines, and J. J. Stachowicz. 2005. Introduction. Pages 1–7 *in* D. F. Sax, J. J. Stachowicz, and S. D. Gaines, editors. Species invasions: insights into ecology, evolution, and biogeography. Sinauer, Sunderland, MA.

Settle, C., T. D. Crocker, and J. F. Shogren. 2002. On the joint determination of biological and economic systems. Ecological Economics 42:301–311.

Settle, C., and J. F. Shogren. 2002. Modeling native-exotic species within Yellowstone Lake. American Journal of Agricultural Economics 84:1323–1328.

Stein, R. A. 1977. Selective predation, optimal foraging, and predator-prey interaction between fish and crayfish. Ecology 58:1237–1253.

Strayer, D. L., V. T. Eviner, J. M. Jeschke, and M. L. Pace. 2006. Understanding the long-term effects of species invasions. Trends in Ecology and Evolution 21:645–651.

U.S. Department of the Interior, Fish and Wildlife Service, U.S. Department of Commerce, and U.S. Census Bureau. 2002. 2001 National survey of fishing, hunting and wildlife-associated recreation. U.S. Fish and Wildlife Service, Washington, DC.

Wilson, K. A., J. J. Magnuson, D. M. Lodge, A. M. Hill, T. K. Kratz, W. L. Perry, and T. V. Willis. 2004. A long-term rusty crayfish (*Orconectes rusticus*) invasion: dispersal patterns and community change in a north temperate lake. Canadian Journal of Fisheries and Aquatic Science 61:2255–2266.

12

Advances in Ecological and Economic Analysis of Invasive Species: Dreissenid Mussels as a Case Study

Jonathan M. Bossenbroek, David C. Finnoff, Jason F. Shogren, and Travis W. Warziniack

In a Clamshell

Since the discovery of zebra mussels and quagga mussels in North America, dreissenid mussels have driven recent U.S. policy development on invasive species. Combining past studies on dreissenid mussels with our current research, we present an unusually (perhaps uniquely) complete synthesis of the entire invasion sequence from both ecological and economic perspectives. This chapter demonstrates models of potential spread, ecological niche models, assessments of the factors that influence establishment, estimates of economic impacts, integrated optimization modeling of the value of slowing the spread, and assessment of the behavior of decision makers. To bring all these ideas into focus, this chapter asks what it is worth to keep dreissenid mussels from becoming established in western states and provinces of North America, as a focal point to our research. Answering this question relies not only on objective ecological and economic estimates of critical variables at the intersection of the intertwined systems, but on the perspectives and attitudes of policy makers about investments in prevention and control. Given limited financial resources, there is no single dollar amount to answer the question; the answer lies in the accuracy of the models and the behavior and priorities of managers and decision makers. This chapter demonstrates a piece-by-piece development of a framework for a comprehensive bioeconomic assessment, which should be useful for assessing the risks of other invasive species.

January 2007 began a new chapter in the invasion of dreissenid mussels in North America. For the first time, dreissenid mussels were not restricted to the eastern United States and Canada. On January 6, 2007, quagga mussels (*Dreissena bugensis* (= *D. rostriformis bugensis* [Andrusov (1897)])) were discovered in a marina in Lake Mead, a large reservoir near Las Vegas, Nevada, on the Colorado River (National Park Service 2007). Preventing this invasion from happening has been the goal of the 100th Meridian Initiative, a cooperative organization including state, federal, and provincial agencies, which was established in 1998. For the past 4 years, we (in collaboration with David Lodge and many of the other contributors in this book) have been working to shed light on two basic questions: What is it worth to keep dreissenid mussels from becoming established in the western states and provinces of North America? How much information and/or modeling is needed to make a policy recommendation given all the uncertainties in the invasion process? Answering these questions requires ecological predictions of dispersal, potential habitat, probability of establishment, and likely abundance and also estimates of direct and indirect economic impacts, the incorporation of policy time horizons, and the behavior of those individuals making decisions. Given these multifaceted and interlinked components, the insights we have generated are incomplete, yet provide a clear example of the research necessary to (eventually) integrate ecology and economics for improved decision making. This integration apparently has been lacking, as agencies of the western United States may now find out sooner than we expected if dreissenid mussels will have the similar economic and ecological impacts as they have had in eastern North America.

In 1986 zebra mussels (*Dreissena polymorpha*) were discovered in Lake St. Clair (Herbert et al. 1989) near Detroit, Michigan, and Windsor, Ontario, and in 1991 quagga mussels were introduced in Lake Ontario (May and Marsden 1992). These introductions occurred even though the dispersal capabilities of dreissenid mussels were known as far back as 1893 and their potential economic impact as early as 1959, including a report for the Environmental Protection Service of Canada in 1981 (for a list of predictions, see Carlton 1991). Despite these warnings, there was no coordinated effort to prevent these invasions from occurring. Dreissenid mussels are estimated to currently cost U.S. industries millions/year (O'Neill 1997). Additionally, they have caused the local extinction of many native mollusks (Strayer and Smith, 1996), have changed the structure of fish communities in the Hudson River (Strayer et al. 2004), and have been implicated in the demise of valuable sport fish populations (Dermott 2001). Given that this invasion was predicted, what would it have been worth to invest to keep dreissenid mussels out of North America?

Early in the North American invasion, Ludyanskiy et al. (1993) went as far as to predict, "Within the next few years, and certainly by the turn of the century, the zebra mussel will be found in almost all parts of the United States and southern Canada." The rapid spread of zebra mussels through the Great Lakes and shipping routes of the United States soon after their introduction in Lake St. Clair supported such predictions (Allen and Ramcharan 2001). However, from 1993 to 2006, the range of dreissenid mussels did not change much (Johnson et al. 2006). The distribution of dreissenid mussels in 2006 was primarily limited to the Great Lakes, rivers with active shipping connected to the Great Lakes (e.g., Mississippi, Kentucky, Tennessee, and Hudson rivers), and inland lakes within 150 km of the distribution as of 1993 (figure 12.1). As of 2007, this distribution has markedly changed with

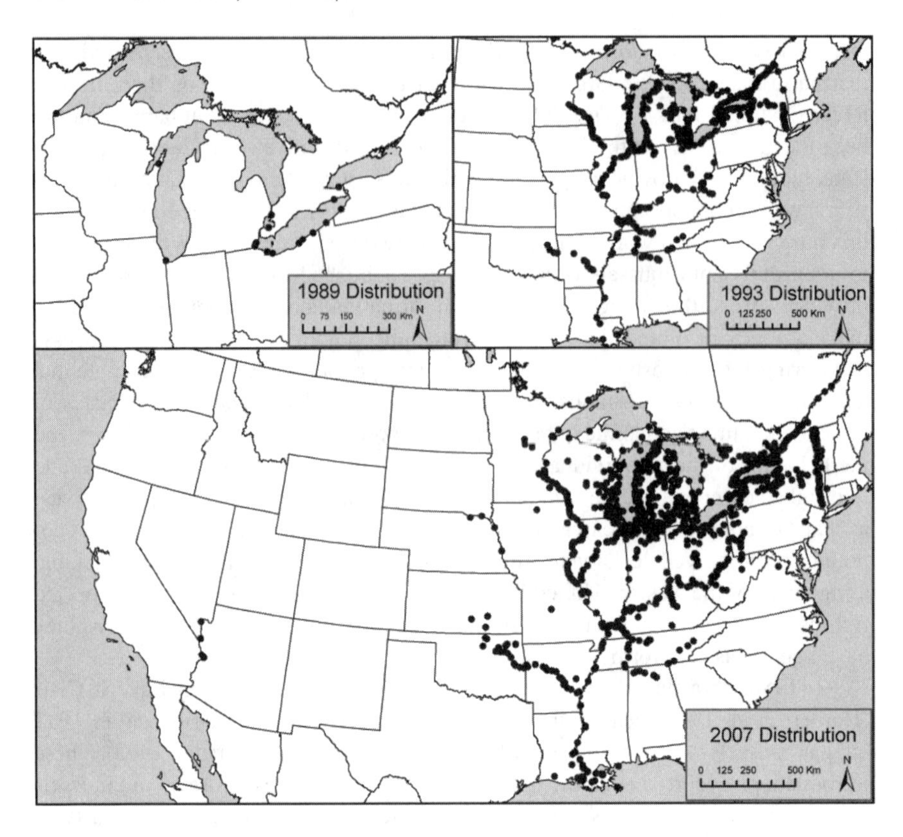

FIGURE 12.1.
The known distribution of dreissenid mussels in North America in 1989, 1993, and 2007. Data from the U.S. Geological Survey Nonindigenous Aquatic Species information resource (http://nas.er.usgs.gov/).

the discovery of quagga mussels in Lake Mead and other reservoirs of the lower Colorado River.

Though the rate of spread had declined, there was still substantial concern recently as to the future distribution of dreissenid mussels in North America. The State of Minnesota has a major public outreach campaign to limit the number of lakes invaded by dreissenid mussels. As of 2007, only three lakes in Minnesota were known to contain dreissenid mussels, despite the presence of dreissenid mussels in the Mississippi River and Lake Superior for previous 15 years. Likewise, the 100th Meridian Initiative focuses on preventing the westward spread of dreissenid mussels and other aquatic nuisance species in North America. Within the context of this concern, the objective of this chapter is to synthesize the research of this poster child of invasive species within the bioeconomic framework discussed in the central portion of this book. We place particular emphasis on how these tools have been used to assess the risk dreissenid mussels pose to the western United States.

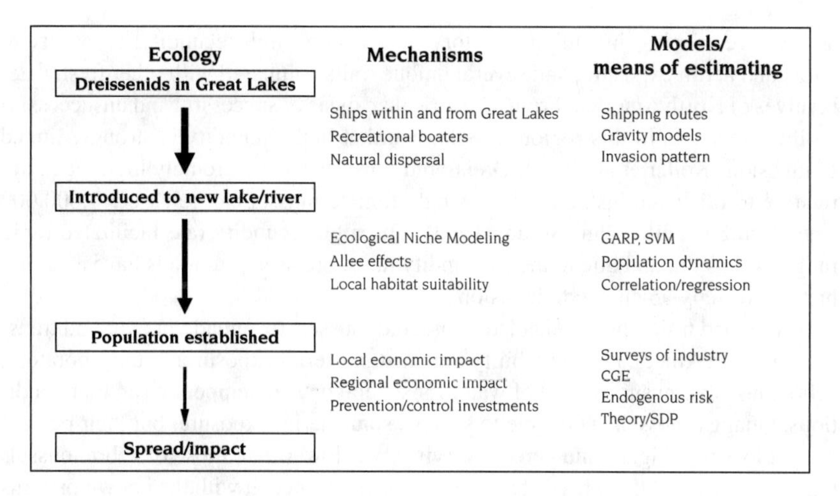

FIGURE 12.2.
General framework of the bioeconomics of the dreissenid invasion of North America, including the mechanisms by which dreissenid mussels transfer between stages and examples of models to predict the success of these transfers. GARP, genetic algorithm for rule-set production; SVM, support vector machine modeling; CGE, computable general equilibrium modeling; SDP, stochastic dynamic programming.

In addition, we hope this chapter provides a synthetic framework by which the bioeconomics of other invasive species can be assessed. This chapter is designed to mimic the overall structure of this book, from predicting the introduction of dreissenid mussels to assessing their regional economic impacts (figure 12.2). Our focus in the chapter is primarily on the later stages of invasion (i.e., secondary spread and impact) and can be compared to the overall invasion process presented in chapter 1 (see figure 1.1). In this chapter, we highlight many of the mechanisms and the tools/models that have been used to understand the dreissenid invasion; specific details of these models can be found in their corresponding chapters in the rest of the book. It should be noted that several stages of the dreissenid mussel invasion have been studied more thoroughly than others, which will be evident in the range and specificity of the research presented in the following sections. This merely suggests that there is plenty of opportunity to further study this process and, more important, the availability of interdisciplinary research opportunities. We expect that through this chapter it becomes apparent that achieving an acceptable answer about the benefits of stopping or slowing the spread of an invasive species requires the integration of ecology, economics, and mathematics at all the stages of an invasion.

PREDICTING SPECIES INTRODUCTIONS

Natural history traits can be used to predict whether a species will become invasive in a certain region. Dreissenid mussels have several traits that make them highly suitable as invasive species, including a history of invasion, high fecundity, an ability to withstand aerial exposure, and several unique traits compared with other freshwater bivalves of North America. From a historical analysis of successful and unsuccessful mollusk invaders in this region, it is known that high fecundity is strongly linked to invasion (Keller et al. 2007). Dreissenid mussels have extremely high fecundity relative to other mollusk species; a single female mussel can release >1,000,000 eggs during her life span (Sprung 1993). The high fecundity rate facilitated rapid range expansion, and due to their fecundity rates, dreissenid mussels have a natural history strongly amenable to invasion.

A second trait that has enabled dreissenid mussels to spread via overland transport in North America is their ability to withstand aerial exposure. In the laboratory, zebra mussels can survive out of water for several days in temperate summer conditions. Quagga mussels were able to survive some aerial exposure, but their percent weight loss was higher and percent survival was lower than those of zebra mussels (Ricciardi et al. 1995). This difference in aerial tolerance fit with the known patterns of zebra and quagga mussels as of 2006. Through 2006, quagga mussels were known only in the Great Lakes, and all the inland lake infestations had been reported as zebra mussels, although we suspect that few of these populations were examined closely to determine if the mussels were zebra mussels or quagga mussels. Contrary to the lower aerial exposure tolerances of quagga mussels, they were the species that made the longest overland jump of more than 1,500 km from the Great Lakes region to the Colorado River. Adult dreissenid mussels, subjected to aerial exposure, have been found on several occasions on boats in many western states (100th Meridian Initiative 2007). It should also be noted that many boats have live wells, bilge tanks, and other compartments that are capable of transporting water, and thus also living mussels or juveniles, eliminating the need for dreissenid mussel to survive aerial exposure. In general, the aerial tolerance of dreissenid mussels has evidently been important in their capabilities for long-distance dispersal.

Beyond their high fecundity and ability to withstand aerial exposure, dreissenid mussels have several unique traits compared with other freshwater bivalves in North America (Mackie 1991). Native freshwater bivalves of North America live their entire adult lives partially buried in sediments in lakes and streams (i.e., infaunal), and the larvae are either brooded (ovoviviparous, Sphaeriidae clams) or parasitic (glochidium, Unionidae clams). The glochidium larvae usually exist on the skin, fins, or gill of fish, and many species are host specific (Parmalee and Bogan 1998). As opposed to native unionids, dreissenids have a pelagic larval stage that enables passive dispersal via water currents. Dreissenid mussels are also epifaunal; that is, they live attached to and above the substrate. The epifaunal nature of dreissenid mussels allows these new species to utilize habitat not used by native bivalves in the freshwaters of North America. Not only do dreissenid mussels colonize natural substrates, such as rock outcroppings, but they also are abundant on human structures, including buoys, boats, docks, and locks. A trait-based assessment of dreissenid mussels would thus have been sufficient to identify dreissenid mussels as being extremely high-risk species long before they became established in the Great Lakes (Carlton 1991).

In addition to species traits increasing the likelihood of invasion success, species that have a history of invasion elsewhere tend to be successful invaders in other similar regions (Kolar and Lodge 2001). Dreissenid mussel are native to the Ponto-Caspian basin in Eastern Europe; during the 1700s and 1800s this region was connected to waterways across Europe by the construction of canals, and by the 1830s zebra mussels had expanded their range to include much of Europe and Britain. Likewise, after their introduction into the Great Lakes, dreissenid mussels quickly expanded their range throughout the navigable waterways that are connected to the Great Lakes. The most recent "island" to become invaded by zebra mussels was Ireland, in 1994, even though Britain has had zebra mussels since 1824 (Pollux et al. 2003). Given their past success as invaders throughout Europe and eastern North America and the similarities of western North America to these previously invaded locations, a clear warning of the potential for dreissenid mussels to become established, and rapidly spread, in western North America should have been heeded.

ESTABLISHMENT SUCCESS AND DISPERSAL

Dreissenid mussels have the right traits to be good invaders, but are all the water bodies across North America equally suitable habitat? Many techniques have been used to model the ecological niches of invasive species (see chapter 4), including several attempts to forecast the potential distribution of zebra mussels in North America, with no specific efforts on predicting the potential habitat of quagga mussels (but see Thorp et al. 2002). These forecasts have included nonspatial models based on water-quality parameters (Ramcharan et al. 1992; Mellina and Rasmussen 1994), regional models that use specific lake and geology parameters (e.g., Neary and Leach 1992; Koutnik and Padilla 1994), and national models based on low-resolution data such as air temperature, frost frequency, and geology (Strayer 1991; Drake and Bossenbroek 2004). The consistent theme through all of these predictions is that water quality or a surrogate, such as bedrock geology, is essential to predicting what bodies of water would allow dreissenids to thrive if they were introduced.

To assess the potential habitat for dreissenid mussels west of the 100th meridian, analyses have been conducted at both regional and local scales. On a national scale, the suitable habitat for zebra mussels has been predicted using a genetic algorithm for rule-set production (GARP; see chapter 4) based on 11 environmental and geologic variables, which were at a resolution of 0.1 decimal degrees (Drake and Bossenbroek 2004), and solely using data on calcium concentration in water (Whittier et al. 2008). The final GARP model was based on five variables thought to be biologically relevant to the distribution of zebra mussels, including frost frequency, maximum annual temperature, slope, bedrock geology, and surface geology. The results suggest that much of the western United States may not be as susceptible to zebra mussel invasion as previously thought, or compared to predictions of previous models based on temperature variables alone (Strayer 1991). The results, however, do predict that certain areas of the Columbia, Colorado, and Sacramento-San Joaquin river basins are at significant risk (figure 4.3; Drake and Bossenbroek 2004). Whittier et al. (2008)

predicted the potential distribution of zebra mussels for each of approximately 60 ecoregions in the United States based on calcium concentrations. The results of this effort were fairly similar to the results of the GARP model, particularly in the Great Plains and Midwest. The differences between the Whittier model and the GARP model were most evident in the Southwest, where the Whittier model predicts higher risk of suitable habitat, and in the Southeast and Northeast, where it predicts a lower risk. The Whittier et al. (2008) results are important because they show that the suitability of Lake Mead and other southwestern reservoirs was predictable.

On a more localized scale, detailed water-quality data were collected to predict the potential densities of zebra mussels if they were introduced in lakes Mead and Roosevelt (Bossenbroek et al. 2007). Population density, the effect of each individual, and the overall range of a species are three key indicators of the overall impact (both ecological and economic) of an invasive species (Parker et al. 1999). Using a previously published model (Ramcharan et al. 1992), predicted abundances of zebra mussels were compared to those of water bodies with reported zebra mussel density. Based on these data and model, density predictions suggest that both Lake Mead (Colorado River) and Roosevelt Lake (Columbia River) could support substantial population densities of zebra mussels (Bossenbroek et al. 2007). Lake Mead is expected to have densities reaching hundreds of thousands per square meter, whereas Roosevelt Lake would most likely maintain more moderate populations in the thousands per square meter. High densities of zebra mussels in lakes Mead and Roosevelt would likely lead to substantial economic impacts because these bodies of water contain more infrastructure than most lakes and rivers in the Midwest, including hydropower dams, municipal water supply systems, and irrigation pumps.

As demonstrated, several tools have now been used to predict the potential distribution and niche boundary of zebra mussels. The compilation of these studies suggests that the areas most at risk to future invasions include the southeastern and southwestern United States and the regions already invaded by dreissenid mussels, that is, the Great Lakes region. These models do acknowledge that suitable habitat for dreissenid mussels exists in many areas throughout the country that have not yet been invaded.

The availability of suitable habitat does not guarantee that a lake will become invaded, which requires the introduction of viable propagules and establishment of a population. Dreissenid mussels have several modes by which they can disperse, both natural and human mediated (Johnson and Padilla 1996). The arrival of dreissenid mussels within North America is believed to be the result of ballast discharge from ships originating from the Ponto-Caspian sea region (Mills et al. 1993), originally in Lake St. Clair. From this initial point of introduction, natural dispersal of dreissenid mussels occurred primarily due to the movement of water, which has been shown in large river systems and coupled lake-stream systems. Coupled lake-stream systems constitute a source-sink model of zebra mussel dispersal (Horvath et al. 1996; Bobeldyk et al. 2005), such that in-stream dreissenid mussel populations are often not self-sustaining but are dependent on continuous recruitment from source populations of the upstream lake. Although populations may not easily establish in moving water, live veligers (the planktonic juvenile stage of dreissenid mussels) can travel more than 300 km in larger rivers, such as the Illinois River (Stoeckel et al. 1997). The downstream dispersal of zebra mussels through streams results in establishment of populations in streams and downstream lakes (Bobeldyk et al. 2005) and is thought to be the source of approximately one-third of all the inland lake invasions (Johnson et al. 2006). The presence of wetlands in streams, however, can inhibit this downstream spread (Bodamer and Bossenbroek 2008). Downstream movement of quagga mussels has already occurred in the lower Colorado River. Lakes Mohave and Havasu have quagga mussels and are downstream of Lake Mead, which is assumed is the initial point of introduction in the Colorado system.

Without the aid of flowing water, long-distance dispersal of dreissenid mussels requires a human vector. After their initial introduction into the Great Lakes, ships and barges played a substantial role in the dispersal of zebra mussels. By 1993, almost all of the navigable waters of eastern North America were invaded with zebra mussels (Allen and Ramcharan 2001). The overland transport to inland lakes also requires an additional human vector, most notably recreational boating (Johnson and Carlton 1996). The spread of zebra mussels via recreational boaters has been modeled on several occasions with the use of gravity models (Schneider et al. 1998; Bossenbroek et al. 2001; Leung et al. 2006; Bossenbroek et al. 2007; for detailed description, see chapters 6 and 7). Linking the spread of invasive species with the movement of people enables integration with geography and economics. Thus far, several invasive species studies have used gravity models, frequently employed by geographers, but there is abundant opportunity to use recreational demand models, which economists use to analyze recreation choices, including consumer behavior.

Based on an understanding of the mechanisms of spread, Bossenbroek et al. (2007) constructed a gravity model to explore the movement patterns of recreational boaters from areas with zebra mussels on a national scale (figure 12.3). The parameters of the model were estimated by comparing model results with survey data collected via the 100th Meridian Initiative. The model results are consistent with the observed slow range expansion of dreissenid mussels in recent years, with the exception of the newly discovered population of quagga mussels in Lake Mead. The model would have predicted the invasion of Lake Mead to be a low probability event, yet more probable than the invasion of most other reservoirs in the western United States and even more probable than many lakes and reservoirs in the eastern United States. The uncertainty involved with these predictions is assessed in detail in chapter 7. Although the Bossenbroek et al. (2007) article was focused on zebra mussels, we believe the model is indicative of the movement patterns that should be exhibited by quagga mussels.

We have updated the national gravity model of Bossenbroek et al. (2007) by considering lakes Mead, Havasu, and Mohave on the Colorado River and the Lake of the Ozarks in Missouri as additional sources of dreissenid mussels (J. M. Bossenbroek unpublished data). Because there are now more sources of dreissenid mussels, the probability of invasion for every lake in the country has increased. The question becomes whether these new sources substantially increase the risk of invasion to noninvaded lakes, such as Roosevelt Lake in the Columbia River. The original gravity model prediction was that 89 boats (on a relative scale) from dreissenid-infested

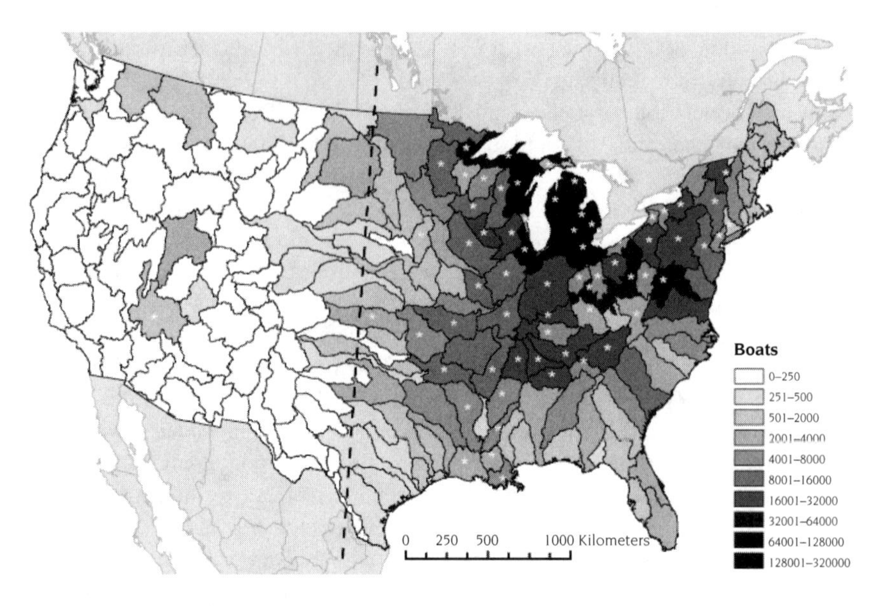

FIGURE 12.3.
Relative predicted number of boats from water bodies invaded by dreissenid mussels traveling to each watershed within the United States as calculated with a gravity model. An asterisk indicates that dreissenid mussels have been observed in the watershed on at least one occasion. The dashed line specifies the 100th meridian. Modified from Bossenbroek et al. (2007).

waters would travel to Roosevelt Lake. The additional sources increased the number of potentially infested boats traveling to Roosevelt Lake by 25% to 114 boats. Thus, Roosevelt Lake is not predicted to have a high risk of introduction compared to many other watersheds in the Great Lakes region (see figure 12.3). The introduction of quagga mussels in Lake Mead was also surprising based on an assessment of the current and past distribution of dreissenid mussels at several scales, which quantified the slow range expansion and the low occurrence of long-overland dispersal events (Kraft and Johnson 2000). As of 2003, only six states had more than 10 inland invasions observed, and they account for 97% of the 293 lakes reported to be invaded in the United States (Johnson et al. 2006). For the four states surrounding Lake Michigan, <8% of suitable lakes (based on Drake and Bossenbroek 2004) larger than 25 hectares had been invaded by 2003. Although the number of invaded lakes has increased over time, the rate of invasions has decreased (figure 12.4). The reasons for this decline in invasion rate in the past several years, despite hundreds of available lakes, could be the result of education efforts, the limited attractiveness to boaters of many lakes, or some indication of a necessary threshold of propagules required for a population to become established.

Little is known about how many dreissenid mussels (juveniles or adults) must be introduced to begin a new population in a previously uninfested and unconnected water body. Dreissenid mussels dispersed by recreational boaters can be transported

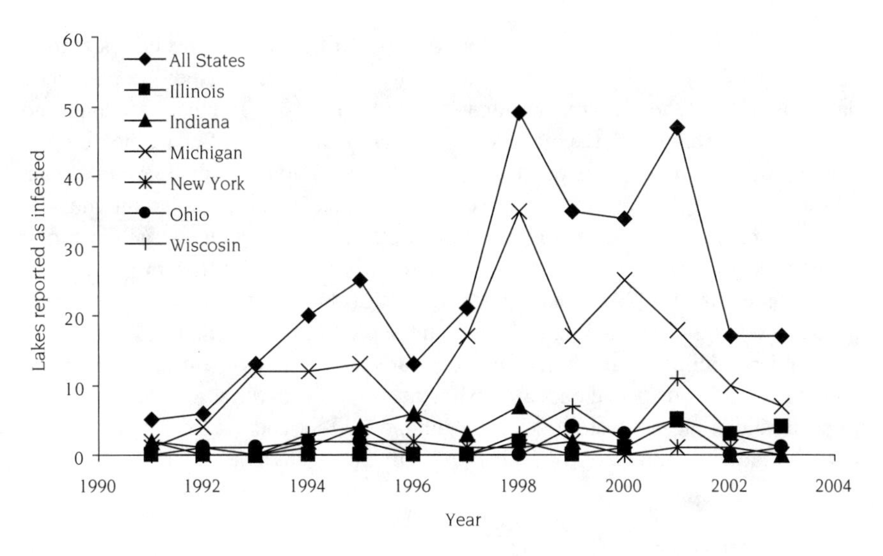

FIGURE 12.4.
Annual numbers of invasions in the six states known to have 10 or more invasions of inland waters and for the entire United States as of 2004. Reproduced from Johnson et al. (2006), with permission of Springer Science and Business Media.

by several mechanisms, including adult mussels attached to parts of boats or attached to macrophytes entangled in a boat or a trailer, and juveniles contained in any pocket of water remaining in a boat, such as a live well or bilge (Johnson and Carlton 1996). In just one example, Johnson and Carlton (1996) examined boats leaving Lake St. Clair and found between 0% and 31% of boats carrying macrophytes with zebra mussels, depending on the day. Depending on the ambient temperature and humidity (Ricciardi et al. 1995) and rate of transport, dreissenid mussels have the ability to survive transport across the entire country. Indeed, quagga mussels in Lake Mead were not the first discovery of dreissenid mussels west of the 100th meridian. On several occasions, boats arrived at Lake Mead with mussels attached (100th Meridian Initiative 2007) but were cleaned prior to launch.

A single boat launched into a lake with dreissenid mussels does not ensure that a population will become established. The probability of a lake becoming invaded with zebra mussels is not linearly related to the number of propagules introduced. Instead, zebra mussels exhibit Allee effects (Courchamp et al. 1999; Leung et al. 2004), which cause inverse density dependence at low densities. Thus, a threshold number of propagules exists above which establishment is much more likely and below which a population is likely to go extinct. There are several possible sources of Allee effects in natural populations, including genetic inbreeding, demographic stochasticity, or a reduction in cooperative interactions or the ability of mates or gametes to encounter. Several aspects of dreissenid mussel life history, including gamete production, gamete release and survival, the external fertilization process, veliger state, and status of settled adults, may be subject to Allee effects. Dreissenid mussels are sexual in their reproduction, thus requiring that gametes be released in close proximity in both space and time. Analyses by A. B. Potapov (unpublished data) suggest that there is a maximum distance of about 10–20 cm between male and female mussels necessary for successful fertilization of gametes. Such a short critical distance thus requires a large population or the introduction of both male and female mussels within close proximity. The existence of Allee effects in dreissenid mussels and the long distance between infested midwestern waters and the large reservoirs of the western United States suggest that the introduction of sufficient numbers of quagga mussels into Lake Mead was a rare event, but one that could be expected based on our gravity modeling described in the preceding section.

Understanding establishment is essential for risk assessment and for policy making. Meshing dispersal and establishment success with the associated economic impacts allows an accurate assessment of the risk invaders pose to society and whether strategies directed toward the problem are worthwhile.

ESTIMATING COSTS OF INVASIONS

The establishment of dreissenid mussels in the water bodies of North America has caused substantial economic impacts in term of both nonmarket and market costs. Nonmarket costs include direct and indirect impacts on the ecology and are not readily quantifiable because they are linked to environmental goods and services not exchanged in the marketplace. The nonmarket effects of dreissenid mussels include changes in water quality and impacts on other species. Dreissenid mussels alter concentrations of nutrients (e.g., Mellina et al. 1995), increase water clarity (e.g., Fahnenstiel et al. 1995), and can negatively affect native mussels by colonizing their shells and inhibiting filter feeding (Schloesser et al. 1996; Ricciardi et al. 1998). The observed nonmarket impacts of dreissenid mussels are one reason the 100th Meridian Initiative is concerned about their western spread.

Despite the paucity of native mussels in the western United States (NatureServe 2006), there are several nonmarket effects that dreissenid mussels could impose on western rivers and reservoirs. One major concern in the Columbia River basin is the potential risk to the native salmonid species that pass through fish ladders. If fish ladders become encrusted with dreissenid mussels, salmonids could be damaged by rubbing against the sharp shells of the mussels (Northwest Natural Resource Group 2003). Dreissenid mussels have also been associated with changes in the distribution of fish communities in river systems, including declines in open-water species and increases in littoral species (Strayer et al. 2004). The lower Colorado River is home to several endangered fish species (Dobson et al. 1997), which are already threatened by several nonindigenous fish species (Stohlgren et al. 2006). The introduction of dreissenid mussels could further imperil these species. Although nonmarket impacts of dreissenid mussels may be high, we concentrate the remainder of the chapter on market impacts given the complete lack of nonmarket impact estimates. Thus, the potential impacts discussed are likely underestimates of the total value of the impacts (if all impacts are assumed to be detrimental).

In relation to the regions of eastern North America currently affected by dreissenid mussel invasions, the West is more dependent on surface-water supplies for power, drinking water, and irrigation. This dependence highlights the importance of understanding the chances of and potential impacts of a dreissenid mussel invasion. While it has been estimated that the one-time cost to install systems to control zebra mussels at Columbia River hydroelectric projects could range from hundreds of thousands to more than a million dollars each (Phillips et al. 2005), a more complete estimate of the consequences has been lacking. Similar to the assessments of O'Neill (1997) and Pimentel et al. (2005), Phillips et al. (2005) assess only the likely direct impacts of dreissenid mussels on an industry, without examining the opportunity costs of the impacts (as described in chapter 8) and how these impacts propagate throughout the entire economy.

To assess the economy wide impacts of dreissenid mussels on the entire economy of the Columbia River basin, we used a computable general equilibrium (CGE) model to capture both primary and secondary (indirect) economic impacts (as described in Warziniack et al. 2008). In the CGE model, the economy consists of households and producing sectors, linked to one another and the rest of the world through commodity and factor markets (figure 12.5). It is through imports into commodity and factor markets that the CGE model can be linked to the transportation of dreissenid

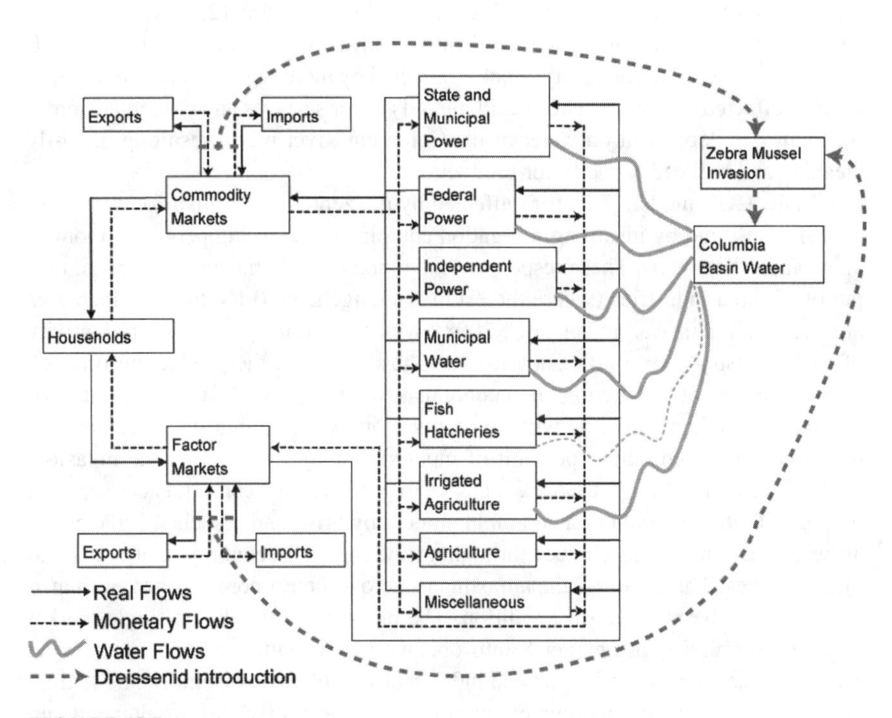

FIGURE 12.5.
Diagram of the Columbia River basin economy used in a computable general equilibrium (CGE) modeling.

mussels, via a gravity model or random-utility model (see chapter 6). Other potential linkages with the biology of dreissenid mussels include habitat suitability models of water sources (e.g., reservoirs), regional spread models, and population dynamics, including growth and abundance models. These explicit linkages will be incorporated into future work on the CGE model. Because the ecological linkages have yet to be made, our analysis was limited to a binary problem: no dreissenid mussels or a "complete" dreissenid mussel invasion. The "complete" scenario is not unfounded, however, because of the dreissenid mussel invasion dynamics that occurred in eastern North America.

In the CGE model, the invasion of dreissenid mussels was incorporated as affecting the production process of waters users. The consequences to these water users were defined based on the observed impacts to independent power companies and municipal water users in the eastern United States (Deng 1996) and estimated impacts on federal hydropower facilities (Phillips et al. 2005). Impacts on irrigated agriculture were also expected and included in the analysis at a similar magnitude as the observed water users. This is a potentially important impact because the Columbia River basin contains more than 5.1 million acres of irrigated farmland and 73.5% ($8.5 billion) of agricultural operating expenditures in the region (U.S. Department of Agriculture 2004). Although impacts on fish hatcheries were also to be expected, no relevant analyses were found to use them in the model, and thus we excluded them (indicated by a dashed gray line in figure 12.5). The industry-specific consequences of a dreissenid mussel invasion are captured in the CGE model by including productivity shocks, reflected by increases in the per-unit output costs of effected industries. Dreissenid mussels cover surfaces and clog intake pipes for industries dependent on water of the Columbia River basin, resulting in costly cleaning and reduced capacity for production.

In the CGE model, industries affected by the zebra mussel invasion were presumed to respond by installing mitigation equipment and hiring people to monitor and control the effects. These responses lead to increases in the cost per unit of output of affected industries (where the estimates range from 0.1% to 0.3% increases and are detailed in Warziniack et al. 2008), which result in declines in productivity of each industry and to efficiency losses. Thus, inputs in the production process, termed "factors of production" by economists, are not as productive in the face of an invasion as they would be without an invasion. Affected industries are not able to produce as much output per unit of input(s) as they were before the invasion, making each unit of output more expensive to produce. The consequences are seen in terms of altered production and input choices by firms and associated changes in all prices, incomes, and choices throughout the regional economy as it adjusts to these changes. The net of all the adjustments across the economy is then evaluated in relation to the noninvaded condition. The procedure extends the single market valuation discussion of chapter 8 into operation in a multimarket, economywide setting that accounts for all direct and indirect effects of the invasion. In this setting, the reciprocal price and income effects from simultaneously shifting demand and supply curves throughout the regional economy are taken into account (although shifts occur in differing degrees and directions). In addition, the method provides aggregate measures of the welfare changes in terms of equivalent variations, or what households would have been willing to pay to not experience the invasion (calculated in relation to the benchmark equilibrium of the noninvaded 2001 data and similar to the consumer surplus measures discussed in chapter 8). The CGE model for the noninvaded economy was based on 2001 IMPLAN data for the Columbia River basin (IMPLAN is the industry standard source of detailed regional economic data developed by the Minnesota IMPLAN Group, Inc.; see www.implan.com).

Given the estimates in changes of unit costs, the potential impacts of dreissenid mussels on the Columbia River basin was estimated to result in an aggregate mean annual welfare loss of roughly $3.31 million, based the CGE model (Warziniack et al. 2008). The standard deviations range from $0.5 million to $1.5 million for that estimate, depending on the precision of the unit-cost impact estimates, where the method of Harrison and Vinod (1992) was employed to account for uncertainty in the unit-cost impact estimates. The method generates unbiased and asymptotically consistent estimators of the welfare change in terms of household equivalent variations, where the only source of uncertainty is the cost impact to industries.

Within the predicted distributions of welfare losses, there are significant differential impacts across households. Households with incomes of $50,000–75,000 bear the largest proportion of mean welfare losses, while those households with incomes of $10,000–15,000 have the smallest mean welfare change (in aggregate). On a per-household basis, welfare impacts range from a low of $1.18 per household (for households with incomes of $25,000–35,000) to a high of $5.22 per household (for households with incomes >$150,000). These results (note that only the mean estimates are reported here) demonstrate a low proportional expected severity of impact from dreissenid mussel invasion of the Columbia River basin and a wide variation in impacts across households. Of course, while a $3 million annual welfare loss may seem substantial to some and not to others, whether it induces a policy response depends not only on what the policy itself might cost and how effective it might be (as noted in chapter 9), but also on how these preferences influence responses of policy makers over time and with differing perceptions of risk.

RISK PERCEPTION AND HUMAN RESPONSES

While the above ecological and economic analyses provide insight into the management of dreissenid mussel invasions, they do not provide specific prescriptions for policy makers. Optimal prescriptions can, however, be provided by merging the population ecology and the potential economic impacts of dreissenid mussels with the economic theory of endogenous risk. Endogenous risk captures the risk-benefit trade-offs created by jointly determined ecosystem conditions, species characteristics, and economic circumstances (as noted in chapter 8 and in Crocker and Tschirhart 1992; Settle et al. 2002). The theory of endogenous risk assumes that people and firms invest scarce resources to change risk. People mitigate risk through prevention (self-protection) efforts to reduce the likelihood of an invasion, and they adapt to risk through control (self-insurance) efforts to reduce the severity of an invasion if it occurs. This framework has been used to examine the risks of biodiversity loss, environmental, and economic damages that dreissenid mussels pose to society using stochastic dynamic programming (SDP; Shogren 2000; Leung et al. 2002).

The optimal decision making of a policy maker (i.e., a resource manager) and (in turn) a private firm (i.e., a power plant) to a potential dreissenid mussel invasion in a Midwest lake was analyzed with an SDP model (Finnoff et al. 2005, 2006, 2007; summarized in part in chapter 9). Optimal decisions were characterized by those of a policy maker that maximized social welfare given the privately optimal choices of the firm (choices that maximize private profit). The manager is the primary decision maker in balancing the risks and benefits of prevention and control, while realizing that private individuals may also make investments to reduce risk. The ecological component of the model incorporates the introduction, establishment, and growth of dreissenid mussels. The SDP model demonstrates (at least) two critical issues relating to human behavior that must be considered in developing integrated optimization models for policy analysis. The first critical issue is to what degree complexity is important for sound policy analysis (as demonstrated in chapter 2). The second critical issue is the attitudes of human decision makers toward time and overt risk, which can have as much influence on optimal decision making as the consequences of the risk.

Variations in a manager's preferences concerning time and risk will influence the choices on the mix of prevention and control. This is a pertinent issue because although scientists have argued that invasive species can be managed most cost-effectively with greater investments in prevention (Leung et al. 2002; Simberloff 2003), investments in prevention are not typically done (Bossenbroek et al. 2005). In many cases, private and public resources are invested primarily to control existing invaders rather than to prevent new invasions. Managers frequently wait until after invaders have arrived and then scramble to limit the damages. For example, farmers often limit investment in prevention efforts because they perceive the introduction of weeds to be outside of their control, while they are comfortable with control methods such as herbicides (Wilson et al. 2008), even though these decisions may not be economically efficient. These paradoxical decisions can be understood by recognizing the link between typical human preferences over time and for risk bearing with the technology of risk reduction (for complete details, see Finnoff et al. 2006, 2007).

The SDP model applied to the aforementioned zebra mussel invasion of a Midwest lake was used to assess a resource manager's preferences over time. Variations in preferences across managers were assessed by varying the discount rate for several types of resource managers differentiated by their level of risk aversion, from risk neutral to highly risk averse. Risk preferences are represented in the SDP model by the curvature of the von Neumann-Morgenstern utility function (Holt and Laury 2002), and risk attitudes were varied from risk neutral to highly risk averse.

The key results of the SDP analysis are somewhat counterintuitive: an increase in the discount rate (i.e., shorter time horizon) causes prevention to fall and control to increase (figure 12.6). Managers with greater preferences toward the present decrease their investment in prevention and increase their investment in control of invasive

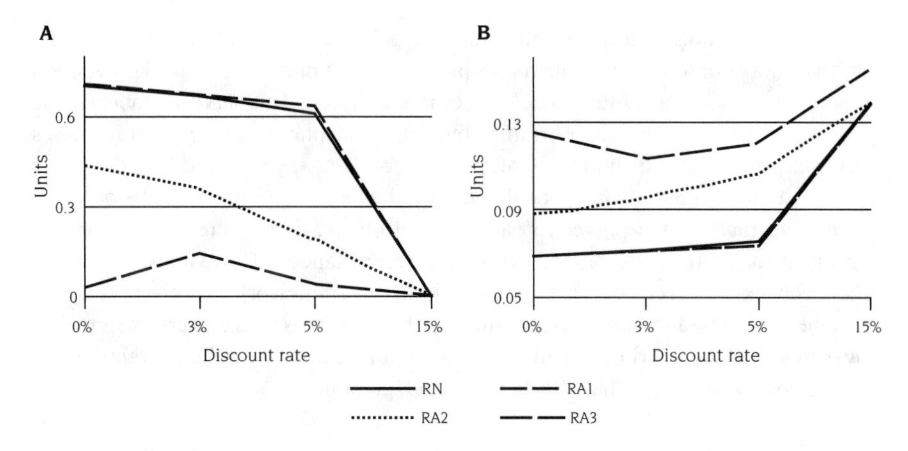

FIGURE 12.6.
An example of the impacts of the discount rate and levels of risk aversion in the endogenous risk framework in mean annual collective prevention (A) and mean annual collective control (B). Units are the average number of prevention (A) or control (B) events that take place on an annual basis. In this example, discount rates were examined at levels of 0%, 3%, 5%, and 15%, with four levels of risk perception, from risk neutral (RN) to very risk averse (RA3).

species. This choice of responses to zebra mussels with less prevention and more control causes the probability of invasion to increase, which is followed by larger populations. The resulting damages require increased levels of adaptation by the power plant, which ultimately lowers overall welfare.

The results for risk attitudes mirror those for the discount rate, where a more risk-averse manager will choose a less risky *alternative* for managing invasive species (Finnoff et al. 2007) and will tend to invest more in control than in prevention (figure 12.6). In theory, greater risk aversion has two effects on the portfolio choice of prevention and control. First, if one is more risk averse, holding onto a dollar is more attractive (i.e., a sure bet) than spending it on either prevention or control, which are both affected by random events, that is, the probability of invasion and stochastic population growth. Second, attenuating indirect effects reflect the technical relationship between the two strategies, but it is ambiguous whether the indirect effect works with or against the direct effect.

In our analysis, risk-averse managers selected *their* less risky combination of strategies, a portfolio with less prevention and more control. Again, this finding seems counterintuitive. But to a more risk-averse manager, a dollar spent on control is worth more than a dollar spent on prevention because control is a sure bet while prevention is risky. The intuition is that control is relatively more attractive because its expected marginal effectiveness exceeds the expected marginal effectiveness of prevention. There is less uncertainty in the application of control—it removes existing invaders from the system; there is more uncertainty in prevention because it only reduces the chance of invasion and does *not* eliminate it. For this reason, the direct effect on prevention dominates the indirect effect; more risk-averse managers use less prevention. Since prevention and control act as substitutes, less prevention implies more control. As with increases in the discount rate, mean annual prevention employment is lower for higher degrees of risk aversion. This serves to increase the probability of invasion, with resulting abundance, adaptation, and control increases that ultimately lower overall welfare.

In summary, the theory of endogenous risk can be used to frame the question of how to manage the prevention and control of invasive species. The approach accounts for both biological and economic circumstances of invasions, as well as the feedbacks between the two systems. Within this framework, one can investigate whether the integration is worth the effort and how changes in managers' preferences over time and for bearing risk influence the optimal mix of public prevention and public control, and how that affects private adaptation.

FUTURE RESEARCH

By integrating the economic and ecological models for dreissenid mussels, reviewed in this chapter, there will be many opportunities to provide recommendations for managers and industries likely to be affected by a dreissenid mussel invasion. In the context of an invasion, many managers and policy makers are often interested in what will be the "cost" of an invasion into a particular location; that is, they want a dollar figure. For the most part, when ecologists have attempted to provide this figure, the exercise has been one of summing up potential (or actual) direct impacts (e.g., Phillips et al. 2005; Pimentel et al. 2005). Unfortunately, as we have shown in this chapter, understanding the bioeconomics of an invasion is much more complex. Defining a dollar figure requires not only ecological predictions of establishment, likely abundance, suitable habitat, and dispersal, but also economic predictions of direct and indirect impacts to the larger economy, the behavior of decision makers, and the incorporation of policy time horizons. Thus, the answer to our first question—what it is worth to keep dreissenid mussels from becoming established in the western states and provinces of North America—is "it depends." As unsatisfactory as this answer may be, we still believe that our bioeconomic assessment of the western spread of dreissenid mussels does provide policy and management prescriptions.

Currently, our recommendations for the Columbia River basin are based on two separate analyses. First, the CGE model suggests that a full-blown invasion of the Columbia River basin would reduce welfare of the regional economy by approximately \$3.3 million/year. Second, gravity model predictions suggest that the probability of dreissenid mussels being introduced into Lake Mead was four times more likely than an introduction into Roosevelt Lake (the largest reservoir on the Columbia River; see chapter 7). Considering that Lake Mead was invaded after dreissenid mussels had been in North America for almost 20 years, the probability of a dreissenid invasion into Roosevelt Lake in any given year is low. The challenge that still remains is to combine these predictions for the Columbia River within an integrated optimization model that also includes an assessment of uncertainty, the discount rate that policy makers in this region work from, and the influence of risk perception.

As with our first question, the answer to our second question—how much information and/or modeling is needed to make a policy recommendation given all the uncertainties in the invasion process—may seem unsatisfactory: at the current state of research into the bioeconomics of invasive species, we do not know exactly how much information or modeling is needed. For most stages of the invasion pathway, only a handful of studies exist on any particular species (even for dreissenid mussels), and even fewer studies exist on their economic impacts, let alone studies that fully integrate economics and ecology in a sophisticated manner. As this chapter shows, those ecological studies that include "economics" do so in an elementary manner, such as simply accounting for direct impacts, and those economic studies that included "ecology" do so in simplistic ways, such as assuming a full invasion or simplifying the movement patterns of human vectors. This chapter, however, does layout the framework and the steps needed to identify satisfactory answers concerning the bioeconomics of invasive species.

The dreissenid invasion of North America has been used time and again as the poster child for aquatic invasive species, as has been demonstrated in this chapter and book. Research on dreissenid mussels has been extensive and has covered many disciplines beyond what has been covered here, including population genetics, bioenergetics, species interactions including competition and facilitation, nutrient dynamics, and so forth. Dreissenid mussels have also been the model organism for efforts to develop or adapt new techniques and models in the field of ecology and economics, including the use of gravity models, niche modeling, and assessments of risk aversion by managers. The integration of invasive species biology and economics has also taken advantage of this base of knowledge and led to understanding the value of spending on prevention and control of invasive species, and dreissenid mussels in particular (Leung et al. 2002; Finnoff et al. 2006, 2007). Our continued goal is to integrate these disciplines to provide advice and quantifiable results to agencies, managers, and other research scientists to efficiently and effectively manage and predict the continuing spread of dreissenid mussels and other invasive species in North America.

Acknowledgments The chapter was substantially improved by the editors of this book and reviews by A. Bobeldyk, C. Jerde, and three anonymous reviewers. This material is based on work supported by the Integrated Systems for Invasive Species project (D. M. Lodge, principal investigator) funded by the National Science Foundation (DEB 02-13698) and by the National Sea Grant. J.M.B. was supported in part by an award from the National Sea Grant and the U.S. Fish and Wildlife Service (awarded to D. M. Lodge). This is publication 2009-004 from the University of Toledo Lake Erie Center.

References

Allen, Y. C., and C. W. Ramcharan. 2001. Dreissena distribution in commercial waterways of the US: using failed invasions to identify limiting factors. Canadian Journal of Fisheries and Aquatic Sciences 58:898–907.

Bobeldyk, A. M., J. M. Bossenbroek, M. A. Evans-White, D. M. Lodge, and G. A. Lamberti. 2005. Secondary spread of zebra mussels (*Dreissena polymorpha*) in coupled lake-stream systems. Ecoscience 12:339–346.

Bodamer, B., and J. M. Bossenbroek. 2008. Wetlands as barriers: effects of vegetated waterways on downstream dispersal of zebra mussels (*Dreissena polymorpha*). Freshwater Biology 53: 2051–2060.

Bossenbroek, J. M., L. E. Johnson, B. Peters, and D. M. Lodge. 2007. Forecasting the expansion of zebra mussels in the United States. Conservation Biology 21:800–810.

Bossenbroek, J. M., C. E. Kraft, and J. C. Nekola. 2001. Prediction of long-distance dispersal using gravity models: zebra mussel invasion of inland lakes. Ecological Applications 11: 1778–1788.

Bossenbroek, J. M., J. McNulty, and R. P. Keller. 2005. Can ecologists heat up the discussion on invasive species risk? Risk Analysis 25:1595–1597.

Carlton, J. T. 1991. Predictions of the arrival of the zebra mussel in North America. Dreissena polymorpha Information Review 2(2):1.

Courchamp, F., T. Clutton-Brock, and B. Grenfell. 1999. Inverse density dependence and the Allee effect. Trends in Ecology and Evolution 14:405–410.

Crocker, T. D., and J. Tschirhart. 1992. Ecosystems, externalities, and economies. Environmental Resource Economics 2:551–567.

Deng, Y. 1996. Present and expected economic costs of zebra mussel damages to water users with Great Lakes water intakes. Ph.D. thesis, Ohio State University.

Dermott, R. 2001. Sudden disappearance of the amphipod *Diporeia* from eastern Lake Ontario, 1993–1995. Journal of Great Lakes Research 27:423–433.

Dobson, A. P., J. P. Rodriguez, W. M. Roberts, and D. S. Wilcove. 1997. Geographic distribution of endangered species in the United States. Science 275:550–553.

Drake, J. M., and J. M. Bossenbroek. 2004. The potential distribution of zebra mussels (*Dreissena polymorpha*) in the U.S.A. BioScience 54:931–941.

Fahnenstiel, G. L., G. A. Lang, T. F. Nalepa, and T. H. Johengen. 1995. Effects of zebra mussel (*Dreissena polymorpha*) colonization on water quality parameters in Saginaw Bay, Lake Huron. Journal of Great Lakes Research 21:435–448.

Finnoff, D., J. F. Shogren, B. Leung, and D. Lodge. 2005. The importance of bioeconomic feedback in invasive species management. Ecological Economics 52:367–381.

Finnoff, D, J. F. Shogren, B. Leung, and D. M. Lodge. 2006. Prevention versus control in invasive species management. Pages 166–202 *in* A. Kontonlean, U. Pascual, and T. Swanson, editors. Biodiversity economics. Cambridge University Press, Cambridge, UK.

Finnoff, D., J. F. Shogren, B. Leung, and D. Lodge. 2007. Take a risk: preferring prevention over control of biological invaders. Ecological Economics 62:216–222.

Harrison, G. W., and H. D. Vinod. 1992. The sensitivity analysis of applied general equilibrium models: completely randomized factorial sampling designs. Review of Economics and Statistics 74:357–362.

Herbert, P. D. N., B. W. Muncaster, and G. L. Mackie. 1989. Ecological and genetic-studies on *Dreissena polymorpha* (Pallas)—a new mollusk in the Great-Lakes. Canadian Journal of Fisheries and Aquatic Sciences 46:1587–1591.

Holt, C. A., and Laury, S. K., 2002. Risk aversion and incentive effects. American Economic Review 92:1644–1655.

Horvath, T. G., G. A. Lamberti, D. M. Lodge, and W. L. Perry. 1996. Zebra mussel dispersal in lake-stream systems: source-sink dynamics? Journal of the North American Benthological Society 15:564–575.

Johnson, L. E., J. M. Bossenbroek, and C. E. Kraft. 2006. Patterns and pathways in the post-establishment spread of non-indigenous aquatic species: the slowing invasion of North American inland lakes by the zebra mussel. Biological Invasions 8:475–489.

Johnson, L. E., and J. T. Carlton. 1996. Post-establishment spread in large-scale invasions: dispersal mechanisms of the zebra mussel *Dreissena polymorpha*. Ecology 77:1686–1690.

Johnson, L. E., and D. K. Padilla. 1996. Geographic spread of exotic species: ecological lessons and opportunities from the invasion of the zebra mussel *Dreissena polymorpha*. Biological Conservation 78:23–33.

Keller, R. P., J. M. Drake, and D. M. Lodge. 2007. Fecundity as a basis for risk assessment of nonindigenous freshwater molluscs. Conservation Biology 21:191–200.

Kolar, C. S., and D. M. Lodge. 2001. Progress in invasion biology: predicting invaders. Trends in Ecology and Evolution 16:199–204.

Koutnik, M. A., and D. K. Padilla. 1994. Predicting the spatial distribution of *Dreissena polymorpha* (zebra mussel) among inland lakes of Wisconsin—modeling with a GIS. Canadian Journal of Fisheries and Aquatic Sciences 51:1189–1196.

Kraft, C. E., and L. E. Johnson. 2000. Regional differences in rates and patterns of North American inland lake invasions by zebra mussels (*Dreissena polymorpha*). Canadian Journal of Fisheries and Aquatic Sciences 57:993–1001.

Leung, B., J. M. Bossenbroek, and D. M. Lodge. 2006. Boats, pathways, and aquatic biological invasions: estimating dispersal potential with gravity models. Biological Invasions 8:241–254.

Leung, B., J. M. Drake, and D. M. Lodge. 2004. Predicting invasions: propagule pressure and the gravity of Allee effects. Ecology 85:1651–1660.

Leung, B., D. M. Lodge, D. Finnoff, J. F. Shogren, M. A. Lewis, and G. Lamberti. 2002. An ounce of prevention or a pound of cure: bioeconomic risk analysis of invasive species. Proceedings of the Royal Society of London Series B Biological Sciences 269:2407–2413.

Ludyanskiy, M. L., D. McDonald, and D. Macneill. 1993. Impact of the zebra mussel, a bivalve invader. BioScience 43:533–544.

Mackie, G. L. 1991. Biology of the exotic zebra mussel, *Dreissena polymorpha*, in relation to native bivalves and its potential impact in Lake St. Clair. Hydrobiologia 219:251–268.

May, B., and J. E. Marsden. 1992. Genetic identification and implications of another invasive species of dreissenid mussel in the Great-Lakes. Canadian Journal of Fisheries and Aquatic Sciences 49:1501–1506.

Mellina, E., and J. B. Rasmussen. 1994. Patterns in the distribution and abundance of zebra mussel (*Dreissena polymorpha*) in rivers and lakes in relation to substrate and other physicochemical factors. Canadian Journal of Fisheries and Aquatic Sciences 51:1024–1036.

Mellina, E., J. B. Rasmussen, and E. L. Mills. 1995. Impact of zebra mussel (*Dreissena polymorpha*) on phosphorus cycling and chlorophyll in lakes. Canadian Journal of Fisheries and Aquatic Sciences 52:2553–2573.

Mills, E. L., R. M. Dermott, E. F. Roseman, D. Dustin, E. Mellina, D. B. Conn, and A. P. Spidle. 1993. Colonization, ecology, and population-structure of the quagga mussel (Bivalvia, Dreissenidae) in the lower Great-Lakes. Canadian Journal of Fisheries and Aquatic Sciences 50:2305–2314.

National Park Service. 2007. Live zebra mussels found at Lake Mead; Resource agencies initiate program to assess extent and prevent spread. Press release. Available at http://www.100thmeridian.org/mead.asp.

NatureServe. 2006. NatureServe Explorer: an online encyclopedia of life [web application]. Version 4.7. NatureServe, Arlington, VA. Available at www.natureserve.org/explorer (accessed June 13, 2006).

Neary, B. P., and J. H. Leach. 1992. Mapping the potential spread of the zebra mussel (*Dreissena-polymorpha*) in Ontario. Canadian Journal of Fisheries and Aquatic Sciences 49:406–415.

Northwest Natural Resource Group. 2003. Preparing to meet the challenge: an assessment of invasive species management in Idaho. Prepared for the Idaho Invasive Species Council. Northwest Natural Resource Group, Boise, Idaho.

100th Meridian Initiative. 2007. News and Announcements. Available at http://100thmeridian.org/news.asp.

O'Neill, C. R. 1997. Economic impact of zebra mussels—results of the 1995 National Zebra Mussel Information Clearinghouse study. Great Lake Research Review 3:35–44.

Parker, I. M., D. Simberloff, W. M. Lonsdale, K. Goodell, M. Wonham, P. M. Kareiva, M. H. Williamson, B. Von Holle, P. B. Moyle, and J. E. Byers. 1999. Impact: toward a framework for understanding the ecological effects of invaders. Biological Invasions 1:3–19.

Parmalee, P. W., and A. E. Bogan. 1998. The freshwater mussels of Tennessee. University of Tennessee Press, Knoxville.

Phillips, S., T. Darland, and M. Sytsma. 2005. Potential economic impacts of zebra mussels on the hydropower facilities in the Columbia River basin. Prepared for the Bonneville Power Administration. Pacific States Marine Fisheries Commission, Portland, OR.

Pimentel, D., R. Zuniga, and D. Monison. 2005. Update on the environmental and economic costs associated with alien-invasive species in the United States. Ecological Economics 52: 273–288.

Pollux, B., D. Minchin, G. Van Der Velde, T. Van Alen, S. Y. Moon-Van der Staay, and J. Hackstein. 2003. Zebra mussels (*Dreissena polymorpha*) in Ireland, AFLP-fingerprinting and boat traffic both indicate an origin from Britain. Freshwater Biology 48:1127–1139.

Ramcharan, C. W., D. K. Padilla, and S. I. Dodson. 1992. Models to predict potential occurrence and density of the zebra mussel, *Dreissena-polymorpha*. Canadian Journal of Fisheries and Aquatic Sciences 49:2611–2620.

Ricciardi, A., R. J. Neves, and J. B. Rasmussen. 1998. Impending extinctions of North American freshwater mussels (Unionoida) following the zebra mussel (*Dreissena polymorpha*) invasion. Journal of Animal Ecology 67:613–619.

Ricciardi, A., R. Serrouya, and F. G. Whoriskey. 1995. Aerial exposure tolerance of zebra and quagga mussels (Bivalvia, Dreissenidae)—implications for overland dispersal. Canadian Journal of Fisheries and Aquatic Sciences 52:470–477.

Schloesser, D. W., T. F. Nalepa, and G. L. Mackie. 1996. Zebra mussel infestation of unionid bivalves (Unionidae) in North America. American Zoologist 36:300–310.

Schneider, D. W., C. D. Ellis, and K. S. Cummings. 1998. A transportation model assessment of the risk to native mussel communities from zebra mussel spread. Conservation Biology 12:788–800.

Settle, C., T. D. Crocker, and J. F. Shogren. 2002. On the joint determination of biological and economic systems. Ecological Economics 42:301–311.

Shogren, J. F. 2000. Risk deductions strategies against the "explosive invader." *In* C. Perrings, M. Williamson, and S. Dalmazzone, editors. The economics of biological invasions. Edward Elgar, Northhampton, MA.

Simberloff, D. 2003. How much information on population biology is needed to manage introduced species? Conservation Biology 17:83–92.

Sprung, M. 1993. The other life: an account of present knowledge of the larval phase of *Dreissena polymorpha*. Pages 39–53 *in* T. F. Nalepa and D. Schloesser, editors. Zebra mussels: biology, impacts, and control. CRC Press, Boca Raton, FL.

Stoeckel, J. A., D. W. Schneider, L. A. Soeken, K. D. Blodgett, and R. E. Sparks. 1997. Larval dynamics of a riverine metapopulation: implications for zebra mussel recruitment, dispersal, and control in a large-river system. Journal of the North American Benthological Society 16: 586–601.

Stohlgren, T. J., D. Barnett, C. Flather, P. Fuller, B. Peterjohn, J. Kartesz, and L. L. Master. 2006. Species richness and patterns of invasion in plants, birds, and fishes in the United States. Biological Invasions 8:427–447.

Strayer, D. L. 1991. Projected distribution of the zebra mussel, *Dreissena polymorpha*, in North America. Canadian Journal of Fisheries and Aquatic Sciences 48:1389–1395.

Strayer, D. L., K. A. Hattala, and A. W. Kahnle. 2004. Effects of an invasive bivalve (*Dreissena polymorpha*) on fish in the Hudson River estuary. Canadian Journal of Fisheries and Aquatic Sciences 61:924–941.

Strayer, D. L., and L. C. Smith. 1996. Relationships between zebra mussels (*Dreissena polymorpha*) and unionid clams during the early stages of the zebra mussel invasion of the Hudson River. Freshwater Biology 36:771–779.

Thorp, J. H., J. E. Alexander, and G. A. Cobbs. 2002. Coping with warmer, large rivers: a field experiment on potential range expansion of northern quagga mussels (*Dreissena bugensis*). Freshwater Biology 47:1779–1790.

U.S. Department of Agriculture. 2004. 2002 Census of Agriculture. National Agricultural Statistics Service. Washington D.C. Available at http://www.agcensus.usda.gov/Publications/2002/index.asp.

Warziniack, T. W., D. Finnoff, and J. F. Shogren. 2008. Evaluating the 100th Meridian Initiative: assessing the impacts of zebra mussel invasion on the Columbia River basin. Working paper, Department of Economics and Finance, University of Wyoming, Laramie.

Whittier, T. R., P. L. Ringold, A. T. Herlihy, and S. M. Pierson. 2008. A calcium-based invasion risk assessment for zebra and quagga mussels (*Dreissena* spp). Frontiers in Ecology and the Environment 6:180–184.

Wilson, R. S., M. Tucker, N. Hooker, J. LeJeune, and D. J. Doohan. 2008. Perceptions and beliefs about weed management: perspectives of Ohio grain and produce farmers. Weed Technology 22:339–350.

13

Putting Bioeconomic Research into Practice

Reuben P. *Keller, Mark* A. *Lewis, David* M. *Lodge, Jason* F. *Shogren, and Martin Krkošek*

In a Clamshell

In this chapter, we step back from the technical issues of bioeconomic modeling to focus on the interactions among science, scientists, policy makers, and managers. The ISIS team has worked throughout the project to advance methods so that we as scientists can better understand bioeconomic systems and provide more useful input to the political decision-making process. Conducting rigorous and cutting-edge work that meets these goals is not always easy within academic settings, but a few relatively simple things can be done when designing, implementing, and communicating research that will improve the chances of that research being appreciated beyond the academic community. We have each conducted projects that we believe have results relevant to policy and management, and we have each met with variable levels of success for different projects. Some of these efforts are described here and range from work conducted by graduate students to senior faculty, and from the local to national scale. The chapter ends with a reexamination of the rationale for bioeconomic approaches in the study of invasive species and other environmental issues.

The Integrated Systems for Invasive Species (ISIS) research team came together nearly a decade ago to explore the contributions that a multidisciplinary bioeconomic approach could make to invasive species science. We are not the first to do this; other researchers cut the path and advanced the field so that our work has been possible (e.g., Gordon 1954; Schaefer 1957; Daly 1968; Hammack and Brown 1974; Clark 1976; Crocker and Tschirhart 1992; Perrings et al. 2000). Each chapter in this book represents the efforts of the ISIS team to add to the foundation provided by previous bioeconomic research.

The field of bioeconomics seeks to explain how economic and biological/ecological systems interact. As discussed throughout this book, the interactions between ecological and economic systems are rarely, if ever, linear or unilateral.

Rather, feedback loops are the norm. The challenge is to create models that capture the essential interactions and that incorporate these feedback loops. To date, feedback loop models have been applied to invasive species when the ecological and economic systems are well known and when abundant data exist to parameterize models. But the potential for these models is enormous. The cost of gathering data and creating models will be dwarfed by the benefits from a better understanding of invasions and how their impacts can be reduced or avoided. As methods improve and more data become available, the development of integrated models is likely to become more common. The demand for integrated models will drive the supply of better data and tools; better tools and data will spur more integration of disciplines.

Our objective in the ISIS project has been to produce methods and results to help decision makers as they confront the issues of invasive species. In this concluding chapter, we discuss our own experiences in integration and in trying to communicate our research to stakeholders, managers, and policy makers. We hope our experience is useful to other scientists who have similar aspirations to provide practical guidance. The remainder of the chapter is divided into four sections. The first three examine whether and how our research has been used by managers and policy makers, and what our roles have been in contributing to the application of the research. In the jargon currently fashionable among scientific administrators, our goal is to assess how "translational" our research has been. First, David Lodge describes his experiences in the policy arena that prompted him to seek collaborators capable of incorporating social and economic data and methods into his work as an ecologist. These collaborative efforts were the genesis of the ISIS project. Lodge also describes more recent work in the U.S. federal policy arena that has been informed by ISIS research. Next, Mark Lewis and Martin Krkošek describe their work on sea lice and wild salmon populations in British Columbia. This research program has contributed to an ongoing debate about appropriate management practices in the salmon farming industry. Reuben Keller describes his dissertation work, how this has led to policy interactions, and the differences between those projects that have so far had a direct influence on policy and those that have not. In the final section, Jason Shogren revisits the rationale for combining ecology and economics to address applied environmental issues. Based on our combined experience, we have put together a short list of actions that have helped us to communicate our science results to policy makers and managers (box 13.1).

ISIS, ECOLOGY, AND U.S. FEDERAL POLICY—DAVID LODGE

Presumably all ecologists were motivated to become scientists because they found the question "how does nature work?" intellectually interesting and challenging. Many ecologists, including me, were also motivated by at least a vague concern that humans were causing changes to the abiotic and biotic environment that negatively affected other species and were possibly counterproductive to long-term human well-being. I therefore believed that the environment should be given more systematic consideration in individual, industry, and policy choices. If such concerns were

BOX 13.1. Communicating Research Results for Better Resource Management and Policy

For researchers who intend to improve resource management or policy, we recommend the following:

- Recognize that science alone does not establish management or policy goals but rather contributes to understanding that informs policy evaluation and change
- Focus on questions, and geographic and temporal scales, that are immediately relevant to timely management or policy issues
- Translate research into short messages in lay language and communicate them effectively and consistently to the media, public, managers, and policy makers
- Repeat a straightforward message of the implications of research results at multiple meetings and workshops that managers and/or policy makers attend
- Create and work with personal networks and face-to-face interactions with stakeholders and agency personnel, which are essential to building trust, opening communication pathways, and delivering a scientific message
- Recognize that changes to management and policy are rarely made quickly but rather occur over years or decades as research results accumulate, withstand scrutiny, and reinforce a direction for policy change

valid—and 30 years later it is now clear they were—knowing how nature worked would certainly be an essential ingredient to improve natural resource management and policy. However, only one of these motivations—the desire to understand the patterns and processes of nature—was honored in the standard scientific training, research funding, and career advancement of 30 years ago. The second motivation—informing wiser individual and social choices—was not, and perhaps still is not given its appropriate due within academia.

When I wrote my first few proposals to the U.S. National Science Foundation (NSF) in the early-mid 1980s, my collaborators and I proposed to use rusty crayfish as a focal species to elucidate how multiple species and trophic levels interacted to create the patterns of ecological communities that we see in lakes. In the northern Wisconsin-Michigan region of our proposed work, rusty crayfish was an "invasive species," but that term was not yet in widespread use. Even if it had been, we would have avoided using it in NSF proposals, because solving problems of societal importance was not central to NSF's mandate and did not impress the academic reviewers of NSF proposals. By couching our research exclusively in terms of "basic" ecological research, we garnered financial support to continue our work. Fortunately, times have changed.

Two decades later, when we wrote the NSF proposal that funded much of the research reviewed in this book, the title of the proposal highlighted a combination of basic and applied themes: "Ecological Forecasting and Risk Analysis of Nonindigenous Species: Strategic Optimization using a Bioeconomic Approach." The "Broader Impacts" section of the proposal—a section required only in recent years—emphasized outreach to the public via collaboration with the John G. Shedd Aquarium in Chicago and workshops with natural resource managers and policy makers in the Great Lakes region. Nevertheless, our collective core purpose and expertise remain scientific research; we are not resource managers or policy makers. In essence, then, this chapter takes stock of whether and how our collective research has been used by managers or policy makers, and what our roles have been in contributing to the application of the research.

The combination of achieving tenure at the University of Notre Dame (in 1992) and a changing attitude at NSF (and other funding agencies) provided my collaborators and me increased freedom to conduct research that both was scientifically innovative *and* provided information or tools that could be usefully employed to better manage invasive species. In hindsight, it was about 10 years before my students, collaborators, and I had honed this collective vision sufficiently to be really useful. And the vision prompted me to take three steps outside of the usual comfort zone of an academic ecologist.

First, I sought explicit training in the appropriate relationships among scientific research, management, and policy. For me, the Aldo Leopold Leadership Program (www.leopoldleadership.org/content/) was an essential source of training, information, and inspiration. Second, I sought opportunities to serve in roles at the nexus of science and policy. Serving as the first chair of the U.S. Invasive Species Advisory Committee opened my eyes to the many perspectives and interests that impinged on the issue of invasive species, and especially to the typically weak influence of science in such discussions and in shaping policy.

Third, the first two experiences prompted me to seek collaborators in other disciplines to better account for multiple influences on decisions about invasive species management. In particular, I wanted to contribute to better answering the set of related questions that I was asked repeatedly at every management or policy-oriented forum: "How much do damages from invasive species cost our city (or state or nation)?" "Given that my agency has a limited budget, should we spend it on control of an existing invader or on preventing new ones from arriving?" As an ecologist, I wasn't trained to answer these questions, although other ecologists and I had some important pieces of the answers. Economists also could not answer those questions alone, at least not without working with ecologists and mathematical biologists—hence, the collaborations reflected in this book.

In my first article aimed directly at policy makers, my coauthors and I (Lodge et al. 2000) made two major recommendations to achieve the then new U.S. federal policy goal (Executive Order 13112 of 1999, which expanded on the Nonindigenous Aquatic Nuisance Prevention and Control Act of 1990 [revised as the National Invasive Species Act in 1996]) of reducing the damage done by invasive species and preventing additional harmful species from entering the country. We focused on crayfishes because the issues were well documented, and the issues generalized widely from crayfishes to many other aquatic and terrestrial taxa. Our two recommendations, paraphrased, were as follows: (1) species proposed for introduction into the nation or a state should be assessed first for risk of harm, and allowed for sale only if the risk is acceptably low; and (2) methods and legal authority or eradication must be developed and employed quickly on newly discovered populations of harmful species. After 8 years and many publications on the same or closely related topics, modest policy change—especially at the local and state level—has occurred. More systematic change at the federal level is poised to happen for the first recommendation.

At the city level, Chicago's list of proscribed species (described by Keller below) was informed to a large extent by three of our collective publications on risk assessment for fishes (Kolar and Lodge 2002), mollusks (Keller et al. 2007a), and other species (Keller and Lodge 2007). The link between our research and Chicago policy is very direct. Working with the horticulture industry and state agencies, we are providing similar direct guidance to the State of Indiana on emerging policy on risk assessment on live plants in commerce (also described by Keller below).

At the federal level, our research publications—contributed to by many other scientists, as well—have motivated congressional interest in seeking policy improvements to reduce the importation of species that have a high probability of damaging human health or U.S. ecosystems and the services they provide society. In congressional hearings before the House Subcommittee on Fisheries, Wildlife and Oceans (resourcescommittee.house.gov), I have relied heavily on our analysis of the failure of the Lacey Act's injurious wildlife provision to prevent the importation of harmful species (Fowler et al. 2007). To replace the current failed policy, I have described risk assessment of species as a potentially important component of better policies; to do so, I have relied heavily on an Ecological Society of America policy paper on invasive species (Lodge et al. 2006), which represents a broad scientific consensus. The conclusion from Keller et al. (2007b) that risk assessment for intentionally imported species brings net economic benefit to society, in addition to environmental benefits, has also been very influential with policy makers. Without these and many other pieces of research by many authors (e.g., National Research Council 2002), arguments to protect the environment would have faltered. Federal legislation to institute risk assessment for imported animals, influenced heavily by this research and our congressional testimony, has been introduced into the House of Representatives, where it will likely gain increasing consideration in the 111th Congress beginning in 2009. Clearly, the federal policy landscape—if not yet legislation passed by Congress—has been heavily influenced by the collective research of the ecological community, including the research from this interdisciplinary project.

On the second of our earlier recommendations—to provide legal authority and resources for eradication of newly discovered species—little impact or policy progress is detectable, especially for aquatic ecosystems. Legal authority remains problematic (Lodge et al. 2006), and little funding is available for the development of more selective toxins or research and development of other control approaches, or even funding for surveillance that would be necessary to discover a species in time to implement a successful eradication effort. This is true not only for crayfishes but also for fishes and most other aquatic invasive species, with the possible exception of plants, which benefit from toxins developed for use on terrestrial plants that are pests in agriculture.

To the extent that our early (Lodge et al. 2000) or more recent (Lodge et al. 2006) policy recommendations have been or will be responsible for policy changes, it is because (1) our recommendations were based on a long history of research published in technical journals by many excellent scientists; (2) our publications for policy audiences were short (Lodge et al. 2000; Fowler et al. 2007) or organized around a small number of recommendations in nontechnical language (Lodge et al. 2006); (3) our technical publications were in journals targeting state and federal natural resource managers and policy makers in addition to academic researchers (*Fisheries*, *Ecological Applications*, and *Frontiers in Ecology and Environment*); (4) our additional publications targeted the public and policy makers more directly (e.g., an op-ed in the *New York Times*); (5) our efforts were not limited to publications but included many meetings, conversations, and presentations in nontechnical language in management and policy settings; and (6) our commitments included extensive cooperation with journalists and filmmakers, which enhanced the coverage of the management and policy implications of our research, including stories in *Science*, *New Scientist*, *Environmental Science and Technology*, *National Geographic*, and the *New York Times* and a short documentary film by the American Museum of Natural History. Our research was conveyed directly to policy makers via our publications and presentations, and indirectly to policy makers via public feedback shaped by media coverage of our research.

SALMON FARMING AND SEA LICE—MARTIN KRKOŠEK AND MARK LEWIS

Although the focus of this book is on invasive species that escape their native range and spread into new environments, another type of "invasive" species is one introduced through agriculture. Such species can greatly affect ecosystem function, even if they do not escape from the agricultural environment to become invasive in the classical sense. For example, the corn and wheat monocultures in the Midwest and prairies replace natural species, dramatically reducing natural biodiversity. Domestic cattle invade areas previously inhabited by deer, bison, and other wild herbivores and introduce new diseases, such as bovine tuberculosis, which then spread to wild hosts in surrounding areas. How should public policy be used to mitigate undesirable impacts of these "invited invaders," particularly in view of their positive economic benefits?

The case study we focus on here is that of our science/policy interactions regarding the aquaculture of an alien species, Atlantic salmon (*Salmo salar*), in the Broughton Archipelago on the West Coast of Canada and its impacts, via disease, on wild salmon stocks. The research was undertaken by two of the coauthors (Krkošek as a Ph.D. student and Lewis as a supervisor), with additional cosupervision by John Volpe and mentorship and collaboration by Alexandra Morton. We do not provide a comprehensive scientific review of the work and subject area (see, e.g., Costello 2006; Harvey 2008). Rather, we give a brief overview and then focus on our experiences with the interactions of science and policy.

Background

Unlike most regions with salmon aquaculture, the West Coast of Canada is home to large and varied populations of wild salmon, including pink (*Oncorhynchus gorbuscha*), chum (*O. keta*), coho (*O. kisutch*), Chinook (*O. tshawytscha*), and sockeye (*O. nerka*). Wild salmon are key to the health of coastal ecosystems. Not only do the fish supply nutrients to marine mammals, bears, and wolves, but also coastal rain forest trees have now been shown to thrive on nutrients left by decomposing salmon after autumn spawning (Reimchen et al. 2003). West Coast First Nations groups have a special relationship with wild salmon, and the salmon are the basis for wilderness tourism operators and fishing lodges.

Salmon farming is a major industry on the West Coast of Canada, with a yearly value of approximately $400 million. Although many farms were originally local enterprises, the majority are now owned by two multinational companies. Regulation of fish farms lies jointly with the Department of Fisheries and Oceans (DFO), whose mandate is to conserve and manage wild fish stocks (as well as to develop aquaculture), and with the provincial Department of Agriculture, which is in charge of site licensing as well as regulation.

Stakeholders in the issue of fish farms and in the preservation of wild salmon include the aquaculture industry, First Nations, environmental organizations, commercial and sport fisheries, and wilderness tourism operators. Conflict regarding the role of fish farms in decline and endangerment of wild fish is intimately tied to differing stakeholder beliefs regarding scientific evidence, and differing values (economic vs. environmental vs. cultural). Historically, stakeholders have had little tolerance for differing perspectives, and this has led to a sometimes strident and acrimonious debate, focused on the fish farms. This is gradually changing, with increased levels of discussion and negotiations among stakeholders.

In our research area, the Broughton Archipelago region of Pacific Canada, salmon farms are found in the nearshore environment, often in long, narrow fjords that are also pathways for wild salmon migrations. Salmon aquaculture typically uses large open-net pens, with each farm about a hectare in size, holding up to one million fish. While some cultured salmon may escape to become feral, the major interaction with wild salmon involves secondary players, fish pathogens that "spill over" and "spill back" between farmed and wild salmon. These pathogens span taxa ranging from viruses to bacteria to crustaceans.

One particular pathogen, the sea louse (*Lepeophtheirus salmonis*), is an ectoparasitic copepod, which is native and common on wild adult salmon, and which can also infect farmed salmon. The very high density of salmon within farms provides conditions for biomagnification (i.e., increase in abundance) of the parasite. Further, because Atlantic salmon are more susceptible to lice than are indigenous Pacific salmon, the potential for biomagnification is increased. Once infected by wild fish, the farmed salmon can "return the favor" by reinfecting the wild salmon. Louse larvae can move through the water and attach to wild salmon in a classic spillover and spillback scenario.

A key issue is the timing of the spillover and spillback of the parasite. In the fall, adult wild salmon return from the open ocean to the nearshore environment, on their way to spawn in rivers and lakes. If these salmon infect the farmed fish, biomagnification occurs during winter among the farmed salmon and the spillback could occur the following spring, when small juvenile salmon migrate seaward, past infected farms. These juvenile salmon are a fraction of their adult size, and this small size is most pronounced in pink and chum salmon. These two species move quickly from hatching to the saltwater environment and are only a few centimeters long, weighing less than a gram, when they first are exposed to the sea lice from fish farms. Such early exposure to lice is rare in unperturbed settings, because young juveniles start their seaward migration when the infected adult population is safely out at sea. In a natural environment, *migratory allopatry* (i.e., migration patterns that keep juveniles and adults apart) separates juvenile and adult cohorts and prevents early pathogen exposure (Krkošek et al. 2007b).

A Story of Science and Policy

In established salmon aquaculture regions, such as Norway and Scotland, sea lice have been associated with fish farms, and there are strict management procedures for their control. Fish farms are relative newcomers to the West Coast of Canada, being first introduced to the Broughton Archipelago region in the 1980s. It was not until 2001 that significant infestations of wild juvenile salmon by sea lice were observed in the Broughton Archipelago. Soon after this, concern about the impact of sea lice on wild salmon led local citizens to become active in the science and policy debate.

One of these local citizens, Alexandra Morton, a biologist who moved to the Broughton Archipelago to study whales, collected data on the sea louse infestations and notified federal authorities. Her efforts and results were disputed and rejected by federal authorities, who claimed that lice are natural and that there was no significant demonstrable impact of salmon farms on wild salmon. Undeterred, Morton continued to collect data, publish studies, and raise public concern on the effect of salmon farms on wild salmon. This work attracted much scientific interest, including our own. Subsequently, much of our research has been in collaboration with Morton, who now directs the Salmon Coast Field Station, a logistical base for our research.

At the start of our research, several key policy-related questions arose: (1) What scientific evidence is there that sea lice spread in significant numbers from farm fish to wild fish? (2) If juvenile salmon are infected with sea lice, how does this affect their survivorship? (3) Do recurrent infestations threaten wild salmon stocks? Can the impact of such infestations be detected in data collected for wild populations?

Early studies showed few sea louse infestations in areas without fish farms relative to those areas with fish farms (Morton et al. 2004). Our field studies measured the level and stage of sea louse infestation on outmigrating juvenile salmon every few kilometers along their 60-km migration route. The analysis of this spatial snapshot shows a clear progression of developmental stages of lice on fish as fish approach a farm and then migrate past the farms. Coupling the large data sets (many tens of thousands of fish sampled) to spatially explicit models allowed us to take the role of "ecological detectives" (Hilborn and Mangel 1997), assessing the level of evidence for competing hypotheses about infection dynamics. Our conclusion was that the infection pattern is consistent with a large point source of infective larvae, located at active farms (Krkošek et al. 2005).

Our further studies showed high mortality of infected fish. Juvenile pink and chum salmon with infections of two adult lice have about a one in three chance of surviving the parasites for a month. Estimates for farm-induced sea louse mortality of outmigrating pink and chum salmon range from 9% to 90%, and commonly exceed 50% (Krkošek et al. 2006a). We also measured the impact of fish farms at the population level, with depressed returns and local extinction risk for salmon in areas where juveniles are exposed to fish farms (Krkošek et al. 2007a). A further, comprehensive meta-analysis shows a 50% reduction in wild salmon survival for many stocks exposed to fish farms around the globe (Ford and Myers 2008).

Although all these studies point toward a serious impact of farm-origin sea lice on wild salmon, the work has not been without critics. We have been involved in a series of critiques and rebuttals of our work by an industry consultant and DFO scientists (Brooks 2005; Brooks and Stucchi 2006; Krkošek et al. 2006b, 2008; Brooks and Jones 2008).[1] Additional critiques have taken the form of Web-based commentaries, posted at salmon farming Web sites and similar sites. We have felt it our responsibility as scientists to respond to some of these (see www.math.ualberta.ca/~mkrkosek/Criticisms&Responses.htm).

Our Policy Interactions

One avenue for informing policy change is increasing public awareness and understanding of science. Our research results have been communicated widely to the public, giving rise to more than 500 news articles in, among others, *Science*, *Nature*, *New York Times*, *Economist*, and *Globe and Mail*, as well as generating several political cartoons. Also, publicity of our work by universities, scientific journals, and marine conservation groups such as the David Suzuki Foundation, SeaWeb, and Lenfest Ocean Program has improved public awareness, motivating some policy initiatives and policy changes. Environmental policy on aquaculture is now a determining electoral issue in some constituencies and features prominently in many government and nongovernment initiatives.

We have engaged in several policy initiatives, often presenting results to groups with widely differing perspectives, some of which were hostile. The British Columbia Pacific Salmon Forum, a provincial initiative to fund and facilitate science that informs policy on aquaculture impacts, met with Krkošek on numerous occasions. The Special Committee on Sustainable Aquaculture of the British Columbia

Legislative Assembly held a day-long debate between disagreeing scientists as part of its mandate to advise policy based on scientific, social, and economic information—on one side of the table were Krkošek, Morton, and colleagues; on the other side were DFO scientists and aquaculture consultants. We have also presented to the National Organic Standards Board of the U.S. Department of Agriculture in a symposium held, in part, to evaluate whether salmon disease issues can be accommodated under organic standards. Additional initiatives, such as the World Wildlife Fund Salmon Aquaculture Dialogue, have brought together stakeholders and scientists to develop guidelines for aquaculture management.

The outcomes of many of the policy initiatives are not yet known. The initiatives have identified a range of possible management actions to protect wild salmon from sea lice: (1) removing farms, (2) relocating farms, (3) reducing salmon density on farms, (4) chemically treating farm fish to remove lice, (5) fallowing (not a viable option for every year), (6) adopting closed containment technology that removes pathogens from farm effluent, and (7) conserving unexploited wild salmon habitats. These options have a range of economic costs in an industry that is global, meaning that local economic sustainability is uncertain. Despite the uncertainties, there are signs of policy change.

Toward Sustainability

Policy that moves toward sustainability in British Columbian aquaculture and fisheries needs to encompass different routes and multiple components. Policy development and change has been influenced by closed-door multistakeholder meetings as well as by public pressure. Behind this public pressure is an ongoing battle between environmental and industrial interests. We have been involved in policy development, participating in meetings, engaging the public, and communicating and clarifying scientific information via news, radio, and internet. At the same time we have witnessed substantial policy changes. One change, aligned with the precautionary principle, is a moratorium on industry expansion in undeveloped habitats in northern coastal British Columbia. This ensures that some regions of British Columbia are unaffected by salmon aquaculture while management actions are developed, implemented, and evaluated.

Although management actions are available, there remains a lack of policy that supports their development and evaluation according to basic scientific principles of control, treatment, replication, and randomization. This has long been a challenge for us and for other scientists. There is a need for policy that provides scientists access to industry data (e.g., louse abundance and stocking density), that allows farm management manipulation in accordance with scientific principles, and that supports long-term funding to evaluate ecological consequences of changes in the management (e.g., fallowing and chemical treatment), changes in the environment (e.g., climate change), and changes in the biology (e.g., louse evolution of chemical resistance). Ongoing support of the science is a crucial element of responsible policy development.

Increasingly, it appears that movement toward sustainability will be based not on federal regulation but on a combination of direct dialogue between industry and stakeholders as well as public opinion and advocacy for management changes.

RISK ASSESSMENT FOR INVASIVE SPECIES IN THE GREAT LAKES—REUBEN KELLER

Many invasive species become established after being intentionally introduced for commerce in the aquarium/pet, horticulture, live food, bait, and other trades. Because these species are specifically identified for commerce, there is great potential for managing the vectors of introduction to prevent the arrival of likely invaders. This in turn requires predictions of which species are likely to become invasive if released. Preventing the intentional introduction of invaders is a natural area of overlap between policy/management and academic ecology because the theory and methods for ecological predictions are still being developed.

Here, I describe two projects that were conduced as part of my Ph.D. dissertation (Keller 2006) and one ongoing project that I am involved in as a result of my Ph.D. research. The first two projects of my dissertation investigated invasion risks from the trades in live aquatic organisms. In one, funded by a federal grant that my adviser, David Lodge, and a previous graduate student, Cindy Kolar, brought into the lab, I spent two summers sampling organisms from the aquarium, water garden, live food, bait, and biological supplies trades. The project was limited geographically to the southern basin of Lake Michigan and involved two summers of visiting stores to purchase and identify samples of all live aquatic species being sold.

Our principal findings were that the trades sell many species that are currently invasive or that appear likely to become invasive in the future. In addition, species identifications used by the trades are often wrong or ambiguous, and aquatic plants sold almost always had "hitchhiker" species (Keller and Lodge 2007). These results are similar to those of other research teams (e.g., Reichard and White 2001; Maki and Galatowitsch 2004; Padilla and Williams 2004; Rixon et al. 2005; Weigle et al. 2005; Cohen et al. 2008).

My second project built upon previous work conducted in the Lodge lab (Kolar and Lodge 2002) and by others (e.g., Richardson et al. 1990; Veltman et al. 1996; Reichard and Hamilton 1997; Pheloung et al. 1999; Champion and Clayton 2000). I gathered trait data for all nonnative mollusk species at two nested geographical scales, the Laurentian Great Lakes and the 48 contiguous United States. We found that annual fecundity is sufficient to accurately discriminate between species that have and have not become invasive (Keller et al. 2007a), and that this relationship holds for both geographic scales. This relationship could be used as part of a decision tool for determining which species pose an acceptable risk in, for example, the aquarium and live food trades.

Although each of these projects has policy implications, only the first has had any direct policy impact to date. This came about when Lodge was invited to meet with the mayor of Chicago. In preparation for the meeting, he and I spent several hours putting together two pages of recommendations for actions the City of Chicago could take to reduce the risks from aquatic invasive species. This included a list of species that we considered high risk, which was drawn partly from the findings of the first project (Keller and Lodge 2007).

Six months after Lodge's meeting, we were contacted by the Chicago Department of Environment and invited to participate in a working group with the aim of identifying aquatic species in trade that pose a high risk to Chicago waterways, including Lake Michigan. Working with other stakeholders, we assembled a list of 13 animals and 13 plants, again drawing in part from my previous work and other work from the Lodge lab (Kolar and Lodge 2002). By action of the Chicago City Council, these species have subsequently been prohibited from live sale within Chicago (Invasive Species Control Ordinance of 2007). In contrast, to the best of our knowledge, the mollusk risk assessment is not being considered directly for policy, although it has contributed to the ongoing federal debate about invasive species policy (described by Lodge, above).

While some of the species identified in my dissertation work are now regulated in Chicago, it is likely that a side project I took on toward the end of my dissertation will have a greater direct impact. The invasive aquatic plant *Hydrilla* was discovered in Indiana in 2006. This prompted the Indiana Aquatic Invasive Species Coordinator to assemble a working group with the aim of determining the invasion risks posed by aquatic plant species in trade. I had previously presented my work at Indiana Department of Natural Resources meetings, and I was invited to be involved. Over the last 18 months I have met regularly with managers, policy makers, nongovernmental organization representatives, and plant retailers to devise a system for assessing risk. This project continues to make progress, and we hope that in the next 12 months we will be able to submit a list of the species in trade, annotated for their invasion risk in Indiana, to the Indiana Department of Natural Resources for consideration.

Conceptually, this final project is similar to my mollusk work. Both projects aim to determine the characteristics of species that make them invasive, and both could be applied to decisions about which species are allowed for import. For two reasons, however, this final project may be more likely to be quickly incorporated into policy and management. First, it was initiated within the Indiana Department of Natural Resources, which is ultimately the agency that has power to regulate species.

Second, it has included a broad range of stakeholders throughout the process. This has forced all members, myself included, to acknowledge and work with the concerns, perceptions, and expertise of others. This will lead to a different, and hopefully more broadly acceptable, outcome than would have occurred if the work had been done entirely within a science lab.

It is worth noting, however, that this final project is being conducted at the scale of a single state, while the mollusk risk assessment produced results applicable to two much larger geographic regions—the Great Lakes region and the 48 contiguous United States. Given the size of these regions and their more complicated political situations, it is reasonable to expect that its influence might be less direct and less rapid.

The Indiana project has also come with drawbacks. The risk assessment model we are pursuing will not advance methods or ecological understanding in the way that my mollusk work did. Although I anticipate that I will eventually spend a similar amount of time working on each project, the final results of the Indiana work will not be publishable in nearly such a high-ranking academic journal as the mollusk work. As an early-career scientist, this is a significant trade-off with respect to a potential academic career.

Through these different experiences, I have been surprised by a number of things about the interaction of science and policy. First, the majority of policy makers, managers, and other stakeholders that I have interacted with do not read scientific publications. To have a direct influence beyond the academic community, it has therefore been necessary to present at management conferences and pursue face-to-face interactions. Second, academic science can, at best, inform decision making. Many other types of knowledge and experience have essential contributions to make. This has been illustrated particularly by my work with aquatic plant retailers and agency scientists in Indiana.

WHY BIOECONOMIC APPROACHES ARE NECESSARY FOR ENVIRONMENTAL ISSUES—JASON SHOGREN

Protecting nature effectively requires scientists and policy makers to integrate economic and ecologic indicators of success and failure. For me, the integration of economics and ecology works on three levels—*models*, *methods*, and *mind-sets*. My perspective on these paths of integration is as an economist who has worked for two decades with other disciplines. This distinction rests on (1) the technical integration of models, (2) the policy integration of methods, and (3) the political integration of mind-sets. While neither simple nor straightforward, integration across disciplinary boundaries matters to better support policy decisions. Integrating models, methods, and mind-sets can help scientists and decision makers address the challenge of balancing rights of self-determination and social preferences to protect nature (see Daly and Cobb 1989).

First, model building is the most straightforward way to integrate economics and ecology. Building a multidisciplinary research team like ISIS to construct a simulation model is a worthy goal in itself. This book focuses on building bioeconomic models that account for the interdependence between economic and biological systems for invasive species. A bioeconomic model can help society design incentive mechanisms that better satisfy political objectives, meet species biological needs, and protect private property concerns. The thought process of integration focuses each scientist's attention on how to create pragmatic positive links between systems. Focusing on how to build the model can improve our understanding of both economic and ecological systems. *Building the model* becomes the primary focus, even though scientists will continue to debate the ethereal normative decisions of morals and policy. The differences and similarities between the disciplines of biology and economics are directly addressed by asking researchers to construct and link the human and natural sectors of the model. Accounting for the joint influences that economic systems and biological systems have on each other improves each discipline's perceptions of the other one.

Second, integrating methods is another path to combine economics and ecology toward achieving a common goal of providing more protection at fewer costs. Economists can propose one method to induce people to protect habitat, say, through voluntary monetary compensation based on economic criteria, whereas biologists might propose another method, say, zoning based on biological criteria. Integration of methods can occur by agreeing on a common goal and working toward finding a common tool. The integration then occurs as we work together to implement and assess the effectiveness of the goal and tool based on sets of both economic and ecological criteria.

We illustrate the integration of methods through how to design a compensation scheme for landowners trying to protect species that meets both economic and biological criteria. Experts judge that compensation of landowners is the key method to achieve environmental objectives on private lands. When compensation is withheld, and private landowners are still required through environmental regulation to protect the attributes of their land by limiting the production of marketable commodities on their land, the landowner pays the cost of environmental protection, yet society as a whole captures the benefits. Landowners avoid the costs of environmental regulation by razing the environmental amenities on their land, thereby eliminating the risk of government regulation prior to federal or local action and oversight (see Brown and Shogren 1998). Approaches exist that offer compensation to landowners for the costs of protecting species on their land, providing incentives to landowners for better practices, relying on the carrot of financial reward rather than the stick of prosecution for violating environmental rules.

Policy makers have addressed the compensation question in part by offering voluntary incentive programs to landowners to increase private species protection and biodiversity conservation. The idea is to transform an environmental liability into a marketable asset. The U.S. Fish and Wildlife Service and more than a thousand nonprofit land trusts promote habitat conservation by using voluntary incentive mechanisms to elicit the cooperation of private landowners. The open question is how to design compensation schemes that satisfy both economic and biological criteria. One idea is the *agglomeration bonus*. With the dual goal of maximizing environmental protection cost-effectively and minimizing private landowner resentment, the agglomeration bonus mechanism pays an extra bonus for every acre a landowner retires that borders on any other retired acre (see Parkhurst et al. 2002). The mechanism provides incentive for landowners to voluntarily create a contiguous reserve across their common border to create a single large habitat, which is usually desired for effective conservation. A government agency's role is to target the critical habitat, to integrate the agglomeration bonus into the compensation package, and to provide landowners the unconditional freedom to choose which acres to retire.

Finally, integration of mind-sets is more challenging. When considering models and methods, integration is mainly a technical matter—researchers recognize why a market failed to protect nature, identify the key feedback loops, and define the appropriate strategy to correct the failure. The key management goal is to find the right tool for the job. But integrating mind-sets is more emotional. Mind-sets reflect the reason you ended up in the discipline you are in—it reflects different principles about how value is created, protected, and enhanced.

Different people have different mind-sets about the proper role of private property and species protection, ranging from anthropocentric to biocentric or ecocentric ethics. The historical dynamics of these differing land conservation mind-sets or ideologies vary across cultures and nations. The challenge is to create a setting in which people with different mind-sets can express their opinions and learn from each other about what matters to them and why (see, e.g., Norton 1991). People appreciate that neither land restrictions nor the call for protection on private land is new. For centuries in Europe and in Euro-America, common-law restrictions have limited what people could do to or with their property (see North and Thomas 1973). In the 1930s, Aldo Leopold argued that the protection of nature "ultimately boil[s] down to reward the private landowner who conserves the public interest" (see Bean 1997).

But environmental regulations have produced a backlash because private property has held special status in the history of many nations, including the United States. Conservatives view laws that restrict private landowner autonomy to protect obscure species as a threat to both the economic system and broader social order (see Epstein 1985; Norton 1991). They view land as capital, albeit natural capital, and argue that capital is the key ingredient that allows people the ability to create, store, and share the wealth necessary for national prosperity. This is the utilitarian mind-set toward nature promoted by John Stuart Mill and Gifford Pinchot—the resource conservation ethic is "the greatest good of the greatest number for the longest time" (Pinchot 1947). This classical liberal viewpoint takes a Hamiltonian perspective—the government should abdicate to market forces that create wealth by allowing for resources to move freely from low valued to high valued uses. They agree with James Madison's argument within the *Federalist Papers* that "the wide diffusion of independent property rights . . . was the essential foundation for stable republican government." Rightly or wrongly, they fear that restricting private land for species protection without just compensation is another step toward collectivism.

A second mind-set toward nature is the romantic-conservation ethic promoted in the United States by Ralph Waldo Emerson, Henry Thoreau, and John Muir. The preservationists, as they are called, believed that nature and land have other uses than just for human financial gain. Landowners would be free to pursue private profits provided they behave as responsible social citizens, too, because by definition land is already in public service. All land uses should be viewed as "harm preventing" rather than as "public good providing." As Sagoff (1997, p. 845) puts it, "the conviction that the freedom to wring the last speculative penny from one's land is of a piece with one's most fundamental civil, political, and personal liberties seems to be grounded less on argument than on assumption."

A third mind-set that emerged in the 1940s, in part from frustration with the other two views, is Leopold's evolutionary-ecological land ethic. Leopold (1949) based his ethic on the scientific notion that nature is not a collection of separate parts but an integrated system of actions, reactions, and feedbacks. This more scientific notion focuses on defining the natural system within the context of human interaction and well-being. Within this mind-set of integrating natural science and social science, one can promote more understanding between mind-sets by working together to define a set of evaluative criteria that reflects the range of ethical views. For the private lands challenge, this set of criteria should address perceived biological needs, landowner interests, and regulatory concerns.

While we appreciate that the process of defining and evaluating the integrated criteria to define the"common good" of society will be challenging given the objectives of different groups, we believe it is worth the effort. Integration of economics and ecology is fundamental both for science and for policy. For science, integration implies more accurate estimates of both economic and ecological phenomena; for policy, integration means a better appreciation of the alternative viewpoints that arise when attempting to address difficult challenges, such as protecting nature on private lands. This book has discussed how we can integrate across disciplines at the modeling, method, and mind-set levels. These three types of integration require explicitly modeling the feedback links between systems, aligning ecosystem goals with personal landowner goals, and accounting for the appropriateness of policy change to a variety of perspectives. Each type of integration sends a similar message—while maintaining a narrow focus when studying a particular discipline is useful, a broader approach matters because it highlights the feedback loops and criteria of success and provides more understanding into how other disciplines approach the issue. Understanding better how to protect nature by appreciating and integrating diverse ideas and tools can move us closer to the objectives we have focused on in this book—how to provide more protection against invasive species at less cost, which will be considered along with other social and political objectives, such as self-determination, species survival, and fairness.

Notes

1. All five articles have been published in *Reviews in Fisheries Science*. Unfortunately (for the science), only one was peer reviewed.

References

Bean, M. 1997. Review of "Private property and the endangered species act: saving habitat, protecting homes." BioScience 49:825.

Brooks, K. M. 2005. The effects of water temperature, salinity, and currents on the survival and distribution of the infective copepodid stage of sea lice (*Lepeophtheirus salmonis*) originating on Atlantic salmon farms in the Broughton Archipelago of British Columbia, Canada. Reviews in Fisheries Science 13:177–204.

Brooks, K. M., and S. R. M. Jones. 2008. Perspectives on pink salmon and sea lice: scientific evidence fails to support the extinction hypothesis. Reviews in Fisheries Science 16:403–412.

Brooks, K. M., and D. J. Stucchi. 2006. The effects of water temperature, salinity and currents on the survival and distribution of the infective copepodid stage of the salmon louse (*Lepeophtheirus salmonis*) originating on Atlantic salmon farms in the Broughton Archipelago of British Columbia, Canada (Brooks, 2005)—a response to the rebuttal of Krkošek et al. (2005a). Reviews in Fisheries Science 14:13–23.

Brown, G., Jr., and J. Shogren. 1998. Economics of the endangered species act. Journal of Economic Perspectives 12:3–20.

Champion, P. D., and J. S. Clayton. 2000. Border control for potential aquatic weeds. Department of Conservation, Wellington, New Zealand.Clark, C., 1976. Mathematical bioeconomics: the optimal management of renewable resources. Wiley, New York.

Cohen, J., N. Mirotchnick, and B. Leung. 2008. Thousands introduced annually: the aquarium pathway for non-indigenous plants to the St Lawrence Seaway. Frontiers in Ecology and the Environment 5:528–532.

Costello, M. 2006. Ecology of sea lice parasitic on farmed and wild fish. Trends in Parasitology 22:475–483.

Crocker, T. D., and J. Tschirhart. 1992. Ecosystems, externalities, and economies. Environmental and Resource Economics 2:551–567.

Daly, H., 1968. On economics as a life science. Journal of Political Economy 76:392–406.

Daly, H., and J. Cobb, Jr. 1989. For the common good: redirecting the economy toward community, the environment, and a sustainable future. Beacon Press, Boston.

Epstein, R. 1985. Takings: private property and power of eminent domain. Harvard University Press, Cambridge, MA.

Ford, J. S., and R. A. Myers. 2008. A global assessment of salmon aquaculture impacts on wild salmonids. PLoS Biology e33 doi:10.1371/journal.pbio.0060033.

Fowler, A. J., D. M. Lodge, and J. F. Hsia. 2007. Failure of the Lacey Act to protect US ecosystems against animal invasions. Frontiers in Ecology and the Environment 5:353–359.

Gordon, H. S. 1954. The economic theory of a common-property resource: the fishery. Journal of Political Economy 62(2):124–142.

Hammack, J., and G. Brown. 1974. Waterfowl and wetlands: toward bioeconomic analysis. Johns Hopkins University Press, Baltimore, MD.

Harvey, B. 2008. Science and sea lice: what do we know? British Columbia Pacific Salmon Forum, Victoria, British Columbia.

Hilborn, R., and M. Mangel. 1997. The ecological detective, confronting models with data. Princeton University Press, Princeton, NJ.

Keller, R. P. 2006. Ecological and bioeconomic risk assessment for invasive species. Ph.D. thesis, Department of Biological Sciences, University of Notre Dame.

Keller, R. P., J. M. Drake, and D. M. Lodge. 2007a. Fecundity as a basis for risk assessment of nonindigenous freshwater molluscs. Conservation Biology 21:191–200.

Keller, R. P., and D. M. Lodge. 2007. Species invasions from commerce in live organisms: problems and possible solutions. BioScience 57:428–436.

Keller, R. P., D. M. Lodge, and D. C. Finnoff. 2007b. Risk assessment for invasive species produces net bioeconomic benefits. Proceedings of the National Academy of Sciences of the United States of America 104:203–207.

Kolar, C. S., and D. M. Lodge. 2002. Ecological predictions and risk assessment for alien fishes in North America. Science 298:1233–1236.

Krkošek, M., J. S. Ford, S. Lele, and M. A. Lewis. 2008. Sea lice and pink salmon declines: a response to Brooks and Jones (2008). Reviews in Fisheries Science 16:413–420.

Krkošek, M., J. S. Ford, A. Morton, S. Lele, R. A. Myers, and M. A. Lewis. 2007a. Declining wild salmon populations in relation to parasites from farm salmon. Science 318:1772–1775.

Krkošek, M., A. Gottesfeld, B. Proctor, D. Rolston, C. Carr-Harris, and M. A. Lewis. 2007b. Effects of host migration, diversity, and aquaculture on sea lice threats to Pacific salmon populations. Proceedings of the Royal Society of London Series B Biological Sciences 274: 3141–3149.

Krkošek, M., M. A. Lewis, A. Morton, L. N. Frazer, and J. P. Volpe. 2006a. Epizootics of wild fish induced by farm fish. Proceedings of the National Academy of Sciences of the United States of America 103:15506–15510.

Krkošek, M., M. A. Lewis, and J. P. Volpe. 2005. Transmission dynamics of parasitic sea lice from farm to wild salmon. Proceedings of the Royal Society of London Series B Biological Sciences 272:689–696.

Krkošek, M., M. A. Lewis, J. P. Volpe, and A. Morton. 2006b. Fish farms and sea lice infestations of wild juvenile salmon in the Broughton Archipelago—a rebuttal to Brooks (2005). Reviews in Fisheries Science 14:1–11.

Leopold, A. 1949. A sand county almanac, and sketches here and there. Oxford University Press, New York.

Lodge, D. M., C. A. Taylor, D. M. Holdich, and J. Skurdal. 2000. Nonindigenous crayfishes threaten North American freshwater biodiversity: lessons from Europe. Fisheries 25:7–20.

Lodge, D. M., S. Williams, H. MacIsaac, K. Hayes, B. Leung, L. Loope, S. Reichard, R. N. Mack, P. B. Moyle, M. Smith, D. A. Andow, J. T. Carlton, and A. McMichael. 2006. Biological invasions: recommendations for policy and management. Ecological Applications 16:2035–2054.

Maki, K., and S. Galatowitsch. 2004. Movement of invasive aquatic plants into Minnesota (USA) through horticultural trade. Biological Conservation 118:389–396.

Morton, A., R. Routledge, C. Peet, and A. Ladwing. 2004. Sea lice (*Lepeophtheirus salmonis*) infection rates on juvenile pink (*Oncorhynchus gorbuscha*) and chum (*Oncorhynchus keta*) salmon in the near shore marine environment of British Columbia, Canada. Canadian Journal of Fisheries and Aquatic Sciences 61:147–157.

National Research Council. 2002. Predicting invasions by nonindigenous plants and plant pests. National Academy of Sciences, Washington, DC.

North, D., and R. Thomas. 1973. The rise of the western world: a new economic history. Cambridge University Press, Cambridge, UK.

Norton, B. 1991. Toward unity among environmentalists. Oxford University Press, New York.

Padilla, D. K., and S. L. Williams. 2004. Beyond ballast water: aquarium and ornamental trades as sources of invasive species in aquatic ecosystems. Frontiers in Ecology and the Environment 2:131–138.

Parkhurst, G., J. Shogren, C. Bastian, P. Kivi, J. Donner, and R. Smith. 2002. Agglomeration bonus: an incentive mechanism to reunite fragmented habitat for biodiversity conservation. Ecological Economics 41:305–328.

Perrings, C., M. Williamson, and S. Dalmazzone, editors. 2000. The economics of biological invasions. Edward Elgar, Cheltenham, UK.

Pheloung, P. C., P. A. Williams, and S. R. Halloy. 1999. A weed risk assessment model for use as a biosecurity tool evaluating plant introductions. Journal of Environmental Management 57: 239–251.

Pinchot, G. 1947. Breaking new ground. Harcourt Brace, New York.

Reichard, S. H., and C. W. Hamilton. 1997. Predicting invasions of woody plants introduced into North America. Conservation Biology 11:193–203.

Reichard, S. H., and P. White. 2001. Horticulture as a pathway of invasive plant introductions in the United States. BioScience 51:103–113.

Reimchen, T. E., D. Mathewson, M. D. Hocking, J. Moran, and D. Harris. 2003. Isotopic evidence for enrichment of salmon-derived nutrients in vegetation, soil and insects in riparian zones in coastal British Columbia. American Fisheries Society Symposium 34:59–69.

Richardson, D. M., R. M. Cowling, and D. C. Le Maitre. 1990. Assessing the risk of invasive success in *Pinus* and *Banksia* in South African mountain fynbos. Journal of Vegetation Science 1:629–642.

Rixon, C. A. M., I. C. Duggan, N. M. N. Bergeron, A. Ricciardi, and H. J. Macisaac. 2005. Invasion risks posed by the aquarium trade and live fish markets on the Laurentian Great Lakes. Biodiversity and Conservation 14:1365–1381.

Sagoff, M. 1997. Muddle or muddle through: takings jurisprudence meets the endangered species act. William and Mary Law Review 38:825–993.

Schaefer, M. B. 1957. Some considerations of population dynamics and economics in relation to the management of marine fishes. Journal of the Fisheries Research Board of Canada 14:669–681.

Veltman, C. J., S. Nee, and M. Crawley. 1996. Correlates of introduction success in exotic New Zealand birds. American Naturalist 147:542–557.

Weigle, S. M., L. D. Smith, J. T. Carlton, and J. Pederson. 2005. Assessing the risk of introducing exotic species via the live marine species trade. Conservation Biology 19:213–223.

Index